Communications in Computer and Information Science 2246

Series Editors

Gang Li⊙, *School of Information Technology, Deakin University, Burwood, VIC, Australia*
Joaquim Filipe⊙, *Polytechnic Institute of Setúbal, Setúbal, Portugal*
Zhiwei Xu, *Chinese Academy of Sciences, Beijing, China*

Rationale
The CCIS series is devoted to the publication of proceedings of computer science conferences. Its aim is to efficiently disseminate original research results in informatics in printed and electronic form. While the focus is on publication of peer-reviewed full papers presenting mature work, inclusion of reviewed short papers reporting on work in progress is welcome, too. Besides globally relevant meetings with internationally representative program committees guaranteeing a strict peer-reviewing and paper selection process, conferences run by societies or of high regional or national relevance are also considered for publication.

Topics
The topical scope of CCIS spans the entire spectrum of informatics ranging from foundational topics in the theory of computing to information and communications science and technology and a broad variety of interdisciplinary application fields.

Information for Volume Editors and Authors
Publication in CCIS is free of charge. No royalties are paid, however, we offer registered conference participants temporary free access to the online version of the conference proceedings on SpringerLink (http://link.springer.com) by means of an http referrer from the conference website and/or a number of complimentary printed copies, as specified in the official acceptance email of the event.

CCIS proceedings can be published in time for distribution at conferences or as post-proceedings, and delivered in the form of printed books and/or electronically as USBs and/or e-content licenses for accessing proceedings at SpringerLink. Furthermore, CCIS proceedings are included in the CCIS electronic book series hosted in the SpringerLink digital library at http://link.springer.com/bookseries/7899. Conferences publishing in CCIS are allowed to use Online Conference Service (OCS) for managing the whole proceedings lifecycle (from submission and reviewing to preparing for publication) free of charge.

Publication process
The language of publication is exclusively English. Authors publishing in CCIS have to sign the Springer CCIS copyright transfer form, however, they are free to use their material published in CCIS for substantially changed, more elaborate subsequent publications elsewhere. For the preparation of the camera-ready papers/files, authors have to strictly adhere to the Springer CCIS Authors' Instructions and are strongly encouraged to use the CCIS LaTeX style files or templates.

Abstracting/Indexing
CCIS is abstracted/indexed in DBLP, Google Scholar, EI-Compendex, Mathematical Reviews, SCImago, Scopus. CCIS volumes are also submitted for the inclusion in ISI Proceedings.

How to start
To start the evaluation of your proposal for inclusion in the CCIS series, please send an e-mail to ccis@springer.com.

Wenjie Zhang · Anthony Tung ·
Zhonglong Zheng · Zhengyi Yang ·
Xiaoyang Wang · Hongjie Guo
Editors

Web and Big Data

APWeb-WAIM 2024
International Workshops

KGMA 2024, SemiBDMA 2024
MADM 2024, AIEDM 2024 and STBDM 2024
Jinhua, China, August 30 – September 1, 2024
Proceedings

Editors
Wenjie Zhang
University of New South Wales
Sydney, NSW, Australia

Zhonglong Zheng
Zhejiang Normal University
Jinhua, China

Xiaoyang Wang
University of New South Wales
Sydney, NSW, Australia

Anthony Tung
National University of Singapore
Singapore, Singapore

Zhengyi Yang
University of New South Wales
Sydney, NSW, Australia

Hongjie Guo
Zhejiang Normal University
Jinhua, China

ISSN 1865-0929   ISSN 1865-0937 (electronic)
Communications in Computer and Information Science
ISBN 978-981-96-0054-0   ISBN 978-981-96-0055-7 (eBook)
https://doi.org/10.1007/978-981-96-0055-7

© The Editor(s) (if applicable) and The Author(s), under exclusive license
to Springer Nature Singapore Pte Ltd. 2025

This work is subject to copyright. All rights are solely and exclusively licensed by the Publisher, whether the whole or part of the material is concerned, specifically the rights of translation, reprinting, reuse of illustrations, recitation, broadcasting, reproduction on microfilms or in any other physical way, and transmission or information storage and retrieval, electronic adaptation, computer software, or by similar or dissimilar methodology now known or hereafter developed.
The use of general descriptive names, registered names, trademarks, service marks, etc. in this publication does not imply, even in the absence of a specific statement, that such names are exempt from the relevant protective laws and regulations and therefore free for general use.
The publisher, the authors and the editors are safe to assume that the advice and information in this book are believed to be true and accurate at the date of publication. Neither the publisher nor the authors or the editors give a warranty, expressed or implied, with respect to the material contained herein or for any errors or omissions that may have been made. The publisher remains neutral with regard to jurisdictional claims in published maps and institutional affiliations.

This Springer imprint is published by the registered company Springer Nature Singapore Pte Ltd.
The registered company address is: 152 Beach Road, #21-01/04 Gateway East, Singapore 189721, Singapore

If disposing of this product, please recycle the paper.

Preface

The Asia Pacific Web (APWeb) and Web-Age Information Management (WAIM) Joint International Conference on Web and Big Data (APWeb-WAIM) is a leading international conference for researchers, practitioners, developers, and users to share and exchange their cutting-edge ideas, results, experiences, techniques, and applications in connection with all aspects of web and big data management. The conference invites original research papers on Web technologies, database systems, information management, software engineering, and big data. As the 8th edition in the increasingly popular series, APWeb-WAIM 2024 was held in Jinhua, China, August 30 – September 1, 2024. Along with the main conference, the APWeb-WAIM workshops provide an international forum for researchers to discuss and share their pioneering work.

This APWeb-WAIM 2024 workshop volume contains the papers accepted by five workshops held in conjunction with APWeb-WAIM 2024. The five workshops were selected after a public call-for-proposal process, and each of them had a focus on a specific area that contributed to the main themes of the APWeb-WAIM conference.

After the single-blinded review process, out of 49 submissions, the five workshops accepted a total of 28 papers, marking an acceptance rate of 57.14%. Each submission was peer-reviewed by 3–4 reviewers. The five workshops were as follows:

- The 7th International Workshop on Knowledge Graph Management and Applications (KGMA 2024)
- The 6th International Workshop on Semi-structured Big Data Management and Applications (SemiBDMA 2024)
- The 1st International Workshop on Mobile Applications and Data Management (MADM 2024)
- The 1st International Workshop on Artificial Intelligence in Education and Educational Data Mining (AIEDM 2024)
- The 1st International Workshop on Spatio-Temporal Big Data Management (STBDM 2024)

As a joint effort, all organizers of APWeb-WAIM conferences and workshops, including this and previous editions, have made APWeb-WAIM a valuable trademark through their work. We would like to express our thanks to all the workshop organizers and Program Committee members for their great efforts in making the APWeb-WAIM 2024 workshops such a great success. Last but not least, we are grateful to the main conference organizers for their leadership and generous support, without which this APWeb-WAIM 2024 workshop volume wouldn't have been possible.

August 2024

Wenjie Zhang
Anthony Tung
Zhonglong Zheng
Zhengyi Yang
Xiaoyang Wang
Hongjie Guo

Organization

APWeb-WAIM 2024 Proceedings Co-chairs

Zhengyi Yang	University of New South Wales, Australia
Xiaoyang Wang	University of New South Wales, Australia
Hongjie Guo	Zhejiang Normal University, China

KGMA 2024

Workshop Co-chairs

Xin Wang	Tianjin University, China
Long Yuan	Nanjing University of Science and Technology, China

Program Committee Members

Peng Cheng	East China Normal University, China
Zi Chen	Nanjing University of Aeronautics and Astronautics, China
Chuan Ma	Chongqing University, China
Yuren Mao	Zhejiang University, China
Kai Wang	Shanghai Jiao Tong University, China
Dong Wen	University of New South Wales, Australia
Jianye Yang	Guangzhou University, China
Shiyu Yang	Guangzhou University, China
Zhengyi Yang	University of New South Wales, Australia
Xuliang Zhu	Shanghai Jiao Tong University, China

SemiBDMA 2024

Workshop Chair

Linlin Ding	Liaoning University, China

Program Committee Members

Ye Yuan	Beijing Institute of Technology, China
Xiangmin Zhou	Royal Melbourne Institute of Technology, Australia
Jianxin Li	Swinburne University of Technology, Australia
Bo Ning	Dalian Maritime University, China
Yongjiao Sun	Northeastern University, China
Yulei Fan	Zhejiang University of Technology, China
Guohui Ding	Shenyang Aerospace University, China
Bo Lu	Dalian Nationalities University, China
Xiaohuan Shan	Liaoning University, China
Yuefeng Du	Liaoning University, China
Tingting Liu	Liaoning University, China
Mo Li	Liaoning University, China

MADM 2024

Workshop Co-chairs

Dong Du	Shanghai Jiao Tong University, China
Mingyu Wu	Shanghai Jiao Tong University, China

Program Committee Members

Xiaocheng Hu	Huawei, China
Changlong Li	East China Normal University, China
Li Li	Beihang University, China
Xin Liu	Lanzhou University, China
Chang Lou	University of Virginia, USA
Bo Wang	Beijing Jiaotong University, China
Chenxi Wang	Chinese Academy of Sciences, China
Chao Wu	Nanjing University of Science and Technology, China
Haoyang Zhuang	Shanghai Jiao Tong University, China

AIEDM 2024

Workshop Co-chairs

Jia Zhu	Zhejiang Normal University, China
Xiang Zhao	National University of Defense Technology, China

Program Committee Members

Ke Deng	Royal Melbourne Institute of Technology, Australia
Zhixu Li	Fudan University, China
Min Yang	Chinese Academy of Sciences, China
Penghe Chen	Beijing Normal University, China
Lingling Zhang	Xi'an Jiaotong University, China
Pasquale De Meo	University of Messina, Italy
Jie Shao	University of Electronic Science and Technology of China, China
Muhammad Umair Hassan	Norwegian University of Science and Technology, Norway
Yu Yang	Hong Kong Polytechnic University, China
Ronghua Lin	South China Normal University, China
Qionghao Huang	Zhejiang Normal University, China

STBDM 2024

Workshop Co-chairs

Kaiqi Zhang	Harbin Institute of Technology, China
Tuo Shi	Aalto University, Finland
Zemin Chao	Harbin Institute of Technology, China

Program Committee Members

Chengliang Chai	Beijing Institute of Technology, China
Quan Chen	Guangdong University of Technology, China
Faming Li	Northeastern University, China
Lei Li	Hong Kong University of Science and Technology, China

Jinfei Liu   Zhejiang University, China
Xueli Liu   Tianjin University, China
Yajun Yang   Tianjin University, China
Ge Yu   Northeastern University, China
Anzhen Zhang   Shenyang Aerospace University, China
Tongxin Zhu   Southeast University, China
Zhaonian Zou   Harbin Institute of Technology, China
Jian Chen   Harbin Institute of Technology, China

Contents

KGMA Workshop

Approximating Temporal Katz Centrality with Monte Carlo Methods 3
  Haonan Yan, Zhengyi Yang, Tianming Zhang, Dong Wen, Qi Luo, and Nimish Ukey

LLM for Uniform Information Extraction Using Multi-task Learning Optimization 17
  Ying Li, Zhen Tan, and Weidong Xiao

ASM: Adaptive Subgraph Matching via Efficient Compression and Label Filter 30
  Yanfeng Chai, Jiashu Li, Qiang Zhang, Jiake Ge, and Xin Wang

SemiBDMA Workshop

Learning Multi-semantic Based on Cross-Attention for Image-Text Retrieval 45
  Bo Lu, Ying Gao, Tianbao Zhao, Xia Yuan, Haibin Zhu, Lin Gan, and Xiaodong Duan

Integrated Global Semantics and Local Details for Image-Text Retrieval 57
  Bo Lu, Lin Gan, Tianbao Zhao, Xiaojin Wu, Tianyuan Zhong, Ying Gao, and Xiaodong Duan

Adversarial Graph Convolutional Network Hashing for Cross-Modal Retrieval 69
  Bo Lu, Tianbao Zhao, Guiyuan Liang, Jiaming Li, and Xiaodong Duan

FS-IGA: Feature Selection Method Based on Improved Genetic Algorithm 81
  Dong Li, Shumei Du, Yong Wei, Lei Qin, and Yuefeng Du

Shearlet Transform Based Multiscale Fusion Network for Image Super Resolution 94
  Wei Wei

Application of Segmental-Based Transfinite Mapping Method for Quadrilateral Mesh Generation of Complex Domain 103
  Dao-Ju Qin and Pei-Pei Shang

Augmenting Knowledge Tracing: Personalized Modeling by Considering
Forgetting Behavior in Learning Process 115
    Hongxin Yang, Yuefeng Du, Tingting Liu, and Linlin Ding

A Time-Aware Sequential Recommendation Based on Attention
Mechanism 123
    *Tingting Liu, Tianrui Li, Baoyan Song, Yuefeng Du, Hongxin Yang,
    and Linlin Ding*

Integrating User Sentiment and Behavior for Explainable Recommendation 135
    *Dong Li, Zhicong Liu, Qingyu Zhang, Yue Kou, Tingting Liu,
    and Haoran Qu*

MADM Workshop

SACC: Secure-Cooperative Adaptive Cruise Control for Unmanned
Vehicles 151
    Wen Ran, Changlong Li, and Edwin H.-M. Sha

A Systematic Mapping Study of LLM Applications in Mobile Device
Research 163
    Chong Chen, Bo Wang, and Youfang Lin

Application Framework for OpenHarmony Distributed Trusted Execution
Environment 175
    Yilong Wang, Yang Yu, and Dong Du

Design and Implementation of a Multi-metric Performance Analysis Tool
for Android System and Its Applications 188
    Zhenyu Yang

Towards Optimal Leakage Assessment of TVLA 200
    Yuanqiao Bi, Weijian Li, and Guiyuan Xie

Joint UAV Trajectory Optimization and Task Offloading in Integrated
Air-Ground Networks 212
    Yuanwei Zhang, Jiaqi Shuai, Weichang Wen, and Haixia Cui

AIEDM Workshop

Research on Ethical Issues of Classroom Dialogue in the Era of Large
Language Model 229
    Jianxia Ling, Jia Zhu, Jianyang Shi, and Pasquale De Meo

Adaptive Exploration: Elevating Educational Impact of Unsupervised
Knowledge Graph Question Answering 239
   Xi Yang, Zhangze Chen, Hanghui Guo, and Tetiana Shestakevych

xLSTM-FER: Enhancing Student Expression Recognition with Extended
Vision Long Short-Term Memory Network 249
   Qionghao Huang and Jili Chen

RDNeRF: Radiance Distribution Guided NeRF for Floaters Removing 260
   Yizhou Chen, Zixuan Huang, Weijun Wu, Weihao Yu, and Jin Huang

Multi-user VR Content Wireless Delivery Using Motion Prediction
and Adaptive Multicasting 271
   Ke Wang, Yuqi Li, Kaikai Chi, and Liang Huang

MEGKT: Multi-edge Features Enhancement for Graph-Based Knowledge
Tracing 281
   Lei Zhang, Linlin Zhao, and Zhenguo Zhang

STBDM Workshop

Collaborative Inference for Adaptive DNN Pipeline-Aware Alignment 295
   Jiankang Ren, Lin Liu, Ran Bi, Simeng Li, Shengyu Li, and Xian Lv

A Unified Framework for Link Prediction on Heterogeneous Temporal
Graph 310
   Chongjian Yue, Qiao Mi, and Lun Du

Massive-Parallel Game of Politicians 323
   Andrew Schumann and Krzysztof Pancerz

Multiscale Convolutional Feature Aggregation for Fine-Grained Image
Retrieval 335
   Caolin Yang, Kaifeng Ding, and Chengzhuan Yang

Author Index 349

KGMA Workshop

Approximating Temporal Katz Centrality with Monte Carlo Methods

Haonan Yan[1], Zhengyi Yang[1(✉)], Tianming Zhang[2], Dong Wen[1], Qi Luo[1], and Nimish Ukey[1]

[1] University of New South Wales, Sydney, NSW, Australia
{haonan.yan,zhengyi.yang,dong.wen,qi.luo1,n.ukey}@unsw.edu.au
[2] Zhejiang University of Technology, Hangzhou, Zhejiang, China
tmzhang@zjut.edu.cn

Abstract. Graphs have long served as fundamental data models across various disciplines such as data mining, social media analysis, and knowledge management. In real-world applications, interactions between nodes often evolve over time, necessitating the use of temporal graphs. Centrality measures are pivotal in graph analysis for identifying key nodes. Specifically, Temporal Katz Centrality (TKC) has garnered significant attention in recent years for its ability to measure node influence in temporal graphs. TKC evaluates the influence of nodes by considering all walks originating from a node, where contributions are weighted based on a user-specified time decay factor. This approach captures temporal dynamics by incorporating both the timing of interactions and the intervals between them, offering a comprehensive assessment of node importance over time. However, computing TKC in large temporal graphs is computationally intensive due to the requirement to traverse all edges and update centrality values along temporal paths. To address this challenge, this paper proposes an efficient Monte Carlo approximation method for TKC. This approach employs random sampling via Simple Random Walk and Alpha Walk to estimate TKC values. The paper rigorously proves the asymptotic consistency of the method using the Law of Large Numbers and Central Limit Theorem, ensuring that estimated TKC values converge to true TKC values as the sample size increases. Experiments conducted on six real-world temporal graph datasets demonstrate the effectiveness of the proposed approximation methods. Under the same running time, the Alpha walk outperforms other methods on large temporal graphs, exhibiting the lowest mean relative error and highest precision.

Keywords: Temproal Graphs · Katz Centrality · Monte Carlo Sampling

1 Introduction

Graphs are commonly used data models in data mining and databases, forming the foundation for social media analysis, epidemiology, knowledge management, and more. However, in many real-world applications, interactions between nodes evolve over time, leading to the emergence of temporal graphs. For example, in social media analysis, temporal graphs track the evolution of interactions, trends, and influencer dynamics over time on social platforms. In epidemiology, they model the spread of diseases by representing how interactions and contacts between individuals change over time, aiding in predicting and controlling outbreaks. In knowledge graph management, temporal graphs facilitate dynamic knowledge representation, capturing updates to facts, relationships, and ontologies in a knowledge base [7,9,13,19]. They enable temporal querying, allowing queries that consider the temporal dimension to find how relationships between entities have changed over specific periods [3,6,26].

Centrality is fundamental in graph analysis for identifying crucial nodes within graphs. In temporal graphs, centrality measures are adapted to reflect evolving node importance over time. In knowledge management, temporal centrality aids in identifying critical entities and relationships within knowledge graphs that change over time, enhancing dynamic knowledge representation and facilitating insights into temporal trends and influential components [18].

Temporal Katz Centrality (TKC) [5] is one of the most fundamental metrics used to evaluate node centrality in temporal graphs. It is a walk-based metric that extends the traditional Katz Centrality [12] to temporal graphs. It measures the influence of nodes in temporal graphs by considering all walks originating from the node, where contributions are weighted based on a user-specified time decay factor. TKC offers advantages in capturing temporal dynamics by incorporating both the timing of interactions and the intervals between them, providing a comprehensive assessment of node importance over time. TKC has demonstrated promising results across various tasks in temporal graph analysis, leading to increased attention in recent years [14,17,24].

Motivation. Despite its broad applicability, computing TKC requires traversing all edges in the entire temporal graph and continuously updating node centrality values along each temporal path. This process results in high computational complexity, which limits the application of TKC to large temporal graphs. Hence, in this paper, we aim to develop an effective and efficient method for approximating TKC in large-scale temporal graphs.

Contributions. In this paper, we explore the use of the Monte Carlo approximation method [1] for approximating TKC. The Monte Carlo method is a commonly used technique that employs random sampling to estimate numerical values of complex mathematical functions. Specifically, we employ random walks, including Simple Random Walk and Alpha Walk [11], to acquire sample walks in temporal graphs and approximate TKC values. We prove that our method is *asymptotically consistent*, namely, the estimated TKC value will converge to the true value as the sample size increases.

- *TKC Approximation with Monte Carlo Methods.* We adopt two sampling methods to approximate TKC: Simple Random Walk and Alpha Walk. For Simple Random Walk, each neighbor of a node has an equal probability of being selected during the walk, ensuring that all neighbors are equally likely

to be visited. In the Alpha Walk method, the continuation of the walk at each step is governed by a fixed probability α. This approach allows the walk to end based on a probability, enabling the approximation method to capture varying lengths of influence paths in the network, thus balancing between immediate and more extended connections.
- *Rigorous Proof on Consistency.* We utilize both the Law of Large Numbers (LLN) and the Central Limit Theorem (CLT) to prove the asymptotic consistency of our method, that is, our algorithm converges to the true centrality measure as the number of samples increases. Specifically, the LLN guarantees that the average of the sample centrality measures will converge to the expected centrality value, while the CLT allows us to make probabilistic statements about the distribution of the centrality estimates as they approach a normal distribution centered around the true value.
- *Extensive Experimental Evaluation.* We conducted extensive experiments on six real-world temporal graphs. The performance of the approximation method is influenced by factors such as the edge-to-vertex ratio of the dataset, the time gap, and data density. The experimental results demonstrate that the Alpha Walk method generally achieves the lowest mean relative error and highest precision than other methods on large temporal graphs.

Outline. The paper is organized as follows. Section 2 provides background and definitions. Section 3 reviews related work. Section 4 presents our Monte Carlo sampling approach for approximating TKC. Section 5 offers a mathematical proof of our method's asymptotic consistency. Section 6 details our experimental evaluation, followed by the conclusion in Sect. 7.

2 Background

This section outlines the definitions of temporal graphs and Temporal Katz Centrality (TKC), and introduces the problem statement.

Definition 1 (Temporal graph). *A temporal graph is represented as $G_t = (V, E_t)$, where $E_t \subseteq V \times V \times T$ and T is the set of time stamps. An edge $e = (u, v, t) \in E_t$ indicates that there is a connection between nodes u and v at time t.*

Definition 2 (Katz Centrality). *Katz Centrality (KC) is a measure of centrality in a network, which takes into account the number of all possible walks between a pair of nodes, with an exponentially decaying weight for longer walks. For a node u in an ordinary graph $G = (V, E)$, the Katz Centrality is defined as:*

$$K_u = \sum_{k=1}^{\infty} \sum_{v \in V} \alpha^k (A^k)_{vu} \qquad (1)$$

where A is the adjacency matrix of the graph, α is a constant ($0 < \alpha < 1$) which ensures the convergence of the series, and k denotes the length of the walk. This centrality measure sums over all paths in the graph, giving more weight to shorter paths.

Definition 3 (Temporal Katz Centrality [5]). *For a node u, the Temporal Katz Centrality (TKC) at time t, denoted as $r_u(t)$, is defined as:*

$$r_u(t) := \sum_v \sum_{\substack{temporal \\ paths\ z}} \Phi(z,t) \qquad (2)$$

where $\Phi(z,t)$ is the temporally weighted sum of all walks from node v to u at time t. This centrality measure respects the temporal ordering of edges, with each walk's contribution diminishing over time based on when its edges were active.

Definition 4 (Time-Dependent Edge Transition Weights). *The weight of a temporal path z at time t, $\Phi(z,t)$, is calculated as the product of edge transition weights. For temporal walks where edges appeared at times $t_1, t_2, \ldots, t_j$, we define:*

$$\Phi(z,t) := \prod_{i=1}^{j} \phi(t_{i+1} - t_i), \qquad (3)$$

where $\phi(t_{i+1} - t_i)$ is the weight of walking at time t, and ϕ is the time-perceived weight function, reflecting the fact that the effect of the interaction diminishes over time.

Definition 5 (Weight Function Specification). *The weight function ϕ is applied to each edge in the path and is defined as an exponentially decaying function:*

$$\phi(t_{i+1} - t_i) = \beta \exp(-c(t_{i+1} - t_i)), \qquad (4)$$

where $t_{i+1} - t_i$ represents the time difference between consecutive interactions within the path, β is the basic weight, and c is the decay constant. The product of edge weight functions on the time path from node v to u is given by:

$$\Phi(z,t) = \beta^{|z|} \exp(-c(t - t_1)), \qquad (5)$$

where $|z|$ represents the length of the path.

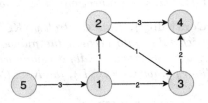

Fig. 1. Example Temporal Graph

Figure 1 we initialized each node centrality to zero and set the factor $\beta = 1.0$ with time decay weight $weight(t) = 0.5^t$. For example, the centrality score for node 4 is computed as follows:

1. At time $t = 2$, edge $(3, 4)$ is processed. The time difference for node 3 and node 4 is 2 since they were last activated at time 0. The decay weights for nodes 3 and 4 are $0.5^2 = 0.25$. Therefore, the updated centrality score for node 4 is:

$$r[4] = 1.0 \times (r[3] \times 0.25 + 1) = 1.0 \times (0 \times 0.25 + 1) = 1.0$$

2. At time $t = 3$, edge $(2, 4)$ is processed. The time difference for node 2 is 2 and for node 4 is 1. The decay weights for nodes 2 and 4 are $0.5^2 = 0.25$ and $0.5^1 = 0.5$, respectively. Therefore, the updated centrality score for node 4 is:

$$r[4] = 1.0 \times (r[2] \times 0.25 + 1) + r[4] \times 0.5$$
$$= 1.0 \times (0 \times 0.25 + 1) + 1.0 \times 0.5 = 2.75$$

Thus, the final TKC score for node 4 is 2.75, and for the other nodes are as follows: Node 1 is 1.0, Node 2 is 0.25, Node 3 is 1.0, and Node 5 is 0.0.

Definition 6 (Truncated Temporal Katz Centrality). *Truncated Temporal Katz Centrality (TTKC) is defined similarly to the temporal Katz centrality, but it restricts the length of the walks considered. This is done to limit the computational complexity and focus on shorter, more relevant paths. Formally, the TTKC is defined as:*

$$r_u^{(K)}(t) := \sum_{v} \sum_{\substack{temporal \\ paths\ z \\ of\ length \leq K}} \Phi(z, t) \tag{6}$$

where K is the maximum length of the paths considered. This restriction helps in managing the computational load while maintaining a reasonable approximation of the node's centrality in temporal graphs.

Problem Statement. This paper aims to approximate the Temporal Katz Centrality values for all nodes in a temporal graph.

3 Related Work

Temporal graph analysis has become crucial due to the need to understand the evolution of complex systems. Various centrality measures has been proposed to capture time-varying importance of nodes in temporal graphs.

Temporal Katz Centrality (TKC) [5] extends classical Katz centrality to temporal graphs by considering temporal random walks and interaction intervals. Efficient TKC algorithms exist for graph streams [2], but scalability in large graphs remains challenging. TATKC [24], a temporal neural graph, approximates TKC rankings using low-dimensional node representations. Betweenness centrality has also been adapted for temporal graphs. Efficient algorithms for temporal betweenness centrality [4,25] highlight its role in facilitating communication over time. Extensions of closeness centrality for temporal settings include algorithms for top-k temporal closeness centrality [15,16], utilizing pruning techniques to improve efficiency. PageRank extends to networks with a temporal dimension [20], incorporating available temporal information judiciously into the model and generalizing it through the concept of temporal random walks. However, those methods require extensive computational resources and hard to stretch on large-scale temporal graphs.

The practical relevance of temporal centrality measures is evident in case studies, such as identifying key influencers using TKC [10] and ranking nodes with approaches like SaPhyRa [21] and TimeSGN [23]. Despite progress, scalability and efficiency remain challenges in large-scale temporal graphs. The application of Temporal Katz Centrality (TKC) to Knowledge Graph (KG) can provide significant insights into the evolving importance of nodes [6,19,22]. Knowledge graphs typically contain large-scale data with complex relationships, and traditional methods may face challenges in terms of scalability and computational efficiency. This paper proposes a Monte Carlo sampling approach for approximating temporal Katz centrality, inspired by fast Monte Carlo algorithms for matrix functions in complex graphs [8].

4 Our Method

In this section, we present our approach to approximate Katz temporal centrality in temporal graphs using Monte Carlo sampling techniques. Our approach leverages random sampling strategies to estimate centrality scores in temporal graphs. We detail two specific sampling method, Simple Random Walk and Alpha Walk, follows by methodology for estimating temporal Katz centrality scores from these samples.

Monte Carlo sampling is a powerful randomized method for approximating Katz temporal centrality in temporal graphs. This method is particularly advantageous when direct computation is not possible due to the complexity of the graph and the need to account for the temporal influence of edges.

Simple Random Walk. In the Simple Random Walk (SRW) approach, each neighbor of a node has an equal probability of being selected during a random walk. This method ensures that all neighbors are equally likely to be chosen, making it a straightforward sampling technique.

For a given node i in the temporal graphs, the probability of selecting a neighbor j at time t is:

$$p_{ij}(t) = \frac{1}{|N(i)|}, \tag{7}$$

where $N(i)$ represents the neighbor set of node i and $|N(i)|$ is the number of neighbors.

Algorithm 1. Alpha Walk

Require: Temporal Edge $< v, u, t >$, initial node i, probability α
Ensure: $C_u(t)$
1: $v \leftarrow i$
2: **while** True **do**
3:    $r \leftarrow$ random number between 0 and 1
4:    **if** $r > \alpha$ **or** no neighbors available **then**
5:       **break**
6:    **end if**
7:    $u \leftarrow$ select neighbor of v uniformly at random
8:    $C_v(t) \leftarrow C_v(t_{vu})\phi(t - t_{vu})$
9:    $C_u(t) \leftarrow C_u(t)\phi(t - t_{vu}) + (C_v(t_{vu}) + 1)\phi(t - t_{vu})$
10:   $v \leftarrow u$
11: **end while**

To perform SRW, the process is as follows:
1. Select a starting node uniformly at random.
2. At each step of the walk, select the next node uniformly at random from the current node's neighbors.
3. Continue the walk until a predefined number of steps is reached or no valid neighbor is available.

The centrality scores are calculated by aggregating the contributions from all random walks performed using this SRW method.

Alpha Walk. In the Alpha Walk sampling approach, the next node in a random walk is selected based on a fixed probability α that determines whether the walk will continue or terminate.

For a given node i in the temporal graph, the probability of selecting a neighbor j at time t and continuing the walk is:

$$p_{\text{continue}} = \alpha, \qquad (8)$$

where α is a fixed probability value between 0 and 1 that controls the termination of the walk.

To perform the Alpha Walk sampling, the process is as follows:
1. Start at the initial node.
2. At each step, generate a random number r between 0 and 1.
3. If $r \leq \alpha$, select the next node uniformly at random from the current node's neighbors.
4. If $r > \alpha$ or no valid neighbor is available, terminate the walk.

The estimated centrality scores $C_u(t)$ are updated based on the time-decayed edge weights along the path traversed during the walk.

Computational Complexity. Our Monte Carlo approximation method for Temporal Katz Centrality (TKC) significantly reduces computational complexity compared to exact methods. By leveraging random walks, the overall complexity is

$O(N \cdot L)$, where N is the number of random walks and L is their average length. In the worst-case scenario, when considering the entire graph, the complexity could be as high as $O(n^2)$, where n represent the number of nodes in temporal graph [12].

Estimation of Centrality Scores. The centrality score for each node is estimated by averaging the contributions from multiple random walks:

$$\tilde{C}_u(t) = \frac{1}{M} \sum_{m=1}^{M} \sum_{n=1}^{L_m} \beta^{n-1} \phi(t - t_{mn}), \qquad (9)$$

where M is the number of random walks, L_m is the length of the m-th walk, β is a base influence parameter, and t_{mn} is the time of the n-th event in walk m.

5 Theoretical Analysis

We use the Law of Large Numbers (LLN) and the Central Limit Theorem (CLT) to prove that our results demonstrate *asymptotic consistency*. These theorems provide the mathematical basis ensuring that our approximation converges to the true centrality measure as the number of samples increases.

Theorem 1 (Monte Carlo Convergence Theorem). *Under certain conditions, the Monte Carlo sampling algorithm for Temporal Katz Centrality converges to the true centrality value $r_u(t)$.*

Proof. Assume N is the number of random walks, and let $X_{u,i}(t)$ represent the value of the centrality measure from the i-th random walk ending at node u at time t. The random walks are independent and identically distributed (i.i.d.) with finite mean and variance. The Monte Carlo estimate of node u's centrality is given by:

$$\hat{C}_u = \frac{1}{N} \sum_{i=1}^{N} X_{u,i}(t) \qquad (10)$$

As N increases, the average $\hat{C}_u$ becomes a more accurate representation of the expected centrality measure $E[X_{u,i}(t)]$. This is ensured by the Law of Large Numbers (LLN), which states:

$$\hat{C}_u \xrightarrow{p} E[X_{u,i}(t)] = r_u(t), \text{ as } N \to \infty \qquad (11)$$

The LLN ensures that:

$$\lim_{N \to \infty} P\left(\left|\hat{C}_u - r_u(t)\right| \geq \epsilon\right) = 0, \text{ for any } \epsilon > 0 \qquad (12)$$

This convergence guarantees that our estimates become accurate as the number of samples increases.

Table 1. Summary of Datasets

Dataset	|V|	|E|	|T|
rating	1,683	100,000	49,282
edit-bnwiktionary	9,803	36,040	35,538
facebook-wosn-wall	46,953	876,993	867,939
slashdot-threads	51,084	140,778	90,345
prosper-loans	89,270	3,394,979	1,259
edit-cswiki	1,030,520	12,998,902	12,464,926

Corollary 1. *The Central Limit Theorem (CLT) allows us to make probabilistic statements about the centrality estimate $\hat{C}_u$:*

$$\hat{C}_u \sim \mathcal{N}\left(E[C_u], \frac{\sigma_u^2}{N}\right) \tag{13}$$

This implies that the distribution of $\hat{C}_u$ approaches a normal distribution centered around $E[C_u]$ as N increases, enabling reliable statistical inferences about the centrality scores.

Thus, using the LLN and CLT, we have proven that our Monte Carlo sampling method for approximating Temporal Katz Centrality demonstrates asymptotic consistency.

6 Experiments

In this section, we present the experimental setup and results obtained from evaluating our proposed methods on real-world network datasets.

6.1 Experimental Setup

Hardware. This experiments are conducted on a standard computing environment with an Apple M1 Max CPU, 64 GB RAM and running in macOS. We implemented our methods in Python.

Datasets. We conducted our experiments using six real-world network datasets from the Network Data Repository [9]. The main datasets used in our experiments are summarized in Table 1. We preprocess the data by normalizing timestamps to the range of $[0, 1]$ using the minimum and maximum timestamps for scaling, ensuring uniformity across datasets.

Metrics. We employ three metrics to evaluate the performance of our approximation methods: Kendall Tau Coefficient, Relative Error, and Precision.

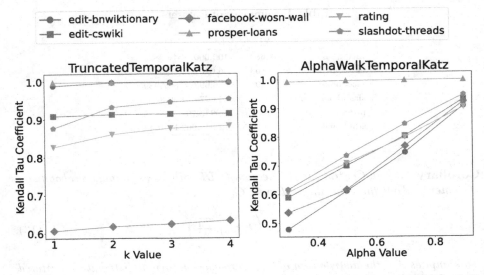

Fig. 2. Left: TTKC: Kendall Tau Coefficient vs. k Value. Right: Alpha Walk: Kendall Tau Coefficient vs. α Value.

6.2 Results and Analysis

Exp1 - Vary Parameter Selection. In this experiment, we aimed to determine the optimal parameter settings for TTKC and Alpha Walk by varying key parameters and observing their impact on performance. We ensured consistency and numerical stability across all methods by using common parameters.

Firstly, we used a normalization factor of 0.01 to scale edge weights uniformly across datasets. The Beta (β) value was set to 1.0 to consistently control the influence of longer paths. Additionally, an exponential weighter with a base of 0.5 was employed to assign decreasing weights to longer temporal paths.

For TTKC, we tested k values of $\{1, 2, 3, 4\}$. The results, as shown in Fig. 2, indicate that the Kendall Tau Coefficient generally increases with higher k values, achieving the highest coefficient at $k = 4$. For Alpha Walk, we varied the restart probability α with values $\{0.3, 0.5, 0.7, 0.9\}$. The runtime for both TTKC with $k = 4$ and Alpha Walk with $\alpha = 0.9$ were nearly identical, demonstrating that both configurations offer similar computational efficiency while providing optimal performance. The runtime for both TTKC with $k = 4$ and Alpha Walk with $\alpha = 0.9$ were nearly identical, demonstrating that both configurations offer similar computational efficiency. Therefore, we chose these two parameter settings for further analysis, as they provide the best balance between performance and runtime efficiency.

Exp2: Mean Relative Error. Table 2 presents the mean relative error of the three approximation methods across different datasets. Overall, Alpha Walk demonstrates the lowest mean relative error across most datasets, indicating its superior accuracy in approximating centrality values compared to TTKC and SRW methods.

Table 2. Mean Relative Error Comparison

Dataset	TTKC	Alpha Walk	SRW
rating	0.7897	**0.6057**	0.9009
edit-bnwiktionary	0.1019	**0.1002**	0.5883
facebook-wosn-wall	**0.2330**	0.3105	0.6525
slashdot-threads	**0.6634**	0.6746	0.8480
prosper-loans	0.1454	**0.1354**	0.5852
edit-cswiki	**0.9223**	0.9287	0.9612

Table 3. Precision for Top-K Query (Top-1%, Top-5%)

Dataset	Top-1%			Top-5%		
	TTKC	Alpha Walk	SRW	TTKC	Alpha Walk	SRW
rating	0.5	**0.75**	0.375	0.5476	**0.7976**	0.4286
edit-bnwiktionary	**0.9691**	0.5773	0.2062	**0.9939**	0.7131	0.3770
facebook-wosn-wall	0.0085	**0.9275**	0.0362	0.0575	**0.9352**	0.0716
slashdot-threads	0.2882	**0.8667**	0.4902	0.5556	**0.9045**	0.6492
prosper-loans	**0.9821**	0.9776	0.9327	0.9982	**0.9984**	0.9857
edit-cswiki	0.4304	**0.8267**	0.4557	0.6090	**0.8594**	0.5066

Exp3: Precision for Top-K Query. Table 3 shows the precision comparison of the three approximation methods. Precision is defined as the proportion of top-1% and top-5% nodes identified by the approximation methods that match the top-1% and top-5% nodes identified by the exact centrality measure. From the result, we can summarize that the Alpha Walk method generally outperforms the TTKC and SRW methods in identifying top-1% and top-5% central nodes across different datasets.

Exp 4: Vary Scales of Datasets. We analyzed the performance of the approximation methods on small-scale, middle-scale, and large-scale datasets, considering factors such as edge-to-vertex ratio, temporal gaps, and data density. Table 4 presents the accuracy comparison (Top-1%) of the methods across these different scales of datasets.

Small-Scale Datasets: For small-scale datasets, we observed distinct performance trends. In 'edit-bnwiktionary', where the number of edges is not significantly

Table 4. Accuracy Comparison across Different Dataset Scales (Top-1%)

Dataset	TTKC	Alpha Walk	SRW
rating	0.5	**0.75**	0.375
edit-bnwiktionary	**0.9691**	0.5773	0.2062
facebook-wosn-wall	0.0085	**0.9275**	0.0362
slashdot-threads	0.2882	**0.8667**	0.4902
prosper-loans	**0.9821**	0.9776	0.9327
edit-cswiki	0.4304	**0.8267**	0.4557

higher than the number of vertices, the TTKC method approximates the true centrality values very closely. In contrast, for 'rating', where the edges vastly outnumber the vertices, Alpha Walk performs better than the TTKC method.

Middle-Scale Datasets: For datasets 'facebook-wosn-wall' and 'slashdot-threads', the performance is influenced by the temporal gaps within the data. The TTKC struggles with datasets that have significant temporal gaps. In these cases, the Alpha Walk method provides a more accurate approximation.

Large-Scale Datasets: For large-scale datasets, the density of the data plays a critical role. Despite 'prosper-loans' having a much larger number of edges compared to vertices, its short temporal intervals make it a dense dataset. Here, both TTKC and Alpha Walk achieve near-perfect accuracy. Conversely, 'edit-cswiki' with its sparse and vast temporal intervals shows better performance with the Alpha Walk method over TTKC.

7 Conclusion

In this paper, we studied Monte Carlo sampling methods for approximating TKC in large temporal graphs. Our methods leverage random sampling strategies including which SRW and Alpha Walk to efficiently estimate centrality scores. We provides mathematical proofs demonstrating the asymptotic consistency of our approach. Extensive experimental evaluation on real-world network datasets showed Alpha Walk demonstrated superior performance in sparse datasets with significant temporal gaps, highlighting its robustness and effectiveness. Future work will focus on developing adaptive sampling methods that adjust the sampling strategy based on the temporal graph. Additionally, we aim to explore more statistical techniques to further improve accuracy and efficiency.

References

1. Bauer, W.F.: The Monte Carlo method. J. Soc. Ind. Appl. Math. **6**(4), 438–451 (1958)
2. Benczúr, A., Béres, F., Kelen, D., Pálovics, R.: Tutorial on graph stream analytics. In: Proceedings of the 15th ACM International Conference on Distributed and Event-Based Systems, DEBS 2021, pp. 168–171 (2021)
3. Bonifati, A., Özsu, M.T., Tian, Y., Voigt, H., Yu, W., Zhang, W.: The future of graph analytics. In: SIGMOD/PODS 2024, pp. 544–545. Association for Computing Machinery, New York (2024)
4. Buß, S., Molter, H., Niedermeier, R., Rymar, M.: Algorithmic aspects of temporal betweenness. In: Proceedings of the 26th ACM SIGKDD International Conference on Knowledge Discovery and Data Mining, KDD 2020, pp. 2084–2092 (2020)
5. Béres, F., Pálovics, R., Oláh, A., Benczúr, A.A.: Temporal walk based centrality metric for graph streams. Appl. Netw. Sci. **3**(1), 1–26 (2018). https://doi.org/10.1007/s41109-018-0080-5

6. Cai, B., Xiang, Y., Gao, L., Zhang, H., Li, Y., Li, J.: Temporal knowledge graph completion: a survey. In: Proceedings of the Thirty-Second International Joint Conference on Artificial Intelligence, IJCAI-2023, International Joint Conferences on Artificial Intelligence Organization (2023)
7. Chen, Z., Yuan, L., Lin, X., Qin, L., Zhang, W.: Balanced clique computation in signed networks: concepts and algorithms. IEEE Trans. Knowl. Data Eng. **35**(11), 11079–11092 (2023)
8. Guidotti, N.L., Acebrón, J.A., Monteiro, J.: A fast Monte Carlo algorithm for evaluating matrix functions with application in complex networks. J. Sci. Comput. **99**, 41 (2024)
9. Kunegis, J.: KONECT – the Koblenz network collection. In: WWW 2013 Companion: Proceedings of the International Conference on World Wide Web Companion, pp. 1343–1350 (2013)
10. Laflin, P., Mantzaris, A.V., Ainley, F., Otley, A., Grindrod, P., Higham, D.J.: Discovering and validating influence in a dynamic online social network. Soc. Netw. Anal. Min. **3**(4), 1311–1323 (2013). https://doi.org/10.1007/s13278-013-0143-7
11. Liao, M., Li, R.H., Dai, Q., Chen, H., Qin, H., Wang, G.: Efficient personalized pagerank computation: the power of variance-reduced Monte Carlo approaches. Proc. ACM Manag. Data **1**(2) (2023)
12. Lin, M., Li, W., Song, L.J., Nguyen, C.T., Wang, X., Lu, S.: SAKE: estimating Katz centrality based on sampling for large-scale social networks. ACM Trans. Knowl. Discov. Data **15**(4), Article no. 66 (2021). https://doi.org/10.1145/3441646
13. Ma, C., Yuan, L., Han, L., Ding, M., Bhaskar, R., Li, J.: Data level privacy preserving: a stochastic perturbation approach based on differential privacy. IEEE Trans. Knowl. Data Eng. **35**(4), 3619–3631 (2023). https://doi.org/10.1109/TKDE.2021.3137047
14. Nathan, E., Bader, D.A.: A dynamic algorithm for updating Katz centrality in graphs. In: Proceedings of the 2017 IEEE/ACM International Conference on Advances in Social Networks Analysis and Mining 2017, ASONAM 2017 (2017)
15. Oettershagen, L., Mutzel, P.: Efficient top-k temporal closeness calculation in temporal networks. In: 2020 IEEE International Conference on Data Mining (ICDM), pp. 402–411 (2020)
16. Oettershagen, L., Mutzel, P.: Computing top-k temporal closeness in temporal networks. Knowl. Inf. Syst. **64** (2022)
17. Ogura, M., Preciado, V.M.: Katz centrality of Markovian temporal networks: analysis and optimization. In: 2017 American Control Conference (ACC), pp. 5001–5006 (2017). https://doi.org/10.23919/ACC.2017.7963730
18. Pan, R.K., Saramäki, J.: Path lengths, correlations, and centrality in temporal networks. Phys. Rev. E **84**, 016105 (2011)
19. Peng, C., Xia, F., Naseriparsa, M., Osborne, F.: Knowledge graphs: opportunities and challenges. Artif. Intell. Rev. **56**, 13071–13102 (2023)
20. Rozenshtein, P., Gionis, A.: Temporal pagerank. In: Frasconi, P., Landwehr, N., Manco, G., Vreeken, J. (eds.) ECML PKDD 2016. LNCS, vol. 9852, pp. 674–689. Springer, Cham (2016). https://doi.org/10.1007/978-3-319-46227-1_42
21. Thai, P., Thai, M.T., Vu, T., Dinh, T.N.: SaPHyRa: a learning theory approach to ranking nodes in large networks (2022)
22. Wu, Y., Xu, Y., Zhang, W., Xu, X., Zhang, Y.: Query2GMM: learning representation with Gaussian mixture model for reasoning over knowledge graphs. In: Proceedings of the ACM on Web Conference 2024, WWW 2024, pp. 2149–2158. Association for Computing Machinery, New York (2024)

23. Xu, Y., Zhang, W., Zhang, Y., Orlowska, M., Lin, X.: TimeSGN: scalable and effective temporal graph neural network (2024)
24. Zhang, T., Fang, J., Yang, Z., Cao, B., Fan, J.: Tatkc: A temporal graph neural network for fast approximate temporal katz centrality ranking. In: Proceedings of the ACM on Web Conference 2024. p. 527-538. WWW '24 (2024)
25. Zhang, T., et al.: Efficient exact and approximate betweenness centrality computation for temporal graphs. In: Proceedings of the ACM on Web Conference 2024, WWW 2024, pp. 2395–2406 (2024)
26. Zhu, C., Chen, M., Fan, C., Cheng, G., Zhan, Y.: Learning from history: modeling temporal knowledge graphs with sequential copy-generation networks (2021)

# LLM for Uniform Information Extraction Using Multi-task Learning Optimization

Ying Li, Zhen Tan[✉], and Weidong Xiao

National Key Laboratory of Information Systems Engineering, National University of Defense Technology, Kaifu, China
tanzhen08a@nudt.edu.cn

Abstract. With the increasing maturity of deep learning technology, large language models have shown excellent performance in the field of natural language processing, but the performance of information extraction is yet to be further improved. In this paper, we use 7B Llama-2 as a base model training to obtain a large language model capable of using natural language to guide information extraction tasks, which can solve the challenges faced by traditional information extraction methods, and also innovatively use multi-task learning optimization to improve the performance of the large model. We performed large-scale pre-training and instruction tuning on the big model LLama-2. Based on the high-quality and richly typed training data automatically constructed by ChatGPT, remote supervision, and other algorithms, which contains a total of 1 million entities, relations, and events, we designed corresponding English templates for instruction tuning. Second, we performed supervised fine-tuning of the model using manually labeled high-quality training sets. To further improve the model performance, we innovatively adopt a multi-task learning optimization strategy—GradNorm, which can dynamically adjust the weights of different tasks, thus balancing the losses among tasks during the training process and reducing the overall training loss. After information extraction experiments, our model is compared with other models to test the performance of uniform information extraction for large models, and the experimental results show that our model performs well.

Keywords: Large Language Model · Uniform Information Extraction · Instruction Fine-tuning · Multi-task Learning Optimization

1 Introduction

In recent years, Large Language Models (LLMs) such as GPT-3 [1], PaLM [2] and Llama [3] have made significant progress in natural language tasks. Ye et al. [4] revealed the performance gap of general LLMs in information extraction tasks. Further, the capabilities of large language models in text comprehension and generation have driven NLP development and stimulated interest in generative information extraction methods [5]. Qi, Guo and Sainz et al. showed [6] that generating structural information using LLMs is more practical than traditional methods [7].

The LLM performs well in tasks such as NER [8], RE [9], and EE [10], and can also model multi-information extraction tasks in a generalized format [11], with command hints to ensure performance consistency. Paolini et al. reveal LLM's ability to generalize [12], to learn from IE data, and to extract information in low/zero-sample scenarios, either through context or commands. All these studies show that LLMs based on instruction tuning are promising in the field of uniform information extraction.

As the amount of data grows and diversifies, it becomes challenging to deal with information from different sources, formats, and complexities. Among them, how to handle information extraction tasks of different types and complexities and ensure their balanced training and learning is key. The multi-task learning optimization method GradNorm [13] provides a powerful tool to solve this problem.

In this paper, we study the UIE-LLM and improve the accuracy of the LLM through pre-training, Instruction Tuning, supervised fine-tuning and multi-task learning optimization. First, the LLama-2 grand model is pre-trained to enhance its ability to handle complex tasks. Then, instruction tuning is performed based on high-quality training data to ensure that the model accurately executes information extraction instructions. Next, supervised fine-tuning of the model is performed using manually labeled data to cover a variety of information extraction scenarios. The GradNorm multi-task learning optimization strategy is used to balance the task loss and improve the model performance. The experiments verify the good performance of the LLM in uniform information extraction, while the ablation experiments conclude that supervised fine-tuning has the greatest impact on the performance of the LLM, followed by instruction tuning, and multi-task learning optimization has the least.

Overall, the main contributions of this paper include:

- We use 7B Llama-2 as a base model training to obtain a large language model that can utilize natural language to guide the information extraction task, and the experimental results show that the large language model performs well after multiple rounds of training.
- We innovatively use the multi-task learning optimization method GradNorm to further improve the performance of the large language model, which provides a new idea for large language model performance improvement.
- The effects of instruction tuning, supervised fine-tuning and multi-task learning optimization on the performance of the large language model are compared, and the experimental results show that supervised fine-tuning has the largest impact, instruction tuning is the second largest, and multi-task learning optimization is the smallest.

2 Methodology

2.1 Supervised Pre-training

Supervised pre-training uses a large labeled dataset related to the final task to train the grand model and enhance the generalization ability. In this paper, the large language model uses the dataset constructed by Zixuan Li et al. [14]. Containing more than 2 million entities, 1 million relationships, and 300,000 events, which is based on the KELM corpus and is rich in content. Supervised pre-training is the basis of the big language model, aiming at extracting linguistic knowledge from text and laying the cornerstone for the unified information extraction large language model.

2.2 Instruction Tuning

Task instructions guide the model to extract information from the text and generate accurate output. It serves as a bridge connecting the input text and structured output, and we have designed instruction templates for different extraction tasks, which are described below one by one.

(1) Relationship extraction (RE) tasks

Table 1. Relationship Extraction Task Instructions.

Instruction_type	Instruction	Output format
zero-shot	Given a phrase that describes the relationship between two words, extract the words and the lexical relationship between them	relation1:word1,word2; relation2: word3, word4
zero-shot	Find the phrases in the following sentences that have a given relationship	relation1:word1,word2; relation2: word3, word4
zero-shot	Given a sentence, please extract the subject and object containing a certain relation in the sentence according to the following relation types	relation1:word1,word2; relation2: word3, word4
zero-shot	Given options, please tell me the relationships of all the listed entity pairs	relation1:word1,word2; relation2: word3, word4

As shown in the table, we designed three instructions for the relation extraction task. The prompt types are all zero-sample, and the output templates are all: relationship 1: entity 1, entity 2; relationship 2: entity 3, entity 4 (Table 1).

(2) Entity extraction (NER) tasks

Table 2. Entity Extraction Task Order Descriptions.

Instruction_type	Instruction	Output format
zero-shot	Please list all entity words in the text that fit the category	type1: word1; type2: word2
zero-shot	Please find all the entity words associated with the category in the given text	type1: word1; type2: word2
zero-shot	Please tell me all the entity words in the text that belong to a given category	type1: word1; type2: word2
zero-shot	Given options, please tell me the categories of all the listed entity words	type1: word1; type2: word2
zero-shot	Please list all entity words in the text that fit the category	word1, word2

(*continued*)

Table 2. (*continued*)

Instruction_type	Instruction	Output format
zero-shot	Please list all entity pairs containing a certain relationship in the given options	word1, word2; word3, word4

As shown in the table, we designed three instructions for the relation extraction task. The prompt types are all zero-sample, and the output templates are all: relationship 1: entity 1, entity 2; relationship 2: entity 3, entity 4 (Table 2).

(3) Event Extraction (EE) Task

Table 3. Event Extraction Task Instruction Description.

Instruction_type	Instruction
zero-shot	Locate the role in the text that participated in the event based on the event type and return it in the event list
zero-shot	Extract the event information in the text and return them in the event list

As shown in the table, for the event extraction task we designed two instructions. The cue types are both zero samples and the outputs are both returned in the event list (Table 3).

2.3 Supervision of Fine-Tuning

Supervised fine-tuning is further training of the large language model on a manually labeled dataset. The dataset used for this step of training is of high quality and can provide a fine-tuning of the large language model. Through this step of training, our large language model can understand the essence of information extraction more deeply and further improve its accuracy and generalization ability of information extraction. This step of training is like a careful polishing and sculpting of the large language model, so that it has a more powerful information extraction ability.

2.4 Multitask Learning Optimization—GradNorm

The core idea of GradNorm is to optimize the performance of multi-task learning by balancing the gradient. Different tasks of information extraction have different loss functions and gradient characteristics. During the training process, if the gradient of a task is significantly larger, it will dominate the model update. GradNorm introduces the concept of gradient paradigm, according to which the weight of each task is adjusted to avoid the model over-learning a task.

The following formulas are the core of Gradient Normalization and aim to represent the Gradient Loss as a function of the loss weights [13].

(1) First defines some variables to measure the magnitude of the task's loss, where:

$$G_W^{(i)}(t) = \|\nabla W w_i(t) L_i(t)\|_2, \tag{1}$$

$$\overline{G_W}(t) = E_{task} G_W^{(i)}(t). \tag{2}$$

- W is a subset of the parameters of the multitask learning network, the last layer of the bottom part of the network share;
- $G_W^{(i)}(t)$ is the value normalized by the gradient of task i, which is the product of the weight $w_i(t)$ of task i and loss $L_i(t)$ The L2 paradigm of the gradient for the parameter W, $G_W^{(i)}(t)$ can measure the magnitude of loss for a given task;
- $\overline{G_W}(t)$ is the mean of the global gradient normalization.

(2) Secondly, some variables are defined to measure the learning speed of the task:

$$\tilde{L}_i(t) = \frac{L_i(t)}{L_i(0)}, \tag{3}$$

$$r_i(t) = \frac{\tilde{L}_i(t)}{E_{task}\left[\tilde{L}_i(t)\right]}. \tag{4}$$

- $L_i(0)$ and $L_i(t)$ represent the loss at step 0 and step t of sub-task i, respectively; $\tilde{L}_i(t)$ measures the training speed of the inverse of task i to a certain extent, and the larger $\tilde{L}_i(t)$ is, the slower the network training;
- $E_{task}[\tilde{L}_i(t)]$ denotes the expectation of the reverse training speed for each task;
- $r_i(t)$ is the relative inverse training speed of the task.

(3) Finally, Gradient Loss (GL) is:

$$L_{grad}(t; w_i(t)) = \sum_i \left| G_W^{(i)}(t) - \overline{G_W}(t)[r_i(t)]^\alpha \right|_1. \tag{5}$$

- $G_W^{(i)}(t) r_i(t)$ denotes the ideal gradient-normalized value;
- $G_W^{(i)}(t)$ and $\overline{G_W}(t)$ are used to balance the magnitude of the loss function for each task;
- $r_i(t)$ is used to balance the speed at which different tasks are trained.

(4) After calculating the Gradient Loss, $w_i(t)$ is updated (GL means Gradient Loss) by the following function:

$$w_i(t+1) = w_i(t) + \lambda * Gradient(GL, w_i(t)). \tag{6}$$

In the large language model training, we design a trainable weight vector (may also be a matrix) w, w to multiply the loss (may also be the loss multiplied by w) and then sum to get a L, this L is optimized by the GradNorm by the loss, with this loss and then go to update the model parameters, after repeated updating the GradNorm can make the loss L minimum.

3 Experiments

In this section, a series of experiments are conducted and relevant experimental results are obtained, which contain the following three main parts: pre-training and instruction tuning only, the former base plus supervised fine-tuning and multi-task learning optimization, and ablation experiments. The ablation experiment compares the effects of instruction tuning, supervised fine-tuning and multi-task learning optimization on the model performance, and also compares with other models to verify the effectiveness of the model. LLMma-2 of 7B is chosen as the base model because previous research [3] demonstrated that the LLama-2 model provides rich background knowledge and contextual comprehension for the UIE task, and the task will be more efficient. Experimental details are described in the following sections.

3.1 Dataset Introduction

(1) Supervised pre-training dataset

This experiment uses the dataset constructed by Zixuan Li et al. [14], which contains 2 million entities, 1 million relations and 300,000 events, based on the KELM corpus. Cleaning the data includes filtering untyped entities, removing "Wikimedia Disambiguation Page" samples, bracketed entities and events with roles such as "of", "follows" and so on. Follows", etc. (Table 4).

Table 4. Supervised pre-training dataset

Task	Data Name	#Types	#Instance	#Tokens	Disk size
NER	KELM	19,009	2,019,990	0.26B	1.15 GB
RE	KELM	810	1,191,199	0.13B	0.54 GB
EE	KELM	499	296,403	0.03B	0.11 GB

(2) Instruction Tuning Dataset

The instruction tuning dataset used in this experiment is a high-quality and richly typed training data containing a total of 1 million entities, relations and events automatically based on the algorithms of ChatGPT, remote supervision and so on (Table 5).

(3) Supervised fine-tuning dataset

Supervised fine-tuning dataset compiled by Xiao WangF et al. [15], containing 32 publicly available datasets covering NER, RE, and EE, from scientific, medical, and social domains, containing news and wiki data, in a unified text-to-text format with manual annotation (Fig. 1).

Table 5. Instruction Tuning Dataset

Task	Data Name	#Types	#Instance	#Tokens	Disk size
NER	UniversalNER	12,072	127,839	0.19B	0.96GB
RE	InstructIE	131	327,984	0.62B	2.61GB
ED	LSEE	20	415,353	0.26B	1.03GB
EAE	LSEE	20	211,635	0.10B	0.50GB

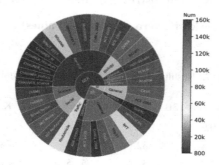

Fig. 1. Data Set of Supervised fine-tuning

The dataset processing includes (1) Uniform label naming to solve the problem of inconsistent labels across tasks; (2) Label conversion to convert complex labels to a natural and understandable format, such as "people_person_place_of_birth" to "place of birth"; (3) Conversion to text-to-text format to ensure consistent representation of input and output.

3.2 Experimental Environment

(1) **Hardware and software environment**

Hard disk: SSD 1.8T
Memory: DDR4 1.0T
GPU: A100-SXM4-40GB * 8
CPU: AMD EPYC 7742 64-Core Processor * 2
OS: Red Hat Enterprise Linux 8.6 (Ootpa)

(2) **Base model and training method**

Base model: llama2
Training method: LoRA fine-tuning (Table 6)

Table 6. Training hyperparameter settings associated with LoRA

Name	Desc	Value
lora_r	The dimension of the LoRA low-rank matrix	64
lora_alpha	Scaling factors for LoRA low-rank matrices	64
lora_dropout	LoRA Layer Discard Rate	0.1

3.3 Results of Pre-training and Instruction Tuning Experiments

(1) **Results of the NER task with zero samples (F1 values)**

Table 7. Results of the NER task with zero samples

Dataset	Vicuna-7B	Vicuna-13B	ChatGPT	UniNER-7B	KnowCoder-7B	Ours
Movie	6.0	0.9	5.3	42.4	50.0	**53.4**
Rest	5.3	0.4	32.8	31.7	**48.2**	47.3
AI	12.8	22.7	52.4	53.5	60.3	**62.1**
Litera	16.1	22.7	39.8	59.4	61.1	**61.5**
Music	17.0	26.6	66.6	65.0	**70.0**	67.9
Politics	20.5	27.2	68.5	60.8	72.2	**75.8**
Science	13.0	22.0	**67.0**	61.1	59.1	60.8

With zero samples, our model generally performs well in the entity extraction task, reaching the highest F1 value for some datasets, but overall still needs to further improve its performance through supervised fine-tuning (Table 7).

(2) **Results of RE task and ED task with zero samples (F1 values)**

Table 8. Results of RE task and ED task with zero samples

Dataset	Ouyang et al. 2022 [109]	Sainz et al. 2023 [110]	KnowCoder-7B	Ours
GIDS(RE)	9.9	-	25.5	**28.7**
CASIE(ED)	-	59.3	58.2	**62.5**

With zero samples, our model obtains the highest F1 values in both relation extraction and event trigger word detection, but the overall performance still needs to be improved by supervised fine-tuning (Table 8).

3.4 Supervised Fine-Tuning and Multi-task Learning to Optimize Experimental Results

(1) Results of the NER task

Table 9. NER mission experiment results

Dataset	InstructUIE	YAYI-UIE	Knowcoder-7B	Ours
ACE05	**86.66**	81.78	86.10	86.04
AnatEM	**90.89**	76.54	86.40	86.29
Broad Twitter	83.14	**83.52**	78.30	81.02
CoNLL03	92.94	**96.77**	95.10	96.09
FabNER	76.20	72.63	**82.90**	80.25
FindVehicle	89.47	98.47	**99.40**	98.46
GENIA	74.71	75.21	**76.70**	76.63
Movie	89.01	70.14	**90.60**	88.21
Rest	**82.55**	79.38	81.30	82.78
MultiNERD	92.32	88.42	**96.10**	93.81
OntoNotes5	**90.19**	87.04	88.20	89.61
bc2gm	**85.16**	82.05	82.00	83.58
bc5cdr	**89.59**	83.67	89.30	89.13
ncbi	**90.23**	87.29	83.80	84.00
Average	86.65	83.07	**86.87**	86.85

For the NER task, our model has competitive F1 scores on most datasets, with average F1 similar to the highest value, and exhibits better generalization, robustness and stability (Table 9).

(2) RE mission results

Table 10. Results of the RE Task Experiment

Dataset	InstructUIE	YAYIUIE	Knowcoder7B	Ours
ADE corpus	82.31	84.14	**84.30**	84.28
CoNLL2004	78.48	79.73	73.30	**79.88**
GIDS	**81.98**	72.36	78.00	81.90

(*continued*)

Table 10. (*continued*)

Dataset	InstructUIE	YAYIUIE	Knowcoder7B	Ours
kbp37	36.14	59.35	**73.20**	73.18
NYT	90.47	89.97	93.70	**93.72**
NYT11 HRL	56.06	**57.53**	-	57.45
SciERC	45.15	40.94	40.00	**45.22**
semeval RE	73.23	61.02	66.30	**73.83**
Avg	67.98	68.13	71.70	**73.68**

For the RE task, our model obtains the highest F1 on half of the datasets, and the other performances are excellent, with the highest average F1 value, showing good generalization, robustness and stability (Table 10).

(3) **EE mission results**

Table 11. EE task experiment results

Dataset	InstructUIE	YAYI-UIE	Knowcoder	Ours
ACE05-EE	**77.13**	65.00	74.20	76.06
ACE05-EAE	72.94	62.71	70.30	**73.08**
Avg	**75.04**	63.86	72.25	74.57

For the EE task, our model obtains the highest F1 on event argument extraction, and performs well on event detection, with an average F1 close to the highest value and more stable, showing good generalization, robustness and stability (Table 11).

3.5 Ablation Experiment

We evaluate the contribution of the three components of supervised fine-tuning, instruction tuning, and multi-task learning optimization to a unified information extraction framework based on a large language model through ablation experiments, the results of which are presented below (Table 12).

Analyzing the experimental results, the model performs differently in the named entity, relation, event and its argument extraction tasks. The absence of supervised fine-tuning results in a significant decrease in the model F1 value, indicating that supervised fine-tuning is crucial for generalization performance. Missing instruction tuning has less impact but is indispensable. Multi-task learning optimization contributes less to large language model performance. It is concluded that supervised fine-tuning contributes the most to model generalization, followed by instruction tuning, and multitask learning optimization has the least impact. These findings inform future work in model optimization and performance improvement.

Table 12. Results of ablation experiments

	NER.avg	RE.avg	EE.avg	EAE.avg
Ours	**86.85**	**73.68**	**76.06**	**73.08**
w/o SFT	81.19	67.98	71.68	66.46
w/o Instr.Tuning	83.59	68.98	72.41	71.72
w/o MultiTask	84.20	72.31	75.67	72.93

4 Related Work

4.1 Instruction Tuning

Large language model instruction tuning improves training efficiency and performance by optimizing the received instructions to reduce training time and cost and obtain better results. Methods include cue word engineering, model fine-tuning, etc. Cue word engineering guides the model to better understand and accomplish the target task by designing high-quality cue words.

4.2 Instruction Tuning

Multi-task learning learns multiple related tasks simultaneously, sharing representations or parameters to improve efficiency and performance. Optimization is achieved by fusing the loss functions of the subtasks with the goal of minimizing the sum of weighted losses. The key is how to improve the performance of the model in handling multi-tasks through strategies and techniques.

4.3 Information Extraction

Information Extraction (IE) extracts event or factual information such as entities, relationships and events from text for automatic classification, extraction and reconstruction of massive content. It is widely used in knowledge graphs, information retrieval, question and answer systems, and other domains. The Unified Information Extraction task includes several key subtasks that work together to achieve information extraction and organization of unstructured text.

5 Conclusion

In this paper, we use 7B LLama-2 as a base model, and after pre-training and instruction tuning, supervised fine-tuning, and multi-task learning optimization, we obtain a large language model capable of guiding information extraction tasks using natural language. We innovatively introduced multi-task learning optimization to further improve the performance of the large language model for unified information extraction tasks. Also after a series of experiments comparing the effects of instruction tuning, supervised fine-tuning and multi-task learning optimization on the performance of the large language model, the experimental results show that supervised fine-tuning has the largest impact, instruction tuning is the second largest, and multi-task learning optimization is the smallest. Our model is also compared with other models, and the experiments show that our model is good at uniform information extraction and achieves the highest F1 value on many datasets. Although the performance improvement of multi-task learning optimization for uniform information extraction for large language models is small, it provides a new idea for improving the performance of large language models for further research.

Acknowledgements. We sincerely thank the reviewers for their valuable comments and suggestions. This work was partially supported by National Key R&D Program of China No. 2022YFB3103600, NSFC under grants Nos. U23A20296, 62272469, 72371245. The Science and Technology Innovation Program of Hunan Province No. 2023RC1007.

References

1. Brown, T.B., Mann, B., et al.: Language models are few-shot learners. Advances in Neural Information Processing Systems 33, pp. 1877–1901 (2020)
2. Chowdhery, A., Narang, S., et al.: PaLM: Scaling Language Modelling with Pathways. arXiv preprint arXiv:2204.02349 (2022)
3. Lu, J., Zhang, H., Li, S., et al.: LLaMA-adapter V2: parameter-efficient visual instruction model. J. Artif. Intell. Res. (2022)
4. Ye, J., Chen, X., Xu, N., et al.: A comprehensive capability analysis of GPT-3 and GPT-3.5 series models. arXiv preprint arXiv:2303.10420 (2023)
5. Chen, W., Zhao, L., Luo, P., et al.: HEProto: a hierarchical enhancing protonet based on multi-task learning for few-shot named entity recognition. In: The 32nd ACM International Conference on Information and Knowledge Management, Birmingham, UK (2023a)
6. Sainz, O., García-Ferrero, I., Agerri, R., et al.: GoLLIE: annotation guidelines improve zero-shot information-extraction. arXiv preprint arXiv:2310.03668 2023
7. Lou, J., Lu, Y., Dai, D., et al.: Universal information extraction as unified semantic matching. In: The Thirty-Seventh AAAI Conference on Artificial Intelligence, Washington, DC, USA (2023)
8. Yuan, S., Yang, D., Liang, J., et al.: Generative entity typing with curriculum learning. In: The 2022 Conference on Empirical Methods in Natural Language Processing, Abu Dhabi, UAE (2022)
9. Wan, Z.: GPT-RE: in-context learning for relation extraction using large language models. In: Proceedings of the 2023 Conference on Empirical Methods in Natural Language Processing, Singapore (2023)
10. Wang, X., Li, S., Ji, H.: Code4Struct: code generation for few-shot event structure prediction. In: The 61st Annual Meeting of the Association for Computational Linguistics, vol. 1: Long Papers, Toronto, Canada (2023d)
11. Lu, Y., Liu, Q., Dai, D., et al.: Unified structure generation for universal information extraction. In: The 60th Annual Meeting of the Association for Computational Linguistics, vol. 1: Long Papers, Dublin, Ireland (2022)
12. Paolini, G., Athiwaratkun, B., Krone, J., et al.: Structured prediction as translation between augmented natural languages. In: The 9th International Conference on Learning Representations, Austrian (2021)

13. Chen, Z., Badrinarayanan, V., Lee, C.-Y., et al.: GradNorm: Gradient Normalization for Adaptive Loss Balancing in Deep Multitask Networks (2018)
14. Li, Z., Zeng, Y., Zuo, Y., et al.: KnowCoder: Coding Structured Knowledge into LLMs for Universal Information Extraction (2024)
15. Wang X., Zhou, W., Zu, C., et al.: InstructUIE: Multi-task Instruction Tuning for Unified Information Extraction (2023)

ASM: Adaptive Subgraph Matching via Efficient Compression and Label Filter

Yanfeng Chai[1(✉)], Jiashu Li[1], Qiang Zhang[2], Jiake Ge[3], and Xin Wang[3(✉)]

[1] Taiyuan University of Science and Technology, Taiyuan 030024, China
yfchai@tyust.edu.cn
[2] North University of China, Taiyuan 030051, China
[3] Tianjin University, Tianjin 300350, China

Abstract. How to efficiently get subgraphs that match the given query graph on large-scale graphs has become a hot topic in both academia and industry. Subgraph matching, as an important research direction of graph algorithms, realizes the basic operation for efficient queries on graph data, and its essence is the subgraph isomorphism problem, which is proved to be an np-complete problem. The filtering phase in the process of subgraph matching is especially critical because it directly affects the efficiency of the overall algorithm. An excellent filtering mechanism can screen out the eligible candidate nodes in a shorter time, thus saving a lot of computing time for the whole process. By analyzing the existing methods for handling the subgraph matching problem, we find that the following problems mainly exist in the existing methods in the filtering stage: (1) the problem of repeated enumeration of equivalent nodes; (2) Incomplete filtering problems with existing structures. These will lead to a large number of redundant validation issues during the validation phase. Then (3) we propose an Adaptive Subgraph Matching (ASM) mechanism to address the aforementioned shortcomings by efficient Compressed Graph Nodes (CGN) and a Label Count Filter (LCF) algorithm to improve the performance. The experiments show that our approach outperforms state-of-the-art subgraph search and matching algorithms by several orders of magnitude in terms of query processing time.

Keywords: Subgraph matching · Subgraph isomorphism · Reducing verification cost

1 Introduction

With the expanding influence of the graph data structure in several modern application domains such as social network analysis [1], bioinformatics [2], and semantic search [3], more effort and cost have been devoted to the study of promoting the searching performance on large-scale graphs. As one of the most crucial issues in graph analysis is subgraph matching, given a data graph G and a query graph q, the subgraph matching problem is to identify all the matches of q in G [4], which is known as the NP-complete problem [5]. That means the time complexity of matching grows exponentially in the worst case. In order to reduce the search space and speed up the query process of subgraph matching, some scholars have proposed different methods to minimize the number of candidate nodes in the data graph, such as constructing indexes that store vertex-adjacent features [6] or designing powerful filtering strategies [7]. However, the cost of substructure enumeration and storage leading to exponential time and space complexity in the computational phase is still relatively high [8]. Therefore, how to make a large data graph enumerate all the query results in an efficient time is a top priority in the research of the subgraph matching problem.

Methods for subgraph matching are broadly categorized into two groups: exploratory backtracking [9-12] and join [4,13]. Each algorithm has its own advantages and disadvantages. For example, Ullmann [10] is the first subgraph isomorphic search algorithm, performing sequential node-depth-first matches without defining the matching order of query vertices. VF2 [11] arbitrarily selects a start node and expands by choosing the next vertex connected to already matched query vertices. RapidMatch [13] for distributed environments performs query operations in main memory without the need to go through an indexing design. According to recent studies, the exploration backtracking method is suitable for large sparse graphs, while the join method is more suitable for small dense graphs [14]. The algorithmic studies mentioned above usually focus on analyzing and comparing the methods and performance of current algorithms [15]. Therefore, instead of trying to identify or compare differences between algorithms, we focus on current problems in subgraph matching and advanced algorithms to solve them.

However, subgraph matching still faces several major challenges: 1) the repeated enumeration of equivalent nodes leads to degradation of search performance, and 2) the other is that for the generation of candidate node sets stage, the filtering conditions are incomplete and there still exists a portion of undesired candidate nodes, which leads to excessive overhead in the validation stage and affects the overall performance. Aiming at the above problems, the main contributions are summarized as follows:

(1) Propose an efficient strict compressed graph node (CGN) algorithm for data and query graphs, which uses the equivalence relationship between the nodes to compress the same equivalence class graph data into a smaller scale.
(2) Propose a label count filter (LCF) technique to cut down the redundant validation cost, which is based on the labeling characteristics of the dataset nodes added in the filtering phase, further reducing the size of the search space and improving the overall performance.
(3) For large-scale graphs, an adaptive subgraph matching (ASM) model is proposed to accelerate the subgraph matching query by reusing previous query results, or judge the overall graph by the situation of some hot nodes, forming an efficient adaptive subgraph matching framework according to the given dataset.
(4) We have conducted a large number of experiments on several real datasets to compare the performance of the ASM algorithm with the other three

subgraph matching algorithms in terms of execution time, and execution algorithm efficiency (the number of callbacks and average callback time). The experiments show that our method achieves better performance on query processing.

2 Background

Previous works show that the preprocessing is crucial for the subgraph matching [16]. Because it significantly reduces memory usage and shortens algorithm running time. By organizing and optimizing data in advance, preprocessing minimizes resource and time consumption during subsequent operations.

Selection of Root Query Node: Choosing the right starting node in subgraph matching is crucial to exclude unmatched nodes early, reducing extended validations. A poor choice leads to redundant enumeration and longer processing times. The ranking rule is determined by the equation $Rank(u) = freq(g, L(u))/d(u)$. Where $freq(g, L)$ denotes the number of nodes labeled L in the graph g, and $d(u)$ denotes the degree of node u.

Determining the Matching (Visit) Order: Usually, when determining the matching order in the subgraph matching process, priority is given to nodes with a smaller number of candidate nodes, i.e., nodes with smaller degrees. The advantage of this approach is that it reduces the search space and decreases the size of the intermediate result set. In addition, if a wrong candidate node is found during traversal, backtracking can also be performed to explore the next node at a lower cost of trial and error.

Generating the Query Tree: When a BFS traversal [17] is performed on the query graph starting from the root query node, thus coming to create the query tree. The edges that appear on the query tree in the query graph are called tree edges (TE). If an edge is on the query graph but not on the BFS tree, it is called a non-tree edge (NTE). BFS is used because in some existing studies, it is shown that BFS minimizes the diameter of the search space [18].

3 Efficient Subgraph Matching Mechanism

To improve performance, the preprocessing enumeration subgraph matching algorithm introduces a preprocessing phase before enumeration. This phase reduces the size of the candidate set for each query vertex and identifies accurate candidate nodes, thereby optimizing the matching order. In order to reduce the time consuming enumeration, it is necessary to use a filtering algorithm to make the candidate set as small as possible while obtaining a complete candidate set.

3.1 Compressed Graph Nodes (CGN) Algorithm

As a large graph, there will be many nodes with the similar structure. How to reduce the number of node one-to-one matches in the subgraph matching plays a crucial role in algorithm efficiency improvement. Therefore, the idea of strict compressed graph node (CGN) algorithm by preprocessing the graph data is proposed to solve the repeated enumeration of equivalent nodes.

Before the matching algorithm starts, many equivalent nodes in the graph are pre-compressed into a single node, which not only reduces the size of the original graph but also reduces the number of one-to-one matches performed during the validation phase.

The graph compression idea is used in the existing $Boost_{ISO}$ [19] algorithm. The algorithm classifies the nodes with equivalence relationships and compresses those nodes from the original graph with the same label into a single node. Then follows the edges between the nodes in the original data graph to determine whether there are any connecting edges between the nodes, and further transforms the nodes with equivalence relationships into an undirected equivalence graph. Although this algorithm avoids unnecessary repetitive enumeration of equivalence nodes in the subgraph isomorphism verification phase. However, there are still some redundant operations in the process of finding the equivalence nodes. For example, in Fig. 1, for the verification process of path V_1-V_2-V_5, it can be concluded that there is no equivalence relationship between V_1 and V_5. However, since the node is not marked as visited, the validation of the relationship between the two nodes V_1 and V_5 is repeated in the path V_1-V_3-V_5.

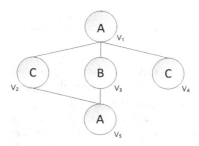

Fig. 1. Example of duplicate lookup of equivalent nodes.

The concept related to equivalent nodes is also proposed by Ren X. et al. [19] to compress the original graph. They propose that two nodes are considered as equivalent nodes if they have the same parent node and the same child node after clipping the edges between them. This is an approximate node-compression technique. Although this approximate node-compression technique has been widely used, its disadvantages still are obvious. Because it is prone to produce results that cannot be returned to the original graph, thus causing the problem of over-compression.

How to determine whether the equivalent node is more appropriate? Our understanding will be given below: (1) Same-labeled nodes with the same father and child nodes in the case of non-prunable existing edges (2) Isolated nodes with the same label (nodes with degree 0).

Algorithm 1: Compressed Graph Nodes (CGN)

Input: data graph g, query graph q
Output: Compressed graph g', q'
1 Create the compressed graph's vertex set V;
2 // Creating a collection of nodes for a compressed graph
3 **for** each $v \in \{v|v \in V(g)\}$ **do**
4 $\quad$ | $nodeExamination(v)$; // Examination of each node
5 **end**
6 $V_0 = \{v|v \in V, d(v) = 0\}$;
7 group $v \in V_0$ by labels
8 update $V = V - V_0$;
9 if(v is equal to the vertex V_n in V)
10 V_n.equal $\leftarrow v$;
11 else $V \leftarrow v$;
12 **for** each $v \in \{v|v \in V(g)\}$ **do**
13 $\quad$ | $edgeExamination(e)$; // Examination of each edge
14 **end**
15 find all edges connect to the equal points of v in g;
16 if(the begin point is in v AND has edges between v and v')
17 create an edge e between v and v';
18 record points in e;
19 update the vertex number and edge number of g';

As shown in Algorithm 1, CGN first the set of nodes of the compressed graph is created (line 1) and each node in the original graph g is examined (line 3). If an isolated node with node degree 0 is found, the processing rules for isolated nodes are used (lines 6–7), at which point the data graph nodes to be found as equivalent nodes are updated (line 8). When the first node V_1 of the original graph arrives, the set V of nodes of the compressed graph is the empty set. So add V_1 directly to the set V. When the second node V_2 of the original graph arrives, go to the collection of nodes of the compressed graph to find out whether V_2 is equivalent to one of the recorded nodes. If V_2 is equivalent to V_1, then V_2 is added to the domain of equivalent nodes of V_1. Otherwise, V_2 will be added directly to V. Scan each node of the original graph in turn. Until all nodes have been visited (line 9–11). At this point, the processing for nodes is finished, and the following processing for edges is carried out, which is based on the compressed node set just obtained. Nodes are selected from V in turn. We select the first equivalent node V_a of V_1. We query the neighbor table of V_a in the original graph, and if we find that there is an edge starting from V_a and its endpoint is not in the domain of equivalent nodes of V_a, we add an edge between V_1 and V_a.

The above Fig. 2 is an example. Suppose we start our examination at node V_1, at which the set of compressed graph nodes $V = \emptyset$. Thus it is straightforward to add V_1 to the node set V, denoted by V_1'. At this point $V = \{V_1'\}$ and $V_1' = \{V_1\}$. For the second node V_2, first determine the equivalence with the nodes in the set V. It is determined that V_2 is not equivalent to V_1', so V_2 is added to V. At this point $V = \{V_1', V_2'\}$. Similarly V_3 is added to V. At this point $V = \{V_1', V_2', V_3'\}$. When examining V_4, it is found that V_4 is equivalent to V_3' in V. Therefore, V_4 is added to the domain of equivalent nodes of the V_3'. At this point V remain unchanged and $V_3' = \{V_3, V_4\}$. Finally, V5 is examined, and since there is no node that is equivalent to it, V_5 is added directly to V. At this point, $V = \{V_1', V_2', V_3', V_4'\}$ and $V_4' = \{V_5\}$.

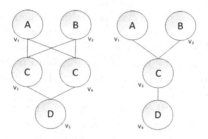

Fig. 2. Strict graph compression techniques

Once the processing of the points is complete, the edges are then processed accordingly. Each node from V is selected in turn for examination. Take out V_1' first and examine each node in V_1'. It is found that $V_1 \in V_1'$, there is an edge from V_1 to V_3 in the original graph and $V_3 \in V_3'$. therefore an edge from V_1' to V_3' is added to the compressed graph, keeping the record as (beginpoint, endpoint). Continuing the examination, it is found that there is an edge from V_1 to V_4 and $V_4 \in V_3'$. However, at this point it will be found that an edge already exists between V_1' and V_3'. So there is no need to repeat the insertion of this edge.

From the above description, it is clear that the process of executing the CGN algorithm requires traversing the original graph. In terms of nodes, every node in the original graph needs to be traversed and examined. Thus the worst case time complexity of compressing the nodes is $O(n^2)$. While working on the edges of the original graph, it is sufficient to explore the edges of the original graph one by one, hence the operation of adding edges is $O(E)$.

3.2 Label Count Filter (LCF) Algorithm

Most of the current algorithms directly use Label and Degree Filter(LDF) or Neighbor Label Count Filter (NLCF) for filtering. And since there are a large number of wrong candidate vertices in the candidate set extracted by using LDF or NLCF directly, we now add a filtering condition based on the dataset features to make the filtering condition more complete and convenient. Since the labels on the data graph and query graph are represented as numbers, this makes the labels not only have the function of identifying nodes, but also increase the computational function. We propose the concept of label count filter (LCF) for this feature, first determining the number of nodes in the query graph, after that adding the label counts of all the nodes and getting the corresponding result, then comparing the filtering with the node labels and of the same number of vertices in the data graph, so that LC in the data graph is equal to the LC in the query graph. Then further use the label sum degree filtering method (LDF) superimposed on the Neighbor Label Counting Filtering method (NLCF) to filter out the nodes whose conditions are not met, which contributes to the extraction of the candidate set in terms of space and time. The main process of extracting the candidate set is given as Algorithm 2:

First, the boundaries of the query are determined, and we generate our bounding set with the concatenated set of the parents of all nodes. For each node, we set up four filters to filter them. The first one is the label count filter, as the name suggests, uses the sum of the numbers of the labels used by the nodes in the query graph to compare with the data graph to find the candidate nodes that satisfy the conditions.

Next, Label Filter (LF) collects neighboring nodes v in v_f with u label, i.e., nodes with different labels are removed for further filtering. Degree Filter (DF) ensures that the degree (number of connected edges) of the data graph node v is greater than or equal to the degree of the query node u. Neighbor Label Count Filter (NLCF) imposes two constraints: one is to satisfy that the number of labels of the neighbors of the current node in the data graph is greater than or equal to the number of neighbor labels of the corresponding node in the query graph; rather, the labels corresponding to the same number of labels need to be the same as the kind as well.

When generating the candidate set through forward neighbors, if tree edge candidates are filtered out, we also remove their parent and child nodes. For non-tree edge candidates that don't meet conditions, we remove the nodes directly. While generating the candidate set with backward neighbors, we refine the set to make it smaller and more accurate. Nodes that pass all the filtering criteria will be stored for subsequent work. The time complexity of Algorithm 2 is $O(|E(q)| \times |E(G)|)$.

3.3 Adaptive Subgraph Matching Algorithm

Through previous experience, we learned that the label features of the dataset could affect the filtering phase's performance, while the density features of the dataset will affect the performance of the sorting phase. So the adaptive algorithm will judge according to the label density, and then adaptively choose the best solution according to the label density characteristics of the dataset. Considering the significant time and space consumption required to import the mega-

Algorithm 2: Extract Candidates Algorithm

Input: data graph g, query graph q
Output: candidate set C

1   forall $u \in T_q$ in BFS order do
2   u_p.frontiers = $\cup u_p$.TE_Candidate
3   forall $v_f \in u_p$.frontiers do
4   LCF(N(v_f),q) // label count filter
5   LF(N(v_f),L_q(u)) // label filter
6   DF(u,v) // degree filter
7   NLCF(u,v) // neighbor label count filter
8   u.TE_candidate[v_f].add(v)
9   //Forward Vertex Neighbor Generation Candidate Set
10   for i ← 2 to $|\varphi'|$ do
11   u ← φ'[i];
12   foreach $v \in \bigcap_{u' \in N_-^{\varphi'}(u)} N(u'.C)$ do
13   //Backward Neighborhood refined Candidate Collection
14   for i ← $|\varphi'|$ to 1 do
15   u ← φ'[i];
16   foreach $v \in$ u.C and $v \notin \bigcap_{u' \in N_+^{\varphi'}(u)} N(u'.C)$ do
17   remove v from u.C

---

graph into memory, we propose a novel concept – local weights. Slicing and dicing the mega-graph allows the candidate set of popular nodes in the query graph to be utilized to locate the approximate region. If isomorphic subgraphs exist, they must exist in the localized subregions, and the isomorphic subgraphs can then be queried directly on these subregions. This can reduce the scale of the subgraph isomorphism query to some extent and improve the query efficiency.

The inputs to the ASM (Algorithm 3) are a data graph G and a query graph q. The output is the result of the isomorphism of all the subgraphs of q in G. A central node that is most closely connected to other nodes is found by comparing the node influence of nodes in the query graph. Local weights consider the importance of a node relative to the entire query graph, including information such as the connectivity between nodes, and the degree of a node.

Firstly, each node is assigned the same value AA_u, and the sum of AA values of all nodes is 1. During a new round of AA value calculation, node u will use the AA_u value of its own node as a weight to equally distribute to the other nodes that are connected. Nodes with no connected edges (degree 0) have unchanged AA_u values; the updated node's AA_u value becomes the sum of the weights of all pointing nodes. Complete the above steps until the AA_u values gradually converge. The stable AA_u value of each node is finally obtained (line 9). Where $N(u)$ is the set of all nodes pointing to node u, node v is a node belonging to the set $N(u)$, and $d(u)$ is the degree of node u. The local influence of a node is measured by comparing its final AA_u value. The labeled density value $score(G)$

Algorithm 3: Adaptive Subgraph Matching (ASM)

Input: data graph g, query graph q
Output: Subgraph isomorphism of all query graphs in a data graph
1　Function CGN(g(v,e)): //preprocessing process of data set
2　**for** *each $v \in \{v|v \in V(g)\}$* **do**
3　　| $nodeExamination(v)$; // Examination of each node
4　　| $edgeExamination(e)$; // Examination of each edge
5　**end**
6　Function
　　ExtractCandidate[LCF(N(v_f),q),LF(N(v_f),L_q(u)),q),DF(u,v),NLCF(u,v)]
7　LD ← GET_ LABEL _DENSITY
8　SCORE ← GET_ GRAPH _DENSITY
9　$AA(u) = \sum_{u \in N(u)} \frac{AA(v)}{d(u)}$
10　//The filtering method of ASM
11　if score(G)>=$score_{threshold}$ then
12　　C ← GQL_ FILTER
13　else
14　　C ← DAF_ FILTER
15　//The ordering method of ASM
16　φ ← GenerateMatchingOrder(q,G,C,score(G));
17　//The enumeration method of ASM
18　SUBGRAPHMATCH(q,G,C,φ ,score(G));
---

of the nodes in the data graph is then computed. When $score(G)$ is greater than or equal to the $score_{threshold}$ (which is set empirically and artificially)(line 11), we recognize the graph as dense and vice versa as sparse. Depending on the characteristics of the dataset, we can choose the appropriate algorithm at each stage, as no algorithm can overwhelm the others on all queries [14].

When subgraph matching queries are performed, a caching approach can also be used to accelerate subgraph matching queries by reusing previous query results [20]. The matching results between the query graph and the data graph are first stored in a cache. These results can be stored in an in-memory data structure, such as a hash table, a tree structure, or a database. When a new subgraph matching query is performed, it first checks whether the matching result corresponding to the current query graph exists in the cache. If so, these results can be retrieved directly from the cache without having to execute the matching algorithm again.

4　Experiments Study

4.1　Experimental Setting

All the source codes are implemented in C++. Experiments are conducted on a PC running ubuntu with Intel i3-6100 3.70 GHz CPUs, 7.7 GB memory, and 1T disk capacity. All evaluation results are averaged over five runs.

DBLP Datasets. A database of scholarly literature in the field of computer science. The dataset contains 317080 nodes and 1049866 edges.

YouTube Datasets. Based on a user's YouTube activity, a relationship graph is built to find and recommend content from users with similar interests. The dataset contains 1134890 nodes and 2987624 edges.

Experiment with ASM we proposed and comparatively analyze the following existing algorithms with different indexing and sorting strategies:

GraphQL [12]: Using Neighbourhood Signature Filter and Left Depth Connection Sorting Strategy.

CFL [18]: One of the most advanced algorithms. Using tree-structured indexing CPI and path-based sorting strategy.

CECI [21]: One of the state-of-the-art algorithms. CECI sorts tree and non-tree edge candidates using forward BFS traversal and reverse BFS refinement.

Comparing with the other three advanced algorithms, we can find that there is a certain progress in terms of the execution time, the number of algorithm callbacks and the average query time in the matching process. Each dataset shows the comparison results by query graph node size 3~7.

4.2 Evaluations on Execution Time

The execution times of the different algorithms on the two datasets are shown in Fig. 3. We notice that all algorithms spend more time on larger query sizes, since larger queries may contain larger sets of labels, which increases the computational complexity of the filtering methods. When the number of node sizes is small, the advantage of ASM is not obvious. However, we see that the effect is more pronounced when the node size is larger.

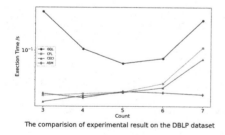
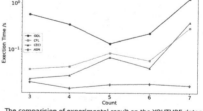

The comparision of experimental result on the DBLP dataset   The comparision of experimental result on the YOUTUBE dataset

Fig. 3. The evaluation of the execution time.

With a node size of 6, ASM at this moment reduces the DBLP and YouTube by 75.34%, 31.27%, 18.35%, 88.87%, 68.56% and 54.48% compared to GQL, CFL, and CECI, respectively. This proves the efficiency and stability of ASM. The GQL algorithm performs BFS tree building and proposes corresponding improvements in filtering and pruning, but the time overhead of building the tree makes the GQL algorithm much slower than several other algorithms in some queries.

4.3 Evaluations on the Callbacks Counts and Average Time

In the algorithm comparison experiments, the lower the number of callbacks or the shorter the search time, the more efficient the algorithm is. In Fig. 4, we note that with the increase in the number of query nodes, the number of callbacks of several other algorithms shows an exponential growth trend, which is very unfavorable for the resource consumption of memory. Whereas ASM grows at a slower rate and can reach a flat trend at later stage. Take node 6 as an example again. ASM reduces the number of callbacks by 71.47%, 73.82%, 75.08%, 79.06%, 90.18% and 88.02% respectively, compared to the other three algorithms.

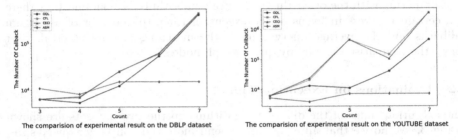

Fig. 4. The evaluation of the counts of callbacks.

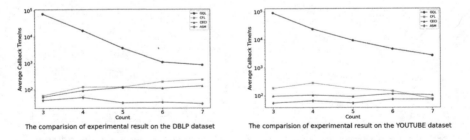

Fig. 5. The evaluation of average callback time.

In terms of average callback time, as shown in Fig. 5. GraphQL can reduce the average search time by increasing the chance of data reuse by adding more nodes at the initial stage. However, when the number of nodes increases to a certain level, the complexity of the data still leads to an increase in the search time of the query due to the need to process more data and more complex queries.

This time also using node 6 as an example, the ASM algorithm reduces 90.33%, 76.5%, 64.54%, 91.38%, 50% and 35.96% compared to the other three algorithms respectively. The other algorithms have a slower increase in average search time, and to our surprise, ASM can maintain a horizontal trend under ideal conditions.

5 Conclusion

In order to accelerate subgraph matching, a strict compressed graph nodes algorithm (CGN) is proposed, which performs both node and edge compression operations on both the data graph and the query graph. Then in the filtering phase, a new label count filter(LCF) based on dataset features is added, and although the time is increased in the filtering phase, the efficiency is improved in terms of the overall time of subgraph matching due to the more fine-grained filtering of the number of candidate nodes. The experiments show that the adaptive subgraph matching algorithm (ASM) has an advantage in matching speed compared to the existing algorithms and has better overall performance.

Acknowledgments. This work is supported by Fundamental Research Program of Shanxi Province (No.202403021211085), Scientific and Technological Innovation Programs of Higher Education Institutions in Shanxi (No. 2022L323), Doctoral Research Start-up Fund of Taiyuan University of Science and Technology (No. 20232003).

References

1. Fan, W.: Graph pattern matching revised for social network analysis. In: Proceedings of the 15th International Conference on Database Theory, pp. 8–21 (2012)
2. Sahoo, T.R., Patra, S., Vipsita, S.: Decision tree classifier based on topological characteristics of subgraph for the mining of protein complexes from large scale PPI networks. Comput. Biol. Chem. **106**, 107935 (2023)
3. Xu, Q., Wang, X., Li, J., Gan, Y., Chai, L., Wang, J.: StarMR: an efficient star-decomposition based query processor for SPARQL basic graph patterns using mapreduce. In: Cai, Y., Ishikawa, Y., Xu, J. (eds.) APWeb-WAIM 2018. LNCS, vol. 10987, pp. 415–430. Springer, Cham (2018). https://doi.org/10.1007/978-3-319-96890-2_34
4. Kim, H., Choi, Y., Park, K., Lin, X., Hong, S.-H., Han, W.-S.: Fast subgraph query processing and subgraph matching via static and dynamic equivalences. VLDB J. **32**(2), 343–368 (2023)
5. Hartmanis, J.: Computers and intractability: a guide to the theory of NP-completeness (Michael R. Garey and David S. Johnson). SIAM Rev. **24**(1), 90 (1982)
6. Sun, Y., Li, G., Du, J., Ning, B., Chen, H.: A subgraph matching algorithm based on subgraph index for knowledge graph. Front. Comp. Sci. **16**(3), 1–18 (2022). https://doi.org/10.1007/s11704-020-0360-y
7. Ba, L.-D., Liang, P., Gu, J.-G.: Subgraph matching algorithm based on preprocessing-enumeration. Comput. Technol. Dev. **33**(12), 85–91 (2023)

8. Zeng, L., Jiang, Y. Lu, W., Zou, L.: Deep analysis on subgraph isomorphism. arXiv preprint arXiv:2012.06802 (2020)
9. Choi, Y., Park, K., Kim, H.: BICE: exploring compact search space by using bipartite matching and cell-wide verification. Proc. VLDB Endow. **16**(9), 2186–2198 (2023)
10. Ullmann, J.R.: An algorithm for subgraph isomorphism. J. ACM (JACM), **23**(1), 31–42 (1976)
11. Cordella, L.P., Foggia, P., Sansone, C., Vento, M.: A (sub) graph isomorphism algorithm for matching large graphs. IEEE Trans. Pattern Anal. Mach. Intell. **26**(10), 1367–1372 (2004)
12. He, H., Singh, A.K.: Graphs-at-a-time: query language and access methods for graph databases. In: Proceedings of the 2008 ACM SIGMOD International Conference on Management of Data, pp. 405–418 (2008)
13. Sun, S., Sun, X., Che, Y., Luo, Q., He, B.: RapidMatch: a holistic approach to subgraph query processing. Proc. VLDB Endow. **14**(2), 176–188 (2020)
14. Sun, S., Luo, Q.: In-memory subgraph matching: an in-depth study. In: Proceedings of the 2020 ACM SIGMOD International Conference on Management of Data, pp. 1083–1098 (2020)
15. Dann, J., Götz, T., Ritter, D., Giceva, J., Fröning, H.: Graphmatch: Subgraph query processing on FPGAs. arXiv preprint arXiv:2402.17559 (2024)
16. Liu, T., Li, D.: EndGraph: an efficient distributed graph preprocessing system. In: 2022 IEEE 42nd International Conference on Distributed Computing Systems (ICDCS), pp. 111–121. IEEE (2022)
17. Bi, F., Chang, L., Lin, X., Qin, L., Zhang, W.: Efficient subgraph matching by postponing cartesian products. In: Proceedings of the 2016 International Conference on Management of Data, pp. 1199–1214 (2016)
18. Gaihre, A., Wu, Z., Yao, F., Liu, H.: XBFS: eXploring runtime optimizations for breadth-first search on GPUs. In: Proceedings of the 28th International Symposium on High-Performance Parallel and Distributed Computing, pp. 121–131 (2019)
19. Ren, X., Wang, J.: Exploiting vertex relationships in speeding up subgraph isomorphism over large graphs. Proc. VLDB Endow. **8**(5), 617–628 (2015)
20. Qin, Y., Wang, X., Hao, W., Liu, P., Song, Y., Zhang, Q.: OntoCA: ontology-aware caching for distributed subgraph matching. In: Li, B., Yue, L., Tao, C., Han, X., Calvanese, D., Amagasa, T. (eds) APWeb-WAIM 2022. LNCS, vol. 13421, pp. 527–535. Springer, Cham (2023). https://doi.org/10.1007/978-3-031-25158-0_42
21. Bhattarai, B., Liu, H., Howie Huang, H.: CECI: compact embedding cluster index for scalable subgraph matching. In: Proceedings of the 2019 International Conference on Management of Data, pp. 1447–1462 (2019)

SemiBDMA Workshop

Learning Multi-semantic Based on Cross-Attention for Image-Text Retrieval

Bo Lu, Ying Gao(✉), Tianbao Zhao, Xia Yuan, Haibin Zhu, Lin Gan, and Xiaodong Duan

Dalian Minzu University, Dalian 116650, China
799835701@qq.com

Abstract. Cross-modal retrieval is a challenging problem for processing multi-modal information, primarily addressing the semantic gap between different modalities. However, current research mainly focuses on aligning image and text regions, which often ignore the importance of contextual background information. Specially, there are uncertainties among heterogeneous modalities which persist in understanding semantic information of cross-modal retrieval. In this paper, we propose a novel Multi-Semantic Based on Cross-Attention Network (MSCA) for cross-modal retrieval tasks. Specifically, a cross-extraction of contextual semantic feature encoder module and a graph structure enhancement module are firstly designed to more effectively address the issue of neglecting contextual background information and to further enhance the fusion of information and features, respectively. Secondly, we effectively solve the cross-modal alignment problem by employing a probability distribution encoder, simultaneously, we integrate three pre-training models, which include image-text contrastive learning, image-text matching loss and masked language model loss for reducing uncertainties among different modalities. We conducted extensive experiments on two large datasets, MS-COCO and Flickr30K. The experimental results show that the MSCA model outperforms other existing methods on both datasets, demonstrating its effectiveness and significant performance advantages.

Keywords: Image-Text Retrieval · Feature Fusion · Coarse-Grained Alignment · Fine-Grained Alignment

1 Introduction

With the proliferation of the internet and social media, information forms have become increasingly diverse. Cross-modal retrieval allows users to conveniently find related forms of information through one modality, addressing the limitations of traditional retrieval methods and becoming a research hotspot. Cross-modal image-text retrieval is a crucial area within cross-modal retrieval, with the core challenge being the effective understanding and alignment of different modalities. Existing methods mainly focus on fine-grained alignment, neglecting the importance of contextual background information, as shown in Fig. 1. Additionally, when dealing with multimodal semantic understanding, it is essential to consider uncertainties both within and between modalities, as illustrated in Fig. 2. This complexity makes understanding and interpretation more challenging, highlighting the need to address these uncertainties In this paper, a Multi-Semantic Cross-Attention Network (MSCA) for image-text retrieval tasks is proposed. Two modules have been designed and three pre-training tasks have been incorporated, resulting in an integrated model that more effectively addresses the issues of ignoring contextual background information and uncertainties within and between modalities. The contributions of our work are presented as follows:

- The limitations of information extraction are addressed by extracting both global and local features of images and utilizing a multi-layer Transformer to enhance contextual background information, which enriches the visual features.
- A semantic relationship graph structure connecting text and image regions is constructed, and the semantic relationship is captured by the Intra-modal Attention layer and the contextual semantic function is learned to enhance the aggregation features. Furthermore, Inter-modal Attention is used to achieve better cross-modal feature fusion.
- The issue of images with multiple entities or texts with multiple images is addressed by employing a PDE probability distribution encoder and incorporating three pre-training tasks: ITC, ITM, and MLM. Both coarse and fine-grained information is handled by the PDE, while cross-modal alignment and inter-modal interaction are achieved by the three tasks, leading to the optimization of downstream task performance.

2 Related Works

We will further introduce related work, including image text retrieval, graph convolutional networks, and pre trained models.

Image-Text Retrieval. VSE++ [6] is a classic model for image and text retrieval, which calculates similarity by encoding images and text. The follow-up SCAN [8] model optimizes graphic and text alignment to improve matching accuracy. PFAN [18] meticulously encodes image information, emphasizing object positional relationships. R-SCAN [7] represents images as scene maps and calculates similarity with text words using attention. Although the above methods have achieved certain results in image and text similarity calculation, they have to some extent ignored the significance of contextual background information.

Graph Convolution Network. VSRN [10] performs graph convolutional networks in image regions. VSRN++ [11] also applies graph convolutional networks in text regions to improve effectiveness. GSMN [12] improves the attention mechanism for image text alignment and constructs a relationship graph for

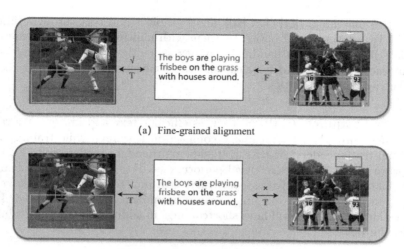

Fig. 1. Provide an example of fine-grained alignment combined with contextual semantic alignment. (a) shows an example of fine-grained alignment, while (b) illustrates an example of contextual semantic alignment. It is evident that incorporating contextual background information in retrieval results in higher accuracy.

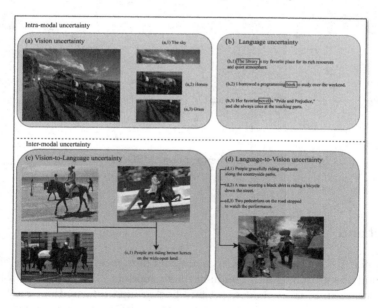

Fig. 2. Multimodal uncertainties. (a) and (b) represent uncertainties within a modality, while (c) and (d) represent uncertainties between modalities.

fine-grained matching. UNITER [20] aligns graphics and text through multiple layers of Transformers and optimizes them with multiple tasks. These methods still have shortcomings in handling complex semantic relationships and image text interactions, failing to fully capture the connections between fine-grained and coarse-grained semantics.

Pre Trained Model. CLIP [9] achieves efficient text and image retrieval by pre training through text and image comparison learning, while training both image and text encoders. ALBEF [14] adds multimodal encoders and ITC tasks to improve image and text retrieval performance. For the Chinese pre training model, Chinese CLIP [19] uses contrastive learning to perform two-stage pre training on Chinese text and image pairs, optimizing the entire model. However, the above methods still have shortcomings in addressing the uncertainty of multimodal tasks.

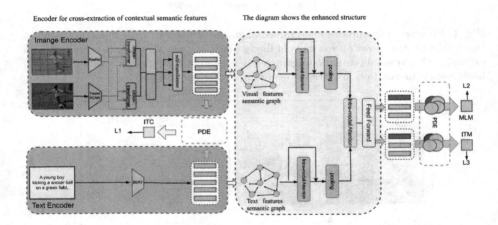

Fig. 3. Overview of the proposed MSCA framework.

3 Method

In this paper, we propose the MSCA model and design two modules, the encoder module and the graph structure enhancement module. They are used to better solve the problem of ignoring context and background information, further better information enhancement and feature fusion. PDE probability distribution encoder is combined with three pre-training tasks, image-text contrast learning (ITC), image-text matching loss (ITM) and mask language model loss (MLM), to solve the cross-modal alignment problem. Reduce uncertainty between different modes. The frame diagram is shown in Fig. 3.

3.1 Encoder for Cross-Extraction of Contextual Semantic Features

Image Feature Extraction. In order to fully capture the rich semantics of the image, we extract the global and local features of the image respectively, and combine them to form a more accurate and comprehensive image representation. (1)Global feature representation: The global features of the image are extracted using the model ResNet152, which has been trained on ImageNet. We delete the last layer of full connection layer and obtain the output features $G = \{g_1, g_2, \ldots, g_m\}, g_i \in R^{D^0}$, which D^0 represent the size of each pixel and the number of feature maps, m is the number of feature maps. Apply the fully connected layer g_i mapping to -dimensional space to get the features: $V_G = W_g G + b_g$, V_G stands for global features, W_g represents the global feature weight matrix and b_g represents the global feature bias.

(2) Local feature representation: we use the model Faster-RCNN [15] trained on Visual Genomes to extract the local features of the image, which is used to recognize the image more accurately to get the feature $L = \{l_1, l_2, \ldots, l_k\}, l_i \in R^{D^0}$, D^0 represents the size of the regional features, k represents the number of regional feature maps, and applying the fully connected layer mapping g_i to the D-dimensional space to get the feature $V_L = W_1 L + b_1$, V_L represents the local features, W_1 represents the local feature weight matrix, and $\mathbf{b}_1$ represents the local feature bias.

(3) Post-fusion feature representation: The global and local features of the image are processed using cross-attention, where the input global and local features are split $V_G \in R^{n \times d_1}$ and $V_L \in R^{n \times d_2}$, Then using one part as the query set and the other part as the key-value set. Its output is a tensor of size $n \times d_2$. For each row vector, its attentional weight for all row vectors is given. Specifically, $Q = V_G W^Q$ and $K = V = V_L W^K$ then the cross attention is computed as follows:

$$CrossAttention(V_G, V_L) = Softmax\left(\frac{Q^T}{\sqrt{d_k}}\right) V \quad (1)$$

where $W^Q \in R^{d_1 \times d_k}, W^K \in R^{d_2 \times d_k}$ is the learned projection matrix and d_k is the dimension of the key-value set (which is also the dimension of the query set). And after the cross-attention aligns the two sets of features, the self-attention is used to make it adaptive to fuse the semantic gaps of the two sets of features, and the obtained features are denoted by V_U.

Text Feature Extraction. Traditional text feature representations often use RNN models, but the BERT [2] model is based on the self-attention mechanism, which can capture semantic relationships more accurately. In this paper, we use BERT as a text encoder to get text features by WordPiece segmentation and extracting word features for enhanced text representation $S = \{s_1, s_2, \ldots, s_l\}$, D^1 denotes the dimension, 1 is the maximum number of words in the word, and then use a fully connected layer $\mathbf{S}_i$ to map to the D-dimensional space to get the feature:$T_S = W_s S + b_s$, T_S represents the weight of the text feature, and b_s represents the bias of the text feature.

3.2 Graph Structure Enhancement Module

To better capture the contextual information, we use the graph enhancement module to further enhance and fuse the obtained image and text features.

Image Area. For images, A semantic relationship graph was created for visual regions, and a module was proposed to understand the context of these regions. The self-attention layer in the graph attention network helps to improve how we understand relationships and represent local features. Through scaled dot-product attention, we establish a semantic relationship graph and obtain new image features V'_U via self-attention learning. Using a weighted sum of the values as the output, and enhanced features by two methods of maximum pooling and average pooling to obtain global embedding v, and control the proportion of representation by parameter β.

$$v = \beta \cdot \text{MaxPool}(V) + (1 - \beta) \cdot \text{AvgPool}(V'_U) \qquad (2)$$

Text Area. For text, the same as for pictures, a semantic relationship graph between text words is constructed to enhance the information of word features. Nodes are word features, and edges represent the relationship between words, just like pictures. Obtain the global embedding u.

Image Text Fusion Area. In the field of multimodal data processing, there are two main types of multi-model converters that have attracted much attention: single-flow model [1,16,22] and dual-flow model [13,17]. The cross attention module takes as input the stacked features of the image region and sentence words. Then, using Scaled Dot-Product Attention, Obtain the fused features of images and text.

$$Y = \begin{pmatrix} v' \\ u' \end{pmatrix} = \begin{pmatrix} Q_R K_R^T V_R + Q_R K_E^T V_E \\ Q_E K_E^T V_E + Q_E K_R^T V_R \end{pmatrix} \qquad (3)$$

where v' represents the feature representation of image fusion, and u' is the feature representation of text fusion. By means of the cross-attention output unit Y.

3.3 Probability Distribution Encoder (PDE) and Three Pre-training Tasks

Probability Distribution Encoder (PDE). The input features are further transformed into a multivariable Gaussian distribution. Utilizing the PDE, With the PDE, we predict the mean vector (μ) and the variance vector ($\sigma 2$) for each input feature. The mean vector can be regarded as the typical position of the feature in the probability space, while the variance vector describes the range of fluctuations of the feature in each dimension. Specifically, we use the multi-head attention mechanism to send the resulting features into two paths $(\mu, \sigma2)$, where $H \in R^{T \times D}$ is divided into K heads, T is the sequence length, and D is the hidden size. The path is represented as follows:

$$\left[Q_\mu^{(i)}, K_\mu^{(i)}, V_\mu^{(i)}\right] = H_\mu^{(i)} W_{qkv}$$
$$\text{Head}_\mu^{(i)} = \text{Act}\left(Q_\mu^{(i)} K_\mu^{(i)} / \sqrt{d_k}\right) V_\mu^{(i)} \quad (4)$$
$$\text{MH}_\mu = \text{concat}_{i \in [k]} \left[\text{Head}_\mu^{(i)}\right] W_O$$

where d_k is set to $D/(2k)$. The weight $W_{qkv} \in R^{d_k \times 3\,d_k}$ is to project the input into the subspace of each head. The weight $W_O \in R^{kd_k \times D}$ projects the cascade of K-head results into the output space.

Coarse-Grained Pre-training. For coarse-grained alignment of the obtained image features and text features without graph structure enhancement, we calculate 2-Wasserstein distances to measure the distance between multivariate Gaussian distributions. For two Gaussian distributions $N(\mu 1, \mu 1)$ and $N(\mu 2, \mu 2)$, their 2-Wasserstein distance is defined as:

$$D_{2\,W} = \|\mu_1 - \mu_2\|_2^2 + \|\sigma_1 - \sigma_2\|_2^2 \quad (5)$$

The similarities between the images and the text can be drawn:

$$s\left(V_U, T_S\right) = a \cdot D_{2w}\left(v_{[CLS]}, w_{[CLS]}\right) + b \quad (6)$$

where a is a negative scale factor and b is a bias. For N image-text pairs in a batch, Loss $\mathscr{L}_{\text{ITC}}$ can be obtained:

$$\mathcal{L}_{\text{ITC}} = \mathcal{L}_{\text{NCE}}^{\text{T2I}}(i) + \mathcal{L}_{\text{NCE}}^{\text{I2T}}(i) \quad (7)$$

Fine-Grained Pre-training. For the image and text features obtained through the cross attention module, we introduce MLM and ITM to perform fine-grained alignment. The MLM is a loss function used in pre-training tasks. By randomly masking certain words in the input text and training the model to predict these words, MLM helps the model learn contextual information and semantic relationships. MLM minimizes the cross entropy loss between these point vectors and other sample points by:

$$\mathcal{L}_{\text{MLM}} = \frac{1}{K+1}\left(\text{CE}(\phi(\mu), y) + \sum_{i=1}^{K} \text{CE}\left(\phi\left(z^{(i)}\right), y\right)\right) \quad (8)$$

K sample point vectors are taken from the Gaussian distribution. These samples are used for training along with the mask word's label y. Where μ is the average vector and $z(i)$ is the random sample point vector. The resulting vectors are fed into the MLM classifier for prediction, followed by average pooling.

ITM method is used to optimize the model's prediction of the matching relationship between image and text. It first extracts the features of both and then calculates the similarity or match score between them. Losses are calculated by comparing the difference between these scores and real labels, and the model is optimized by backpropagation.

$$\mathcal{L}_{\text{ITM}} = \frac{1}{K+1} \left(\text{CE} \left(\phi \left(\text{concat} \left[v_\mu, w_\mu \right] \right), y \right) + \sum_{i=1}^{K} \text{CE} \left(\phi \left(\text{concat} \left[v^{(i)}, w^{(i)} \right] \right), y \right) \right) \quad (9)$$

where V_μ, W_μ are the average vector of the picture and text distribution, $v^{(i)}$ and $w^{(i)}$ are the sample points. Additionally, regularization loss is added to maintain the distribution's uncertainty at an appropriate level.

$$\mathcal{L}_{\text{reg}} = \max \left(0, \gamma - h \left(\mathcal{N} \left(\mu, \sigma^2 \right) \right) \right) \quad (10)$$

where γ is a set threshold that affects the level of uncertainty of the distribution, and $h\left(\mathcal{N}\left(\mu, \sigma^2\right)\right)$ is the entropy of a multivariate Gaussian distribution.

Optimize. To sum up, the total loss we can get is:

$$\mathcal{L}_{\text{pre}} = \mathcal{L}_{\text{MLM}} + \mathcal{L}_{\text{ITM}} + \mathcal{L}_{\text{ITC}} + \alpha \mathcal{L}_{\text{reg}} \quad (11)$$

where α is the weight.

4 Experiments

In order to evaluate the image-text matching task of our MSCA, we have compared it with existing technical methods as shown in Table 1.

4.1 Datasets

We used two publicly available datasets: MS-COCO and Flickr30K. Flickr 30K contains 31,783 images, each with 5 titles, divided into 29,000 for training, 1,000 for verification, and 1,000 for testing. The MS-COCO dataset contains 123,287 images, each with five descriptive titles, of which 113,287 were used for training, 5,000 for validation, and another 5,000 for testing.

4.2 Evaluation Indicators and Results

we used R@ K(K = 1, 5, 10) recall rate to evaluate model performance, the proportion of correct results in the first K results. In addition, the comprehensive index Rsum is also used to evaluate the overall performance of the model. The following latest methods are compared with: SCAN [2], VSRN [10], IMRAM

[5], SGRAF [3], NAAF [21], HREM [4], Table 1 shows the quantification of our method on two datasets. Our MSCA significantly outperforms state-of-the-art methods on most assessment metrics. Specifically, MSCA performance on MS-COCO dataset I2T retrieval and T2I retrieval R@1 is significantly improved by 1.1% and 1.2%, respectively, compared with the existing best local matching method HREM [4]. Second, the RSUM on the datasets MS-COCO and Flickr30K were improved by 1.6% and 1.1%, respectively.

Table 1. The results of our method on two datasets. The bests are in bold.

Datasets	Methods	Image-to-Text			Text-to-Image			
		R@1	R@5	R@10	R@1	R@5	R@10	RSUM
MS-COCO	SCAN [8]	72.7	94.8	98.2	58.8	88.4	94.6	507.5
	VSRN [10]	76.2	94.8	98.2	62.8	89.7	95.0	516.7
	IMRAM [5]	76.7	95.6	98.5	61.7	89.1	95.0	516.6
	SGRAF [3]	79.6	96.2	98.5	63.2	90.7	96.1	524.3
	NAAF [21]	80.5	96.5	98.8	64.1	90.7	96.5	527.1
	HREM [4]	82.9	**96.9**	**99.0**	67.1	92.0	96.6	534.5
	MSCA(Ours)	**84.0**	96.0	98.8	**68.3**	**92.3**	**96.7**	**536.1**
Flickr30K	SCAN [8]	67.4	90.3	95.8	48.6	77.7	85.2	465.0
	VSRN [10]	71.3	90.6	96.0	54.7	81.7	88.2	482.5
	IMRAM [5]	74.1	93.0	96.6	53.9	79.4	87.2	484.2
	SGRAF [3]	77.8	94.1	97.4	58.5	83.0	88.8	499.6
	NAAF [21]	81.9	96.1	98.3	61.0	85.3	90.6	513.2
	HREM [4]	**84.0**	96.1	98.4	**64.4**	88.0	93.1	524.0
	MSCA(Ours)	83.6	**96.2**	**98.8**	63.3	**89.2**	**94.0**	**525.1**

Precision-Recall Curve. The Precision-recall curve is drawn to show the variation of accuracy with recall under different methods. As can be seen from Fig. 4, the area under the PR curve of the proposed method is larger than that of other comparison methods. This proves that our model is more optimized.

4.3 Ablation Study

MSCA is composed of three modules, and ablation experiments were conducted on two datasets for each module, as shown in Table 2.

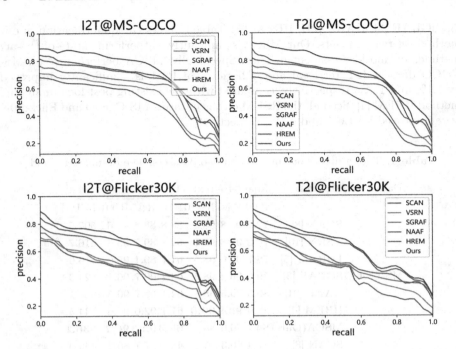

Fig. 4. PR curve of cross-modal retrieval task based on two data sets

Table 2. Ablation Study Results on Two Datasets.

Datasets	Methods				Image-to-Text		Text-to-Image	
	Gobal feature	Local feature	Graph structure	Pre training	R@1	R@5	R@1	R@5
MS-COCO	✓		✓	✓	82.3	94.1	89.1	93.3
		✓	✓	✓	83.1	94.3	91.2	94.2
	✓	✓		✓	82.5	93.9	92.4	94.5
	✓	✓	✓		81.6	95.2	91.5	95.6
	✓	✓	✓	✓	**83.7**	**95.7**	**92.6**	**96.8**
Flickr30K	✓		✓	✓	81.9	96.3	87.3	92.8
		✓	✓	✓	82.4	95.5	87.9	93.5
	✓	✓		✓	83.6	96.6	88.4	92.5
	✓	✓	✓		84.2	97.1	88.1	93.7
	✓	✓	✓	✓	**85.0**	**97.7**	**88.7**	**94.2**

4.4 Visualization of Retrieval Results

To validate the multi-semantic matching performance leveraging cross-attention, we benchmarked our approach against MS-COCO using experimental HERM. We exhibited two scenarios: I2T and T2I retrieval, highlighting the top 3 results for each. Correct matches were visually identified with checkmarks for texts and red borders for images. As shown in Fig. 5.

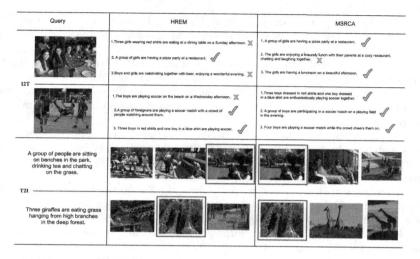

Fig. 5. Comparison of visualization results between HREM and MSCA is shown (Color figure online)

5 Conclusion

In this paper, we propose a new fusion model that contains a contextual semantic feature cross-extraction encoder and a graph structure enhancement module that utilizes a multi-layer transformer to enhance visual features. The model combines PDE with ITC, ITM, and MLM pre-training tasks to optimize parameters and reduce cross-modal uncertainty. Experiments show that it outperforms existing methods on both datasets. Future plans are to expand to more types of multimedia data to study the similarities.

Acknowledgements. This work is supported by the National Natural Science Foundation of China (Grant No. 61602085).

References

1. Chen, Y.C., Yu, L., Kholy, A.E., Ahmed, F., Cheng, Y., Liu, J.: Supplementary material UNITER: universal image-text representation learning, pp. 104–120 (2020)
2. Devlin, J., Chang, M.W., Lee, K., Toutanova, K.: BERT: pre-training of deep bidirectional transformers for language understanding abs/1810.04805, pp. 4171–4186 (2018)
3. Diao, H., Zhang, Y., Ma, L., et al.: Similarity reasoning and filtration for image-text matching. Cornell University - arXiv (2021)
4. Fu, Z., Mao, Z., Song, Y., Zhang, Y.: Learning semantic relationship among instances for image-text matching, pp. 15159–15168 (2023)

5. Hui, C., Guiguang, D., Xudong, L., Zijia, L., Ji, L., Jungong, H.: IMRAM: iterative matching with recurrent attention memory for cross-modal image-text retrieval, pp. 12652–12660 (2020)
6. Kalantidis, Y., Sariyildiz, M.B., Pion, N., Weinzaepfel, P., Larlus, D.: Hard negative mixing for contrastive learning, vol. 33, pp. 21798–21809 (2020)
7. Kuang-Huei, L., Hamid, P., Xi, C., Houdong, H., Jianfeng, G.: Learning visual relation priors for image-text matching and image captioning with neural scene graph generators (2019)
8. Lee, K.-H., Chen, X., Hua, G., Hu, H., He, X.: Stacked cross attention for image-text matching. In: Ferrari, V., Hebert, M., Sminchisescu, C., Weiss, Y. (eds.) ECCV 2018. LNCS, vol. 11208, pp. 212–228. Springer, Cham (2018). https://doi.org/10.1007/978-3-030-01225-0_13
9. Li, J., Selvaraju, R.R., Gotmare, A.D., Joty, S., Xiong, C., Hoi, S.: Align before fuse: vision and language representation learning with momentum distillation. abs/2107.07651, pp. 9694–9705 (2021)
10. Li, K., Zhang, Y., Li, K., Li, Y., Fu, Y.: Visual semantic reasoning for image-text matching, vol. 2019, pp. 4653–4661 (2019)
11. Li, K., Zhang, Y., Li, K., et al.: Image-text embedding learning via visual and textual semantic reasoning (2023)
12. Liu, C., Mao, Z., Zhang, T., Xie, H., Wang, B., Zhang, Y.: Graph structured network for image-text matching abs/2004.00277, pp. 10918–10927 (2020)
13. Lu, J., Batra, D., Parikh, D., Lee, S.: ViLBERT: pretraining task-agnostic visiolinguistic representations for vision-and-language tasks, vol. 32, pp. 13–23 (2019)
14. Radford, A., et al.: Learning transferable visual models from natural language supervision, vol. 139, pp. 8748–8763 (2021)
15. Schuster, M., Paliwal, K.K.: Bidirectional recurrent neural networks. IEEE Trans. Signal Process. **45**, 2673–2681 (1997). https://api.semanticscholar.org/CorpusID:18375389
16. Su, W., Zhu, X., Cao, Y., Li, B., Lu, L., Wei, F., Dai, J.: VL-BERT: pre-training of generic visual-linguistic representations. abs/1908.08530 (2020)
17. Tan, H., Bansal, M.: LXMERT: learning cross-modality encoder representations from transformers, vol. D19-1, pp. 5100–5111 (2019)
18. Wang, Y., et al.: Position focused attention network for image-text matching. abs/1907.09748, pp. 3792–3798 (2019)
19. Yang, A., et al.: Chinese CLIP: contrastive vision-language pretraining in Chinese (2022)
20. Chen, Y.-C., et al.: UNITER: universal image-text representation learning. In: Vedaldi, A., Bischof, H., Brox, T., Frahm, J.-M. (eds.) ECCV 2020. LNCS, vol. 12375, pp. 104–120. Springer, Cham (2020). https://doi.org/10.1007/978-3-030-58577-8_7
21. Zhang, K., Mao, Z., Wang, Q., Zhang, Y.: Negative-aware attention framework for image-text matching, vol. 2022, pp. 15640–15649 (2022)
22. Zhang, P., et al.: VinVL: revisiting visual representations in vision-language models, pp. 5575–5584 (2021)

Integrated Global Semantics and Local Details for Image-Text Retrieval

Bo Lu, Lin Gan(✉), Tianbao Zhao, Xiaojin Wu, Tianyuan Zhong, Ying Gao, and Xiaodong Duan

Dalian Minzu University, Dalian 116650, China
ganlin923@126.com

Abstract. Image-text retrieval is a critical challenge for understanding the semantic relationship between vision and language domains. Previous studies have focused on analyzing either global or local features, neglecting the intrinsic connections between these two levels of granularity. In addition, current mainstream methods attempt to construct a unified semantic space by aggregating the weighted features of different segments, aiming to enhance the interaction with different granularities of information. However, these methods may be disrupted by irrelevant segments, leading to semantic misalignment. To address these challenges, we propose a Bidirectional Focused Attention and Global-Local Matching Fusion Network (BFAGL). The proposed method well integrates the similarity matching of global semantic contexts with the precision of local detail analysis, ensuring that global semantic alignment is achieved without compromising critical local information. Furthermore, it incorporates bi-directional focal attention into the local matching process to promote a nuanced understanding of the contextual semantic relationships of images and text. Experiments on the Flickr30K and MS-COCO datasets demonstrate the state-of-the-art performance of BFAGL in image-text retrieval tasks.

Keywords: Image-Text Retrieval · Global-Local Matching · Bidirectional Focused Attention

1 Introduction

In the era of big data, the proliferation of multimodal data has accelerated exponentially, thereby amplifying the significance of multimodal retrieval. Traditional retrieval methods, which are confined to a single modality, are increasingly inadequate when confronted with the vast and varied landscape of multimodal data. Therefore, cross-modal retrieval has emerged as a burgeoning field of study. Cross-modal retrieval offers more convenience and flexibility than traditional unimodal methods, providing related results across modalities for any query. The semantic gap, where image and text representations differ, is a key issue in image-text retrieval. Innovative strategies to bridge this gap are now central to research, focusing on precision and efficiency in retrieval, gaining academic attention.

Previous research has focused on global and local matching. The former involves summarising the overall information inherent in images and text. The latter involves establishing specific links between discrete image regions and corresponding regions of text. However, these studies have been the underappreciation of the imperative to forge interconnections of global and local features. The methods currently in vogue largely depend on pre-established feature representations to quantify the degree of similarity with images and texts. However, these representations frequently fall short of fully encapsulating the contextual and semantic nuances present within the two modalities. In some cases, they may bring interference from irrelevant segments, which hinders the efficiency of the retrieval process. To address these challenges, this paper introduces an innovative framework known as the Bidirectional Focused Attention and Global-Local Matching Fusion Network (BFAGL). This network framework rapidly narrows down the search scope through global semantic matching and incorporates local detail analysis to pinpoint the most relevant images or texts to the query text or image. In addition, BFAGL integrates a bi-directional focused attention mechanism in the local matching process to effectively capture contextual and semantic information in images and texts. This method not only improves the retrieval speed but also ensures the accuracy of the retrieval results. Our contributions can be summarized as follows:

- An innovative architecture is proposed that combines global and local matching methods, which utilizes a combination of global semantics and local details to pinpoint the most relevant image (or text) to the query text (or image), thus improving the accuracy and efficiency of matching.
- A bi-directional focused attention mechanism is introduced that utilizes the effects of image-to-text attention and text-to-image attention to infer the similarity of images and texts.
- Extensive experiments on Flickr30K and MS-COCO have been conducted to verify the efficiency of our methods. The experimental results show that BFAGL outperforms the state-of-the-art methods.

The rest of this paper is organized as follows. Section 2 reviews related works. In Sect. 3, the method of BFAGL is described in detail. Section 4 shows the experimental results and a conclusion is drawn in Sect. 5.

2 Related Works

2.1 Global Matching

Global matching methods aim to deepen the semantic connections between images and texts by integrating their global features, ensuring the consistency of contextual semantics throughout the matching process. For instance, Sarafianos et al. [2] have delved into the application of adversarial representation learning in text-to-image matching, demonstrating the potential of adversarial training to enhance the model's sensitivity to semantic consistency. Subsequently, Zheng et al. [3] introduced an innovative dual-path convolutional image-text embedding model that employs an instance loss mechanism to more efficiently learn global feature representations of images and text. Global matching methods, despite their progress, have limitations in processing local semantic details, potentially hindering the model's fine-grained semantic capture in images and text.

2.2 Local Matching

Local matching methods provide an intricate understanding of the semantic interplay between images and text by precisely correlating specific image regions with particular textual phrases. Such as Chen et al. [6] introduced the methods that employ a recurrent attention mechanism to iteratively refine cross-modal image-text retrieval. Liu et al. [7] developed a graph-structured network that adeptly captures the nuanced relationships across images and text, subsequently boosting the precision of the matching process. Zhang et al. [8] bolstered cross-modal retrieval performance by embedding deep relational insights within their model, which learns and leverages correlations of two modalities to elevate retrieval accuracy. In contrast to the above, Pan et al. [11] focusing on information coding, effectively discerned shared semantics between images and text by implementing a coding framework and a hard-assigned coding scheme. In addition, Li et al. [5] improve accuracy through visual semantic reasoning that delves into the semantic depth of visual contexts. Liu et al. [10] introduced a method that ensures semantic importance across modalities through consistency constraints, addressing traditional approaches' shortcomings in handling semantic variability and alignment inconsistency. Studies show local matching excels at capturing detailed semantics, but over-focusing on details can miss broader context, leading to incomplete semantic understanding.

2.3 Attention Mechanism

Despite significant advancements in global and local matching methods for image-text retrieval, these approaches often rely on initial feature representations to measure the similarity in different modalities, which presents limitations in capturing contextual and semantic information of images and text. To address this challenge, researchers have begun to explore the application of attention mechanisms, aiming to enhance the matching performance of models within complex contexts. Lee et al. [14] proposed the stacked cross-attention network significantly enhances the accuracy of matching by deepening the association with image and text features. Wang et al. [15] optimized the capture of complex relationships of the two modalities through learning a dual-branch neural network. Liu et al. [16] proposed the bidirectional focal attention network introduced further improves matching precision by focusing on key regions in both images and text. Wang et al. [17] developed a position-focused attention network that enhances the capability of bidirectional retrieval. He et al. [18] employed a cross-graph attention mechanism to improve the performance of fine-grained image-text retrieval. Zhang et al. [19] proposed an image-text matching method based on negative perceived attention, which utilizes the positive effects of matched fragments and the negative effects of unmatched fragments to infer image-text similarity together. These studies demonstrate that attention mechanisms significantly enhance the performance of image-text retrieval, especially when dealing with images and text rich in contextual information.

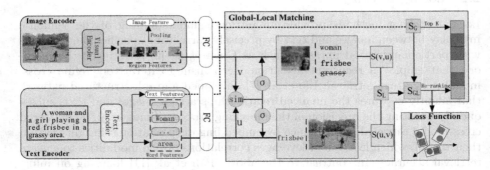

Fig. 1. Overview of the proposed BFAGL framework.

3 Our Methodology

This section presents the Bidirectional Focused Attention and Global-Local Matching for Fusion Networks (BFAGL) as shown in Fig. 1. Section 3.1 explains how to extract features from images and texts. Section 3.2 introduces the fusion network with bidirectional focal attention and local-global matching. Finally, Sect. 3.3 discusses the objective function of training.

3.1 Text Representation and Image Representation

Text Representation. To process the input text, we use a pre-trained BERT model to extract the word-level features and sentence-level features of the text T. Finally, we obtain $T = \{u_1, u_2, ..., u_n, u_g\}$, where $u_1, u_2, ..., u_n$ are the word-level feature representations of the text, n represents the number of word-level features, and u_g is the sentence-level feature representation of the text.

Image Representation. We use the Faster R-CNN to extract local features from the image, and then average pooling is used to get the global feature representation of image I. Finally, we obtain $I = \{v_1, v_2, ..., v_m, v_g\}$, where m represents the number of extracted local features and v_g is the global feature representation of the image I:

$$v_g = \frac{1}{m}\sum_{i=1}^{m} v_i \tag{1}$$

3.2 Fusion Network

Global Matching. Only the global features of both modalities are used for similarity computation during global matching, and here we use cosine similarity to measure the similarity between them:

$$S_G = \frac{v_g^T \times u_g}{\|v_g\| \times \|u_g\|} \tag{2}$$

Local Matching. In local matching, we further enhance the matching effect by introducing a two-way focused attention mechanism.

In the direction of text-to-image, text words are considered as shared semantics, and the task is to identify the image region associated with each text word. This process involves three core steps: pre-allocating attention, identifying the associated image regions, and then reallocating attention based on the results. Specifically, we first assign attention scores to each region in advance, which is achieved by calculating the cosine similarity with the region and the word:

$$w_{ij} = \sigma\left(\alpha \frac{u_i^T \times v_j}{\|u_i\| \times \|v_j\|}\right), i \in [1,m], j \in [1,n] \tag{3}$$

where σ represents the softmax activation function, α is the scaling factor that increases the gap between correlated and uncorrelated regions.

Secondly, when identifying the relevant regions, we distinguish each region based on its attention score about the other regions. If a region's score was above zero, it was considered textually relevant; otherwise, it was deemed textually irrelevant:

$$F(v_{ij}) = \sum_{t=1}^{n} f(v_{ij}, v_{it}) g(u_{it}), i \in [1,m], j \in [1,n] \tag{4}$$

To determine the relative attention between the j-th region and the t-th region, we calculate $f(v_{ij}, v_{it})$ as the difference in their pre-assigned attention scores. We also define the confidence score of the t-th region as the correlation with that region and the i-th query item. This enables us to measure the variation in attention levels across different regions and assess the degree of correlation of them and the query item.

Then, We identify relevant image regions associated with the text and adjust attentional weights for these regions through a renormalization process. This ensures focused attention on the relevant parts of the image:

$$w'_{ij} = \frac{w_{ij} H(v_{ij})}{\sum_{j=1}^{n} w_{ij} H(v_{ij})}, i \in [1,m], j \in [1,n] \tag{5}$$

The shared semantics based on the i-th word selected from the image is computed as $v'_i = \sum_{j=1}^{n} w'_{ij} v_j$. Subsequently, the text-to-image local correlation formula is derived as:

$$S(u, v) = \frac{1}{m} \sum_{i=1}^{m} S(u_i, v'_i) \tag{6}$$

In the image-to-text direction, each image region is associated with a shared semantic feature. The objective is to identify the textual vocabulary that is most relevant to each image region. This process ensures that text words that are highly relevant to the image region receive greater attention, while even those that are not directly relevant contribute to the semantic sharing with the image region and the target text. Subsequently, we score the text words based on these pre-assigned attentional weights, that is:

$$F(u_{ij}) = \sum_{t=1}^{n} f(u_{ji}, u_{it}) g(v_{jt}), i \in [1, m], j \in [1, n] \tag{7}$$

The indicator function $H(u_{ji})$ is then utilized to establish the text vocabulary linked to the image region, based on the calculated score. If a vocabulary is deemed to be an associated word, its value is assigned as 1; otherwise, it is assigned as 0. The attention to the related words will be reallocated:

$$w'_{ji} = \frac{w_{ji} H(u_{ji})}{\sum_{i=1}^{m} w_{ji} H(u_{ji})}, i \in [1, m], j \in [1, n] \tag{8}$$

The attention reallocation will focus on all relevant words using the element-level product and its representation in d-dimensional space. The shared semantics of the j-th region from the text is selected and computed as a weighted combination of related words $u'_j = \sum_{i=1}^{m} w'_{ji} u_i$, where the learned focus attention determines the weights. Subsequently, the local relevance score from image to text is computed:

$$S(v, u) = \frac{1}{n} \sum_{j=1}^{n} S(v_j, u'_j) \tag{9}$$

The concept involves two processes: Text-to-Image Focused Attention and Image-to-Text Focused Attention. The first identifies similar image regions to each text word, while the second selects text words corresponding to each image region's meaning. Considering the intersection and semantic overlap among images and texts, the overall relevance score is computed by integrating the focused attention from both directions. Specifically, the local relevance scores of images and texts in both directions are computed separately and then integrated by summing them to obtain the final local similarity of the image-text pair:

$$S_L = S(\mathbf{u}, \mathbf{v}) + S(\mathbf{v}, \mathbf{u}) \tag{10}$$

Finally, we combine the global similarity with the local similarity obtained above to obtain the hybrid similarity:

$$S_{GL}(I, T) = (1 - \theta) S_G + \theta S_L \tag{11}$$

where θ is the hyperparameter used to adjust the ratio of the two similarities.

Subsequently, the top K candidate samples with the largest global matching similarity are selected, and the top K samples are reordered by applying the hybrid similarity based on global and local matching to obtain the final result.

3.3 Objective Function

To further improve the performance of the designed network, which combines the intra-modal and inter-modal ordering losses during training. The inter-modal loss here takes the triplet loss function commonly used for image-text retrieval:

$$\begin{aligned} L_1 = &\max\left(0, \beta - S(I,T) + S\left(I,T_{l^-}\right)\right) \\ &+ \max\left(0, \beta - S(I,T) + S\left(I_{v^-},T\right)\right) \end{aligned} \tag{12}$$

where β is a hyperparameter used to ensure that the cross-modal distance between negative samples is at least one margin larger than the distance with positive samples. $S(I_{v^-}, T_{v^-})$ and $S(I_{l^-}, T_{l^-})$ are two hard-negative sample pairs, and v^- and l^- are the indexes of samples, respectively.

While this loss pulls mismatched samples of different modalities apart, it lacks constraints between samples of the same modality. As a result, the distances between mismatched samples of the same modality may be very close. Therefore, the loss of contrast within the same modality is defined as:

$$\begin{aligned} L_2 = &\max\left(0, |S\left(I,I_{l^-}\right) - S\left(T,T_{l^-}\right)| - \gamma\right) \\ &+ \max\left(0, |S\left(I,I_{v^-}\right) - S\left(T,T_{v^-}\right)| - \gamma\right) \end{aligned} \tag{13}$$

where γ is a slack variable that ensures that the distances between samples of different modalities can have specific gaps and may not be identical, to relax the constraints on the distances with samples.

The total loss incurred by the model is obtained by adding the inter-modal contrast loss and the intra-modal contrast loss:

$$L = L_1 + L_2 \tag{14}$$

4 Experiments

4.1 Datasets

The **Flickr30K** dataset comprises 31,783 images, each with five corresponding text descriptions. The dataset was split into 29,783 for training, 1,000 for validation, and 1,000 for testing [20]. The **MSCOCO** dataset contains 123,287 images divided into 113,287 for training, 5,000 for validation, and 5,000 for testing. Each image has five corresponding texts, with an average text length of 8.7 [21].

4.2 Evaluation Metric and Results

R@K and RSUM. We will use R@K and RSUM to evaluate the performance of the BFAGL model concerning other methods in the field of image-text retrieval. R@K is defined as the percentage of correctly matched queries in the most recent K retrieval instance. RSUM is the sum of all R@K. The methods we compare are VSE++ [1], VSRN++ [12], HREM [13], and SGRAF [9], SCAN [14], BFAN [16], and NAAF [19]. Table 1 displays the experimental results of our method on the two datasets. Our method outperforms the most advanced methods in most evaluation indicators. Compared to HREM and NAAF, our RSUM on Flickr30k increased by 3.5% and 4.5% respectively. Similarly, the RSUM on MSCOCO increased by 3.5% and 3.7% respectively.

Table 1. The results of our method on two datasets. The bests are in bold.

Datasets	Methods	Image-to-Text			Text-to-Image			
		R@1	R@5	R@10	R@1	R@5	R@10	RSUM
MS-COCO	VSE++ [1]	64.6	90.0	95.7	52.0	84.3	92.0	478.6
	SCAN [14]	70.9	94.5	97.8	56.4	87.0	93.9	500.5
	BFAN [16]	74.9	95.2	-	59.4	88.4	-	317.9
	VSRN++ [12]	77.9	96.0	98.5	64.1	91.0	96.1	523.6
	SGRAF [9]	79.6	96.2	98.5	63.2	90.7	97.5	525.7
	NAAF [19]	80.5	96.5	98.5	64.1	90.7	**96.5**	526.8
	HREM [13]	81.2	96.5	**98.9**	63.7	90.7	96.0	527.0
	BFAGL	**82.6**	**96.9**	98.2	**64.8**	**92.1**	96.3	**530.9**
Flickr30K	VSE++ [1]	52.9	80.5	87.2	39.6	70.1	79.5	409.8
	SCAN [14]	67.9	89.0	94.4	43.9	74.2	82.8	452.2
	BFAN [16]	68.1	91.4	-	50.8	78.4	-	288.7
	VSRN++ [12]	79.2	94.6	97.5	60.6	**85.6**	91.4	508.9
	SGRAF [9]	77.8	94.1	97.4	58.5	83.0	**88.8**	499.6
	NAAF [19]	81.9	96.1	98.3	61.0	85.3	90.6	513.2
	HREM [13]	81.4	96.5	**98.5**	60.9	85.6	91.3	514.2
	BFAGL	**83.1**	**97.2**	98.4	**62.3**	85.2	**91.5**	**517.7**

Precision-Recall Curve. We draw the precision-recall (PR) curve to show the accuracy rate changing with the recall rate using different methods. Figure 2 shows that the proposed method has a larger area under the PR curve than other comparison methods.

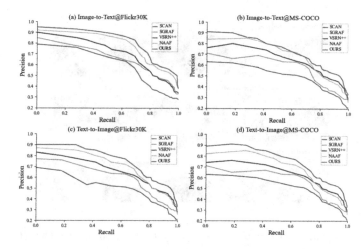

Fig. 2. PR curves for cross-modal retrieval tasks based on two datasets.

Time Complexity. Global methods offer higher efficiency but lower accuracy due to less focus on detailed features. In contrast, local methods improve accuracy by emphasizing context and local semantics, though at a higher time cost. While BFAGL strikes a satisfactory balance inter accuracy and efficiency. To comprehensively evaluate the performance of different methods, a comprehensive comparison was performed on the Flickr30K test set, as shown in Fig. 3. The time complexity of BFAGL was compared with that of the existing algorithms VSRN++ [12], HREM [13], SCAN [14], SGRAF [9], and NAAF [19]. The results demonstrated that BFAGL outperforms the other models in balancing accuracy and efficiency.

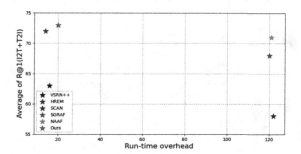

Fig. 3. The comparison between accuracy and speed for cross-modal retrieval.

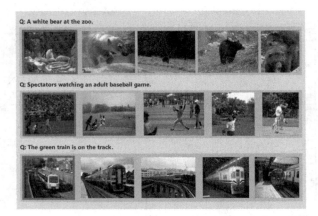

Fig. 4. The retrieval results of text-to-image by BFAGL on MS-COCO. (Color figure online)

4.3 Visualization of Results

To visually demonstrate the retrieval performance of the proposed BFAGL, we present several retrieval examples on the MS-COCO dataset. The results are shown in Figs. 4 and 5. In Fig. 4, the top-5 retrieved images are arranged from left to right, with the images in bold red boxes being the ones matched with the sentence queries. The top-1 retrieved images indicate that the model integrates both global and local features, combining information to ensure comprehensive matching, even if there is partial similarity in regional features. Figure 5 shows the retrieval results for the image-to-text task, with the top-5 retrieved sentences listed in descending order. It can be seen that almost all matching sentences were retrieved. Bidirectional focused attention in complex scene matching effectively targets relevant areas, minimizes interference, and notably enhances performance.

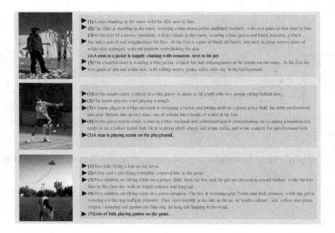

Fig. 5. The retrieval results of image-to-text by BFAGL on MS-COCO.

5 Conclusion

In this study, an innovative fusion model known as the Bidirectional Focused Attention and Global-Local Matching Fusion Network (BFAGL) is introduced. This model is designed to balance accuracy and efficiency in the retrieval process by integrating both global and local matching methods. Additionally, an attention mechanism is incorporated within the local matching process, further enhancing the model's performance. Inter-modal loss and intra-modal loss are designed to maintain high-level semantic feature similarity among different modalities. Experiments on two datasets show that our method outperforms state-of-the-art methods. Future research will extend the network to more multimedia data types to better explore their semantic similarity while balancing accuracy and efficiency.

Acknowledgments. This work is supported by the National Natural Science Foundation of China (Grant No. 61602085).

References

1. Faghri, F., Fleet, D.J., Kiros, J.R., Fidler, S.: VSE++: improving visual-semantic embeddings with hard negatives. arXiv preprint arXiv:1707.05612 (2017)
2. Sarafianos, N., Xu, X., Kakadiaris, I.A.: Adversarial representation learning for text-to-image matching. In: Proceedings of the IEEE/CVF International Conference on Computer Vision, pp. 5814–5824 (2019)
3. Zheng, Z., Zheng, L., Garrett, M., Yang, Y., Xu, M., Shen, Y.-D.: Dual-path convolutional image-text embeddings with instance loss. ACM Trans. Multimedia Comput. Commun. Appl. (TOMM) **16**(2), 1–23 (2020)
4. Ji, Z., Wang, H., Han, J., Pang, Y.: Saliency-guided attention network for image-sentence matching. In: Proceedings of the IEEE/CVF International Conference on Computer Vision, pp. 5754–5763 (2019)
5. Li, K., Zhang, Y., Li, K., Li, Y., Fu, Y.: Visual semantic reasoning for image-text matching. In: Proceedings of the IEEE/CVF International Conference on Computer Vision, pp. 4654–4662 (2019)
6. Chen, H., Ding, G., Liu, X., Lin, Z., Liu, J., Han, J.: IMRAM: iterative matching with recurrent attention memory for cross-modal image-text retrieval. In: Proceedings of the IEEE/CVF Conference on Computer Vision and Pattern Recognition, pp. 12655–12663 (2020)
7. Liu, C., Mao, Z., Zhang, T., Xie, H., Wang, B., Zhang, Y.: Graph structured network for image-text matching. In: Proceedings of the IEEE/CVF Conference on Computer Vision and Pattern Recognition, pp. 10921–10930 (2020)
8. Zhang, Y., Zhou, W., Wang, M., Tian, Q., Li, H.: Deep relation embedding for cross-modal retrieval. IEEE Trans. Image Process. **30**, 617–627 (2020)
9. Diao, H., Zhang, Y., Ma, L., Lu, H.: Similarity reasoning and filtration for image-text matching. In: Proceedings of the AAAI Conference on Artificial Intelligence, vol. 35, no. 2, pp. 1218–1226 (2021)
10. Liu, Z., Chen, F., Xu, J., Pei, W., Lu, G.: Image-text retrieval with cross-modal semantic importance consistency. IEEE Trans. Circuits Syst. Video Technol. **33**, 2465–2476 (2022)

11. Pan, Z., Wu, F., Zhang, B.: Fine-grained image-text matching by cross-modal hard aligning network. In: Proceedings of the IEEE/CVF Conference on Computer Vision and Pattern Recognition, pp. 19275–19284 (2023)
12. Li, K., Zhang, Y., Li, K., Li, Y., Fu, Y.: Image-text embedding learning via visual and textual semantic reasoning. IEEE Trans. Pattern Anal. Mach. Intell. **45**(1), 641–656 (2022)
13. Fu, Z., Mao, Z., Song, Y., Zhang, Y.: Learning semantic relationship among instances for image-text matching. In: Proceedings of the IEEE/CVF Conference on Computer Vision and Pattern Recognition, pp. 15159–15168 (2023)
14. Lee, K.-H., Chen, X., Hua, G., Hu, H., He, X.: Stacked cross attention for image-text matching. In: Ferrari, V., Hebert, M., Sminchisescu, C., Weiss, Y. (eds.) ECCV 2018. LNCS, vol. 11208, pp. 212–228. Springer, Cham (2018). https://doi.org/10.1007/978-3-030-01225-0_13
15. Wang, L., Li, Y., Huang, J., Lazebnik, S.: Learning two-branch neural networks for image-text matching tasks. IEEE Trans. Pattern Anal. Mach. Intell. **41**(2), 394–407 (2018)
16. Liu, C., Mao, Z., Liu, A., Zhang, T., Wang, B., Zhang, Y.: Focus your attention: a bidirectional focal attention network for image-text matching. In: Proceedings of the 27th ACM International Conference on Multimedia, pp. 3–11 (2019)
17. Wang, Y., et al.: PFAN++: bi-directional image-text retrieval with position focused attention network. IEEE Trans. Multimed. **23**, 3362–3376 (2020)
18. He, Y., Liu, X., Cheung, Y.-M., Peng, S.-J., Yi, J., Fan, W.: Cross-graph attention enhanced multi-modal correlation learning for fine-grained image-text retrieval. In: Proceedings of the 44th International ACM SIGIR Conference on Research and Development in Information Retrieval, pp. 1865–1869 (2021)
19. Zhang, K., Mao, Z., Wang, Q., Zhang, Y.: Negative-aware attention framework for image-text matching. In: Proceedings of the IEEE/CVF Conference on Computer Vision and Pattern Recognition, pp. 15661–15670 (2022)
20. Young, P., Lai, A., Hodosh, M., Hockenmaier, J.: From image descriptions to visual denotations: new similarity metrics for semantic inference over event descriptions. Trans. Assoc. Comput. Linguist. **2**, 67–78 (2014)
21. Lin, T.-Y., et al.: Microsoft COCO: common objects in context. In: Fleet, D., Pajdla, T., Schiele, B., Tuytelaars, T. (eds.) ECCV 2014. LNCS, vol. 8693, pp. 740–755. Springer, Cham (2014). https://doi.org/10.1007/978-3-319-10602-1_48

Adversarial Graph Convolutional Network Hashing for Cross-Modal Retrieval

Bo Lu, Tianbao Zhao(✉), Guiyuan Liang, Jiaming Li, and Xiaodong Duan

Dalian Minzu University, Dalian 116650, China
838401449@qq.com

Abstract. Graph-based hashing has recently emerged as a prominent research area due to the rapid expansion of multimodal datasets. Most existing graph-based methods rely on Graph Convolutional Networks (GCNs) as a guiding mechanism, primarily because of the unique characteristics of cross-modal retrieval tasks. However, these methods often encounter several challenges: 1) They struggle to preserve the inherent similarities within the original datasets. 2) They frequently employ intractable binary quadratic programs, which limits their scalability when dealing with extensive data volumes. In this paper, we introduce an innovative framework termed Adversarial Graph Convolutional Network Hashing (AGCNH), which leverages dual-cycle Generative Adversarial Networks (GANs) alongside GCNs. This framework incorporates two asymmetric graph layers and a standard graph layer, effectively addressing the issue of excessive parameterization in GCNs and enabling the direct generation of binary codes. To demonstrate the effectiveness of our proposed method, we have conducted experiments on four widely used datasets, thereby validating the robustness and applicability of our approach.

Keywords: Cross-Modal Retrieval Adversarial · Graph Convolutional Network · Hashing

1 Introduction

In the current landscape, there is a growing emphasis on developing effective and efficient search strategies, largely in response to the proliferation of multimedia data. Among these strategies, cross-modal retrieval has emerged as a notable area of exploration, aiming to identify images (or texts) that are relevant to a given text (or image) query. The primary objective of cross-modal retrieval is to extract specific subjects from diverse modal data sources, with text and image modalities being the most commonly encountered. As depicted in Fig. 1, an image can be utilized to retrieve pertinent text within the same category or topic, while conversely, text can be leveraged to retrieve relevant images within the same category or topic. Furthermore, the hashing method has garnered significant interest due to its ability to convert high-dimensional, multi-modal data points into a single Hamming space, thereby assigning identical hash codes to similar cross-modal data points. This method is particularly appealing due to its efficient storage requirements and swift query times.

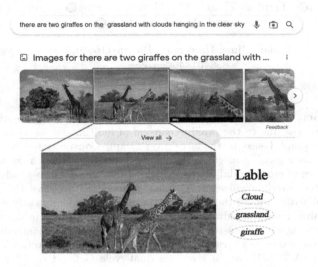

Fig. 1. The example of cross-modal retrieval in common search engine

Many cross-modal hashing techniques have been successfully used up to this point to solve or facilitate real-world issues. Supervised hashing methods [20,24] learn hash codes under the supervision of semantic labels or their own generation. Unsupervised hashing methods [15,21] usually depends on the advanced potential semantic information between training data. The convolutional neural network model's potent feature representations are used by the majority of these techniques to create their hash functions. However, it is inevitable that each model will include features generated from multi-modal data. Therefore, accurate cross-modal hashing continues to face difficulties due to the modality divide.

Adversarial networks have received a lot of interest because they have the potential to bridge the modality gap and further reduce the distance between various modalities. For instance, Unsupervised Generative Adversarial Crossmodal Hashing (UGACH) [17] uses the GAN and unsupervised learning method to learn similar structural information between cross-modal data. Adversary Guided Asymmetric Hashing (AGAH) [4] uses the multi-label semantics guided by confrontation learning to enhance the feature learning.

A great number of approaches for retrieval using graph representations have been proposed in recent years, such as [2,3,10,11]. Among them, graph-base hashing has become very successful at learning highly discriminative features in cross-modal retrieval, Inter-Media Hashing [13] revealing the semantic relevance among various media types through manifold learning, Local Graph Convolutional Networks Hashing(LGCNH) [1] builds local graph in accordance with the neighborhood relationship between samples in the depth feature space, Aggregation-Based Graph Convolutional Hashing [19] using multiple similarity measures to construct more accurate affinity matrix for learning.

While numerous graph-based hashing techniques for cross-modal retrieval have shown promising results, the majority are limited to supervised data. This paper proposes a novel approach aiming to generate binary code using a graph convolution layer. However, due to intractable binary quadratic programs, scalability to massive datasets remains a challenge for graph-based hashing methods. Furthermore, integrating different model features into the graph convolution framework requires mapping them to the same representation. Additionally, scalability to large datasets poses a significant hurdle for graph-based hashing, and addressing the issue of out-of-sample extension remains a challenging problem.

- We propose a novel AGCNH framework, which combines dual adversarial neural networks with a graph convolution layer. Powerful common representation can be learned in the adversarial neural networks, and output reliable hash codes after graph convolution layer.
- This is the first work that use graph convolution layer to direct generate binary code. In addition, we merge the asymmetric graph convolutional to simultaneously convolve the anchor graph, through which the problem of large parameter quantity of graph convolution network is well solved.
- Tests on four benchmarks show that our proposed AGCNH significantly outperforms the most advanced cross-modal hashing techniques currently available, including both conventional and deep learning based techniques.

The rest of this paper is organized as follows. We will introduce the related work in Sect. 2.

2 Related Work

In this section, we introduce the related cross-modal work, including Hashing and Graph Convolutional Networks.

According to whether supervised information is used, the cross-modal hashing methods can be divided into two categories: supervised and unsupervised methods. For unsupervised cross-modal hashing methods, Unsupervised coupled Cycle generative adversarial Hashing networks (UCH) [9] design an unsupervised coupled cycle generative adversarial hashing network, powerful common representation and reliable hash codes can be learned simultaneously in an unified framework.

Supervised methods directly utilize label information to supervise the learning procedure. As the first CMH method to integrate feature learning and hash code learning into the same deep learning framework, Deep cross-modal hashing [5] forces hash codes to keep semantic relevance for similar data points using similarity matrix in an obvious way, without paying too much attention to latent structure of cross-modal data. Self-Supervised Adversarial Hashing [8] is also learn hash codes by preserving label similarity correlation, but the algorithm is very time-consuming.

With the development of graph, Graph neural networks have attracted high attention in various research areas. Graph Convolutional Networks (GCNs) [6] is based on an efficient variant of convolutional neural networks which integrates both local graph structure and features of nodes. Graph Convolutional Hashing [16] learns modality-unified binary codes via an affinity graph, it notably outperforms other cross-modal hashing methods at that time. Aggregation-based Graph Convolutional Hashing [18] is the first work that applies GCNs in unsupervised hashing learning which can fully considers the neighborhood relevance in the learning procedure. There are two difficulties in the application of GCNs, it can hardly scale to massive datasets in the main scenario of applying hashing techniques and the issue of out-of-sample extension. Graph Convolutional Networks Hashing (GCNH) [22] introduce a novel asymmetric graph convolutional layer which convolves the anchor graph and well solved the above problems in image retrieval. In this paper, our work utilizes generative adversarial network to map different modality features to common representations, then employs GCNH to direct generate binary code.

3 Notation and Problem Definition

Suppose that the training set $O = \{o_i\}_{i=1}^n$ contains n data points, $o_i = (x_i, y_i, l_i)$, where x_i an y_i are original image and text of the i-th data point, and $l_i = [l_{i1}, ..., l_{ic}]$ is the ground belongs to the j-th class, $l_{ij} = 1$, otherwise $l_{ij} = 0$. In addition, we utilize multi-label similarity matrix S to denote similarity of two data points, if they belong to at least one same class, $S_{mn} = 1$, otherwise $S_{mn} = 0$. The goal of cross-modal hashing is to generate reliable hash codes for image and text: $B \in \{-1, 1\}^K$, where K is the length of binary codes. $L \in \{1, 0\}^{n1 \times c}$ denotes the corresponding labels of the labeled data from l class. The outputs of GCN layers are defined as $H^*, * \in \{v, t\}$ for image and text respectively, binary hash code B^* are generated by applying a sign function to H^*:

$$B^* = sign(H^*), * \in \{v, t\} \tag{1}$$

4 Methodology

In this study, we concentrate on retrieval between the two modalities that are utilized the most frequently in daily life: text and image. The suggested AGCNH model's flowchart is depicted in Fig. 2. It consists of two main components: Dual circulation generator adversarial networks and Graph Convolutional Networks, which will both be specifically discussed in the framework chapter.

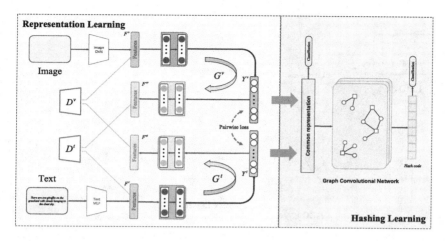

Fig. 2. The flowchart of our AGCNH method. First, input the image and text into CNN and MLP respectively to obtain the initial feature representation called f^v and f^t. In the dual cycle generative adversarial networks, the image feature f^v generates a fake text feature $f^{v'}$ through the generator G^v. Similarly, the text feature f^t generates a fake text feature $f^{t'}$ through the generator G^t. The middle representation is input into the hash learning module as a public representation, and the corresponding hash code is finally generated through GCN

4.1 Representation Learning

Representation Learning include a cycle generator adversarial layer which has two generator networks for image and text features, it goal is to obtain reliable common representation for image and text. By training these two networks, powerful individual representation can be learned for different modalities and thus modality gap can be bridged effectively.

Input image-text pair (x, y), though the feature encoding functions $F^* = E^*(*, \theta^*), * \in \{v, t\}$ with convolutional neural network for image and Multi-Layer Perception for text, where θ^* are network parameters. Within the cycle GAN, the generation can be formulated as:

$$F^{*'} = G^*(F^*, \eta^*), * \in \{v, t\} \qquad (2)$$

where $F^{*'}$ are fake image and fake text representation, G^* are two generation functions, η^* are network parameters. Moreover, to judge the quality of the generative data, we further devise two discriminators $D^*(\cdot, \mu^*), * \in \{v, t\}$ to provide adversarial loss for generators, where μ^* are network parameters. Finally, we hope to generate similar representations Y^v and Y^t output for common representations.

In addition, we construct a classification layer to supervise the generation of common representation, it consists of fully connected layer.

To enable GCN module to learn more multidimensional structural features, we need to fuse common representation without losing too much semantic

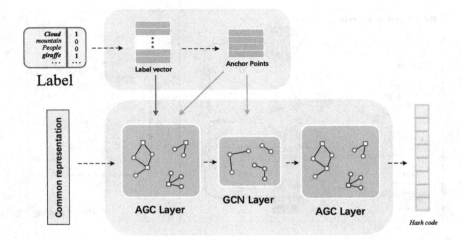

Fig. 3. The flowchart of the Graph Convolutional Layer. During the training phase, we sample a portion of the training set's data and choose a random subset of anchor points. The GCN is simultaneously fed the data batch, the asymmetric graph between the sampled data and anchor points, and the anchor graph.

relevance. This work use self-attention for structural-preserving fusing method and it can be formulated as:

$$F_g = norm(y^t \oplus y^v)$$
$$y^t = Y^t \times (Y^v \times Y^{t\top}) \qquad (3)$$
$$y^t = Y^v \times (Y^v \times Y^{t\top})$$

Here, $norm$ is matrix normalization, $\oplus$ denotes matrix concat and $\times$ denotes matrix inner product.

4.2 Hashing Learning

Hashing Learning include graph convolutional layers and classification layer. Figure 3 is the flowchart of the Graph Convolutional Layer, it consists of three parts: two asymmetric graph convolutional layer and a graph convolutional layer.

Common representation is fed into graph convolutional layers along with the asymmetric adjacency matrix Z of $n \times p$, where p is the randomly selected anchor points. Here, Z is denote similarity of two data points, if they belong to at least one same class, $Z_{ij} = 1$, otherwise $Z_{ij} = 0$.

Suggested by GCNH, the graph convolutional layers in our AGCNH can take the following form:

$$H^{(1)} = \sigma(\tilde{D}^{\top -\frac{1}{2}} \tilde{Z}^\top \tilde{D}^{\top -\frac{1}{2}} H^{(0)} W^{(1)}) \qquad (4)$$

$$H^{(2)} = \sigma(\tilde{D'}^{-\frac{1}{2}} \tilde{A} \tilde{D'}^{-\frac{1}{2}} H^{(1)} W^{(2)}) \qquad (5)$$

$$H^{(3)} = \sigma(\tilde{D}^{-\frac{1}{2}}\tilde{Z}\tilde{D}^{-\frac{1}{2}}H^{(2)}W^{(3)}) \qquad (6)$$

Here, $H^{(0)} = F_g$ and $\tilde{Z} = Z + I_N$ is the normalized adjacency matrix of the asymmetric graph. $\tilde{A} = A + I_N$ is the normalized adjacency matrix of the second graph, it can be computed as $A(i,j) = l_i \times l_j$, here l_i is the ground truth label of data point i. I_N is the identity matrix, indicating every node is connected to itself. $\tilde{D}_{ii} = \sum_j \tilde{Z}_{ij}$ is the degree matrix of $\tilde{Z}$ and $\tilde{D}'_{ii} = \sum_j \tilde{A}_{ij}$ is the degree matrix of $\tilde{A}$. $\sigma(\cdot)$ denotes an activation function. $H^{(l-1)}$ represents the features learned by the preceding layers and $W^{(l)}$ denotes the convolutional filter.

Though the graph convolutional layers, the soft binary codes can be learned: $U = H^{(3)}$. Final binary codes is $B = sign(U)$, where $sign(\cdot)$ is an element-wise sign function which is defined as follows: $sign(\cdot) = 1$ if $x > 0$, otherwise $sign(\cdot) = -1$.

In this paper, out-of-sample data's features will feed into graph convolutional layers with classification results from the Representation Learning layer. Through this and asymmetric layer, we can overcome the out-of-sample problem.

Algorithm 1. Adversarial Graph Convolutional Network Hashing for Cross-Modal Retrieval

---
Input: The training data: $O = \{o_i\}_{i=1}^n$, hash code length: c, balance parameters: α, β.
Output: Learned network parameters $\theta^{v,t}, \eta^{v,t}$; the parameters $W(l)$ of all layers of graph convolutional layers; Binary codes B.
Initialize: Initialize network parameters $\theta^{v,t}, \eta^{v,t}, \mu^{v,t}, W(l)$.
repeat: Update $\eta^{v,t}$ with BP algorithm;
  Update $\theta^{v,t}$ and $\mu^{v,t}$ with BP algorithm;
  Update $W(l)$ with BP algorithm;
  Update hash codes matrix B;
until convergent.

4.3 Objective Function

Our goal is to obtain reliable hash codes for image and text. In Representation Learning, we can learn the common representation of different modalities via minimizing the classification error and maximizing the dissimilarity of samples from different categories. In addition the modality gap will be broken by the interaction between generators and discriminators.

The generative adversarial losses can be written as:

$$\mathcal{L}_{gan} = \mathcal{L}_{ad}^v + \mathcal{L}_{ad}^t + \mathcal{L}_{re}^v + \mathcal{L}_{re}^t + \mathcal{L}_{co} \qquad (7)$$

where $\mathcal{L}_{ad}^v, \mathcal{L}_{ad}^t, \mathcal{L}_{re}^v$ and $\mathcal{L}_{re}^t$ are formulated as follows:

$$\mathcal{L}_{ad}^v = \min\max \ \log D^v(F^v) + \log(1 - D^v(F'^v)) \qquad (8)$$

$$\mathcal{L}_{co} = \min \sum \|Y^v - Y^t\|_2^2 \qquad (9)$$

Finally, The following objective function is defined to measure the classification loss:
$$\mathcal{L}_{cl} = \min ||L - Y^v W^{(4)}||_2^2 + ||L - Y^t W^{(5)}||_2^2 \tag{10}$$
where $W^{(4)}$ and $W^{(5)}$ retains the parameters of the fully connected layer which works for classifies the common representation to corresponding classes. As to the representation learning, the whole loss function can be defined as:
$$\mathcal{L}_1 = \alpha \mathcal{L}_{gan} + \mathcal{L}_{cl} \tag{11}$$

In Hashing Learning, We expect the learned binary codes to preserve the original semantic/similarity structure. For this purpose, the binary codes of images in the same class or similar images in terms of the Euclidean distance should be as close as possible, while the binary codes of dissimilar images should be far away. We propose to minimize the classification error over them by considering the binary code learning as a linear classification problem, leading to the following objective function:
$$\mathcal{L}_2 = \min \sum ||L - BW^{(6)}||_2^2 \tag{12}$$
Combining the two aforementioned objective functions, the overall objective function of AGCNH can be formulated as:
$$\mathcal{L} = \beta \mathcal{L}_1 + \mathcal{L}_2 \tag{13}$$

5 Experiments

In order to evaluate the effectiveness of our AGCNH, the experiment included four commonly used cross modal datasets: MIRFlickr25K and NUS-WIDE-10K. Each data set is divided into training and test data, which are described in detail as follows:

MIRFlickr25K is a benchmark dataset collected from Flickr. It has 25000 pictures, and each picture has corresponding tags and annotations. Tags can be used as text descriptions, of which 1386 tags appear in at least 20 pictures; As labels, there are 24 annotations in total. We select 2000 pairs of image-text pairs as the test set and other 18015 pairs as training set.

NUS-WIDE-10K dataset is the largest multi label dataset in the field of cross modal retrieval. This dataset contains 269648 samples, which can be divided into 81 classes. We use the most commonly used 20 tags as the dataset for retrieval. The first step is to divide the data set into training set and test set; In the test set, each class selects 1000 image-text pairs to form a query set of 10000 images.

Pascal Sentence contains 1000 pairs of images and texts from VOC 2008, most of which have 5 descriptions. It has 20 categories, each with 50 image text pairs. 60% of the data pairs are selected as training set, and 40% of the data pairs are selected as test set.

Wikipedia is the dataset used for retrieval, it contains 2866 samples, 10 classes, and two models of image and text. According to the order of arrangement, we selected the first 2173 as training sets and the last 693 as test sets.

Table 1. MAP comparison results in MIRFlickr25K and NUS-WIDE-10K, where the best performance is boldfaced.

Method	MIRFlickr25K						NUS-WIDE-10K					
	I2T			T2I			I2T			T2I		
	16	32	64	16	32	64	16	32	64	16	32	64
DAGNN	0.583	0.585	0.591	0.579	0.582	0.589	0.581	0.586	0.588	0.584	0.587	0.592
AGCH	0.612	0.615	0.630	0.610	0.617	0.629	0.615	0.621	0.634	0.619	0.620	0.636
DCMH	0.486	0.510	0.518	0.491	0.513	0.521	0.479	0.493	0.501	0.487	0.508	0.512
SSAH	0.497	0.516	0.523	0.505	0.517	0.518	0.488	0.502	0.514	0.499	0.516	0.523
UCH	0.501	0.528	0.537	0.513	0.536	0.534	0.511	0.525	0.528	0.503	0.522	0.541
GCNH	0.532	0.539	0.540	0.535	0.523	0.547	0.527	0.539	0.535	0.530	0.543	0.550
Bi-CMR	0.623	0.635	0.638	0.613	0.632	0.636	0.628	0.629	0.633	0.622	0.643	0.639
AGCNH	**0.664**	**0.673**	**0.686**	**0.658**	**0.668**	**0.670**	**0.660**	**0.677**	**0.682**	**0.665**	**0.684**	**0.679**

Table 2. MAP comparison results in Pascal Sentence and Wikipedia, where the best performance is boldfaced.

Method	Pascal Sentence						Wikipedia					
	I2T			T2I			I2T			T2I		
	16	32	64	16	32	64	16	32	64	16	32	64
DAGNN	0.643	0.647	0.651	0.650	0.654	0.670	0.573	0.579	0.591	0.655	0.656	0.660
AGCH	0.677	0.678	0.683	0.669	0.680	0.693	0.611	0.619	0.620	0.683	0.688	0.692
DCMH	0.513	0.525	0.536	0.482	0.488	0.494	0.511	0.511	0.519	0.483	0.484	0.490
SSAH	0.523	0.525	0.525	0.495	0.509	0.511	0.510	0.521	0.526	0.489	0.497	0.509
UCH	0.564	0.568	0.567	0.563	0.571	0.576	0.562	0.567	0.576	0.531	0.542	0.541
GCNH	0.560	0.572	0.581	0.563	0.564	0.572	0.546	0.552	0.557	0.588	0.589	0.589
Bi-CMR	0.665	0.671	0.672	0.658	0.661	0.665	0.633	0.641	0.644	0.679	0.677	0.682
AGCNH	**0.685**	**0.690**	**0.693**	**0.688**	**0.689**	**0.696**	**0.668**	**0.671**	**0.672**	**0.691**	**0.701**	**0.704**

5.1 Baselines and Evaluation

We compare with two adversarial cross-modal retrieval methods and five state-of-the-art deep hashing cross-modal retrieval methods to illustrate the effectiveness of our AGCNH, including DAGNN [12], AGCN [2], DCMH [5], SSAH [8], AGAH [18], and Bi-CMR. We adopt two common used protocols Mean Average Precision (MAP) and Precision-Recall curves (PR curves), to evaluate the retrieval performance of all methods in our experiments. We compare the proposed AGCNH with all baselines in the Image-to-Text and Text-to-Image search tasks with code length varying from 16 bits to 64 bits.

5.2 Experiment Results

Table 1 and 2 reports the MAP results for our proposed AGCNH and the compared the-state-of-art methods on four benchmark datasets. The results show that on all four datasets, our suggested AGCNH has two significant benefits: 1) AGCN performs appreciably better than the compared methods; 2) Within a certain range, the larger the data set is, the more substantial the advancement

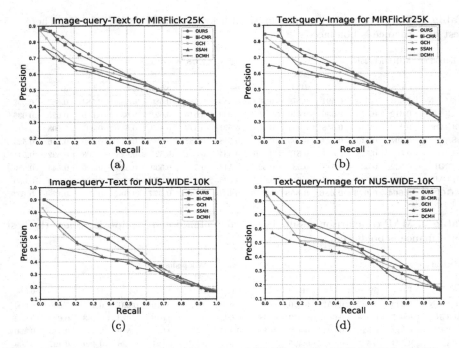

Fig. 4. Precision-recall curves evaluated on MIRFlickr25k, NUS-WIDE-10K, Pascal Sentence and Wikipedia. The code length is 64.

of the model we supply. In particular, while on NUS-WIDE, it surpasses the second best competitor by 4.1%–4.9% in the task.

Figure 4 show the precision-recall curves evaluated on MIRFlickr25k, NUS-WIDE-10K, Pascal Sentence and Wikipedia. We adjusted the length of cross-modal hashing methods to 64 bits in order to ensure the comparability of experimental outcomes because most hashing methods often produce satisfactory results when the hash code length is 64 bits.

5.3 Ablation Studies

We define five alternatives, AGCNH1, AGCNH2, AGCNH3, AGCNH without GAN, and AGCNH without GCN, to study the impact of different components and independently training strategy. Here, AGCNH1 does not consider forward learning. AGCNH2 trains multilabel network without considering $\mathcal{L}_{cl}$. AGCNH3 does not consider $\mathcal{L}_{co}$. For fair comparison, all of these variants use the same network architecture settings and evaluation indicators.

In order to evaluate the influence of hyper-parameters in Eq. 13 and Eq. 15, we separately tune α in $\{10, 100, 1000\}$ and β in $\{0.01, 0.1, 1\}$. The variety of MAP can be seen from Fig. 2.

6 Conclusion

In this paper, a graph convolutional hashing network with generative adversarial network is proposed for cross-modal retrieval. First, to break the modality gap we proposed a cycle generator adversarial layer which has two generator networks for image and text features. Then we use the features input to a graph convolutional Network and get their corresponding hashing code. We construct datasets with MIRFlickr25K and extensive experiment results have proved the high effectiveness of our ACGNH.

Acknowledgements. This work is supported by the National Natural Science Foundation of China (Grant No. 61602085).

References

1. Chen, Y., Wang, S., Lu, J., Chen, Z., Zhang, Z., Huang, Z.: Local graph convolutional networks for cross-modal hashing. In: Proceedings of the 29th ACM International Conference on Multimedia, pp. 1921–1928 (2021)
2. Dong, X., Liu, L., Zhu, L., Nie, L., Zhang, H.: Adversarial graph convolutional network for cross-modal retrieval. IEEE Trans. Circuits Syst. Video Technol. **32**(3), 1634–1645 (2021)
3. Duvenaud, D.K., et al.: Convolutional networks on graphs for learning molecular fingerprints. In: Advances in Neural Information Processing Systems, vol. 28 (2015)
4. Gu, W., Gu, X., Gu, J., Li, B., Xiong, Z., Wang, W.: Adversary guided asymmetric hashing for cross-modal retrieval. In: Proceedings of the 2019 on International Conference on Multimedia Retrieval, pp. 159–167 (2019)
5. Jiang, Q.Y., Li, W.J.: Deep cross-modal hashing. In: Proceedings of the IEEE Conference on Computer Vision and Pattern Recognition, pp. 3232–3240 (2017)
6. Kipf, T.N., Welling, M.: Semi-supervised classification with graph convolutional networks (2016)
7. Kipf, T.N., Welling, M.: Semi-supervised classification with graph convolutional networks. arXiv preprint arXiv:1609.02907 (2016)
8. Li, C., Deng, C., Li, N., Liu, W., Gao, X., Tao, D.: Self-supervised adversarial hashing networks for cross-modal retrieval. In: 2018 IEEE/CVF Conference on Computer Vision and Pattern Recognition (2018)
9. Li, C., Deng, C., Wang, L., Xie, D., Liu, X.: Coupled cycleGAN: unsupervised hashing network for cross-modal retrieval (2019)
10. Niepert, M., Ahmed, M., Kutzkov, K.: Learning convolutional neural networks for graphs. In: International Conference on Machine Learning, pp. 2014–2023. PMLR (2016)
11. Qian, S., Xue, D., Zhang, H., Fang, Q., Xu, C.: Dual adversarial graph neural networks for multi-label cross-modal retrieval. In: Proceedings of the AAAI Conference on Artificial Intelligence, vol. 35, pp. 2440–2448 (2021)
12. Qian, S., Xue, D., Zhang, H., Fang, Q., Xu, C.: Dual adversarial graph neural networks for multi-label cross-modal retrieval. In: Thirty-Fifth AAAI Conference on Artificial Intelligence, AAAI 2021, Thirty-Third Conference on Innovative Applications of Artificial Intelligence, IAAI 2021, The Eleventh Symposium on Educational Advances in Artificial Intelligence, EAAI 2021, Virtual Event, 2–9 February 2021, pp. 2440–2448. AAAI Press (2021). https://ojs.aaai.org/index.php/AAAI/article/view/16345

13. Song, J., Yang, Y., Yang, Y., Huang, Z., Shen, H.T.: Inter-media hashing for large-scale retrieval from heterogeneous data sources. In: Proceedings of the 2013 ACM SIGMOD International Conference on Management of Data, pp. 785–796 (2013)
14. Wang, D., Cui, P., Ou, M., Zhu, W.: Deep multimodal hashing with orthogonal regularization. In: Twenty-Fourth International Joint Conference on Artificial Intelligence (2015)
15. Wu, B., Yang, Q., Zheng, W.S., Wang, Y., Wang, J.: Quantized correlation hashing for fast cross-modal search. In: Twenty-Fourth International Joint Conference on Artificial Intelligence (2015)
16. Xu, R., Li, C., Yan, J., Deng, C., Liu, X.: Graph convolutional network hashing for cross-modal retrieval. In: Twenty-Eighth International Joint Conference on Artificial Intelligence IJCAI-19 (2019)
17. Zhang, J., Peng, Y., Yuan, M.: Unsupervised generative adversarial cross-modal hashing. In: Proceedings of the AAAI Conference on Artificial Intelligence, vol. 32 (2018)
18. Zhang, P.F., Li, Y., Huang, Z., Xu, X.S.: Aggregation-based graph convolutional hashing for unsupervised cross-modal retrieval. IEEE Trans. Multimed. 1 (2021)
19. Zhang, P.F., Li, Y., Huang, Z., Xu, X.S.: Aggregation-based graph convolutional hashing for unsupervised cross-modal retrieval. IEEE Trans. Multimed. **24**, 466–479 (2021)
20. Zhen, L., Hu, P., Wang, X., Peng, D.: Deep supervised cross-modal retrieval. In: Proceedings of the IEEE/CVF Conference on Computer Vision and Pattern Recognition, pp. 10394–10403 (2019)
21. Zhou, J., Ding, G., Guo, Y.: Latent semantic sparse hashing for cross-modal similarity search. In: Proceedings of the 37th International ACM SIGIR Conference on Research & Development in Information Retrieval, pp. 415–424 (2014)
22. Zhou, X., Shen, F., Liu, L., Liu, W., Nie, L., Yang, Y., Shen, H.T.: Graph convolutional network hashing. IEEE Trans. Cybern. 1460–1472 (2018)
23. Zhu, X., Huang, Z., Cheng, H., Cui, J., Shen, H.T.: Sparse hashing for fast multimedia search. ACM Trans. Inf. Syst. (TOIS) **31**(2), 1–24 (2013)
24. Zhu, X., Huang, Z., Shen, H.T., Zhao, X.: Linear cross-modal hashing for efficient multimedia search. In: Proceedings of the 21st ACM International Conference on Multimedia, pp. 143–152 (2013)

FS-IGA: Feature Selection Method Based on Improved Genetic Algorithm

Dong Li, Shumei Du, Yong Wei, Lei Qin, and Yuefeng Du(✉)

Liaoning University, Shenyang 110036, China
dongli@lnu.edu.cn

Abstract. Feature selection is one of the important techniques of machine learning and data mining. However, with the continuous growth of data scale and application scenarios, the traditional feature selection methods have some defects. In this paper, a Feature Selection method based on Improved Genetic Algorithm (FS-IGA) is proposed. Firstly, the fitness function is improved by introducing the separability criterion based on the distance between classes. Secondly, the decision tree is introduced as the classifier in the genetic algorithm to optimize the performance of the algorithm. Finally, experimental results show that our proposed method can effectively improve the accuracy and adaptability of feature selection. The method presented in this paper boasts of a unique advantage in that it transcends reliance on any specific model framework, thereby ensuring broader applicability. As such, this paper provides a more profound reference value for the research on feature selection in the field of machine learning and data mining.

Keywords: Genetic Algorithm · Feature Selection · Decision Tree · Machine Learning · Data Mining

1 Introduction

The purpose of feature selection is to extract some useful features from the original features, so as to optimize model performance [1], improve learning efficiency and reduce model complexity. The traditional feature selection methods mainly including filter, wrapper and embedded method, but these methods have some defects. The filtering feature selection method only consider characteristic and the target variable correlation of the two, without considering the relationship between characteristics, unable to make full use of the characteristic collection of the information. The wrapper feature selection method faces challenges of high computational complexity and the risk of overfitting. The embedded feature selection method combines feature selection with specific models, which may cause the selected features to only target a specific model, and reduce the generalization of the model.

Aiming at the limitations of traditional methods, this paper proposes a Feature Selection method based on Improved Genetic Algorithm (FS-IGA). Genetic algorithm [2] is a kind of natural selection, crossover and mutation operation strategy of optimization of intelligent algorithm. Genetic calculation method of the fitness function can be adjusted according to the different needs of the user to modify that of individuals and populations through crossover and mutation operations such as degree of optimization. The improved method proposed in this paper introduces a separability criterion, which is used as the fitness function of the genetic algorithm. This enhances the performance of the genetic algorithm and subsequently integrates a decision tree classifier into the genetic algorithm framework. In order to explore the adaptation degree and the effective degree of the improved method are tested on real data sets. Experiments show the results show that the improved method has better accuracy and adaptability, can in a relatively short period of time to find a more optimal feature subset, and improve the accuracy of the classifier and generalization performance. Compared with other existing algorithms, the improved method has better effect of feature selection.

2 Related Works

Feature selection [1] is the process of choosing a subset of relevant features to represent a problem, improving model simplicity and performance by eliminating irrelevant characteristics from the dataset.

Feature selection algorithms are techniques used in machine learning and data mining to identify the most crucial features for prediction. The goal is to improve model performance, reduce overfitting, speed up training, and enhance interpretability by removing irrelevant or redundant features. Chandrashekar and Sahin [3] summarized that there are mainly three methods for feature selection algorithms: filter [4], wrapper [5] and embedded method [6].

2.1 Filter Feature Selection

The filter feature selection algorithm is a commonly used feature selection method, which main idea is to utilize the statistical information of the features and the relationship between the features and the target variable for feature selection. This method is primarily applied in the data preprocessing stage, where all features are ranked from high to low according to their scores, eliminating those with lower scores. The advantages lie in its simplicity and speed, unconstrained by algorithmic complexity, making it suitable for feature selection tasks on large-scale datasets.

Early feature selection algorithms were mostly filter methods, which require feature selection before model training. These two processes are independent. Filter methods evaluate the predictive power of selected features using heuristic criteria based on information statistics. The chosen feature subset varies with different evaluation criteria but can quickly eliminate noisy features. These methods are computationally efficient and widely applicable. The inconsistency measure criterion was first introduced by Almualliam et al. [7], who subsequently proposed the FOCUS feature selection algorithm. This algorithm has relatively high computational complexity and is overly reliant on the search process. The paper [8] used average normalized mutual information as a measure of redundancy between features and introduced a filter feature selection method based on mutual information. They combined this method with genetic algorithms to develop a hybrid filter approach called GAMIFS. Brunato et al. [9] introduced a filter feature selection method based on precise mutual information. The research shows that using precise mutual information to identify feature sets in binary and multiclass classification tasks yields significant performance advantages.

2.2 Wrapper Feature Selection

Wrapper feature selection algorithms are model-evaluation-based methods that use the performance of a feature subset as the objective function, searching for the optimal feature subset through search algorithms. Wrapper methods are typically employed with specific machine learning models, such as Support Vector Machines (SVM) and Logistic Regression (LR). The advantage of this approach is its ability to more accurately assess the contribution of feature subsets to model performance, thereby selecting the best feature subset.

Wrapper methods rely on the learning algorithm chosen, using the performance of the learner that will be employed as the evaluation criterion for the feature subset. The wrapper method was first introduced [6]. The paper [10] proposed a wrapper feature selection method based on a dual strategy of interdependence. In paper [11] a wrapper feature selection method for Multi-Layer Perceptron (MLP) neural networks with random perturbation is introduced, which has particularly good representational capabilities. The paper [12] developed a wrapper feature selection method for classifying missing data, which starts with the missing data in the dataset to classify the filtered missing data, significantly improving the classifier's test accuracy. Wrapper methods require multiple training rounds of the learner and have relatively higher computational costs compared to filter methods.

2.3 Embedded Feature Selection

The embedded feature selection algorithm is a method that combines feature selection with the model training process, where the optimal feature subset is directly learned during model training. In embedded feature selection, the feature selection process is integrated into model training via adjusting the optimization process of the model to select features that contribute most to model performance. Compared to filter and wrapper methods, embedded feature selection algorithms can better consider the relationship between the features and the model, resulting in more accurate and stable feature subsets.

In practical applications, embedded feature selection algorithms are often used with models such as linear regression, logistic regression and Support Vector Machines (SVM), achieving feature selection by adjusting the model's regularization terms. Since embedded methods can consider both feature correlations and their relationship with the model, they perform well when dealing with high-dimensional data and situations where the number of features far exceeds the number of samples. The embedding method is a widely used feature selection algorithm. The paper [13] proposed an embedded minimum-maximum feature selection algorithm that optimizes kernel functions through search strategies, forming an optimal feature subset. Numerical experiments show that the algorithm proposed in this paper is effective for binary classification problems. The paper [14] introduced two types of penalty functions and proposed improved optimization algorithms based on Newton's method and line search for optimizing concave functions. The integration of this method with the SVM approach significantly enhances model performance on the majority of datasets, outperforming the use of all available features. The paper [15] considered a special type of Universum to enhance the classification ability of classical support vector machines, further enabling it to play a role in useful feature identification and separation hyperplane construction, thereby improving the feature selection capability and classification performance of Universum-supported vector machines.

3 FS-IGA Model

3.1 Model Overview

In this paper, we propose a novel Feature Selection method based on Improved Genetic Algorithm (FS-IGA). Firstly, the fitness function is improved by introducing the separability criterion based on the distance between classes. Secondly, the decision tree is introduced as the classifier in the genetic algorithm to optimize the performance of the algorithm.

The genetic algorithm, modeled after natural evolution, optimizes solutions through genetic processes but often gets trapped in local optima and struggles with high-dimensional feature spaces. Therefore, some improvements have been introduced to improve the performance of genetic algorithms in feature selection problems, and here are some advantages of improving the genetic algorithm feature selection methods: improve search speed, avoid falling into local optimum, dealing with high-dimensional feature space, strong adaptability to abnormal data.

3.2 Adaptability

3.2.1 Inter-class and Intra-class Distance

The separability criterion based on intra-class distance refers to judging whether the data set can be effectively divided into different classes by calculating the distance between different classes.

When using the separability criterion based on the distance between classes within a class, the following two indicators are usually used to evaluate the validity of data set classification.

(1) Inter-class distance: indicates the relative distance between classes. The farther the distance indicates the greater the difference between classes, indicating the better the classification effect.
(2) Intra-class distance: indicates the distance of different data points in the same class. The smaller the distance, the higher the similarity of data points in the class, the better the classification effect.

When feature selection is carried out, the first thing to do is to determine the selection criteria, and then the relevant feature selection problem is transformed into the selection of M features that can make the criterion function be called the optimal value, but in fact, the number of m is far less than the number of M.

Fisher's linear test is actually an intuitive criterion for class separability, which can be expressed as the average of the distance between any two pairs of samples in the two classes, and can now be derived from this test for multiple classes.

By calculating the average interval distance δ of M-dimensional feature vectors in two different classes $x_k^{(i)}$ and $x_l^{(j)}$, the relative positional relationship between two different M-dimensional classes feature vectors can be obtained, so as to determine the relative relationship between them. The average interval distance of various feature vectors is shown in Eq. (1).

$$J_d x = \frac{1}{2} \sum_{i=1}^{c} P_i \sum_{j=1}^{c} P_j \frac{1}{n_i n_j} \sum_{k=1}^{n_i} \sum_{l=1}^{n_j} \delta x_k^{(i)}, x_l^{(j)} \tag{1}$$

where c is the number of population classes, n_i is the number of samples in the first M-dimensional class, n_j is the number of samples in the second M-dimensional class, and P_i and P_j are the prior probabilities of corresponding classes.

In a multidimensional space, the distance between two vectors can be expressed by Euclidean distance theorem, and its expression can be shown in Eq. (2).

$$\delta(x_k^{(i)}, x_l^{(j)}) = \left(x_k^{(i)} - x_l^{(j)}\right)^T (x_k^{(i)} - x_l^{(j)}) \tag{2}$$

m_i is used to represent the mean vector of the class i sample set, then the m_i expression is Eq. (3).

$$m_i = \frac{1}{n_i} \sum_{k=1}^{n_i} x_k^{(i)} \tag{3}$$

m is used to describe the population average for all different types of sample sets, which is shown in Eq. (4).

$$m = \sum_{i=1}^{c} P_i m_i \tag{4}$$

Combine all the formulas into the mean distance formula to Eq. (5).

$$J_d(x) = \sum_{i=1}^{c} P_i \left[\frac{1}{n_i} \sum_{k=1}^{n_i} \left(x_k^{(i)} - x_l^{(j)}\right)^T (x_k^{(i)} - x_l^{(j)}) + (m_i - m)^T (m_i - m) \right] \tag{5}$$

By using the example defined below, we can build an expression for $J_d(x)$ and get the result as Eq. (6) and (7).

$$\widetilde{S_b} = \sum_{i=1}^{c} P_i (m_i - m)(m_i - m)^T \tag{6}$$

$$\widetilde{S_w} = \sum_{i=1}^{c} P_i \frac{1}{n_i} \sum_{k=1}^{n_i} (x_k^{(i)} - m_i)(x_k^{(i)} - m_i)^T \tag{7}$$

Because it is necessary to ensure that the intra-class dispersion is small enough and the inter-class dispersion is large enough, the following Eq. (8) is defined as the distance criterion.

$$J_d(x) = \frac{tr S_b}{tr S_w} \tag{8}$$

By using $J_d(x)$ as a separable criterion, similarity between features can be evaluated more effectively, resulting in more accurate model construction.

3.3 Fitness Function

Genetic algorithm is to use a specific coding form of the candidate target region, encode it into a chromosome, so as to select the relatively optimal D features in d features, and combine them into a string represented by 0/1 characters, 0 represents not selected, 1 represents selected, this string is called chromosome, referred to as m. Obviously, what is needed is a chromosome with d 1 genomes.

By introducing the concepts in 3.2.1, the purpose of optimization will be transformed into a function with fitness, which value $f(m)$ represents an eigenvalue, and $J_d(x)$ represents an eigenvalue, which represents an eigenvalue and is used to measure whether a eigenvalue conforms to the eigenvalue.

3.4 Improved Genetic Algorithm

By introducing $J_d(x)$ as fitness, the genetic algorithm can be adjusted according to the specific conditions within and between classes, as shown in Fig. 1.

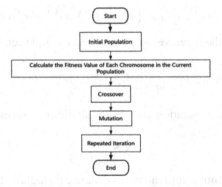

Fig. 1. Flowchart of the improved genetic algorithm.

3.5 Population Initialization

Table 1. Chromosome sequences.

Random replacement of dimensions	Random chromosome sequence
30	[1. 1. 0. 0. 1. 0. 0. 0. 0. 1. 1. 1. 0. 0. 0 0. 1. 0. 0. 0. 1. 1. 1. 0. 1. 1. 1. 1. 0. 0 1. 0. 1. 0. 1. 0. 0. 1. 1. 0. 1. 0. 1. 0. 1 1. 0. 0. 0. 1. 1. 1. 0. 1. 1. 0. 1. 0. 0. 1.]

Using sonar dataset, we can create a 1*60 dimensional dye body where 60 features can be regarded as 30 valid references, so by converting 30 bits of a zero vector to 1, we can effectively extract 30 references and build a qualified chromosome sequence from them. Then, one of the initialized chromosomes changes the random 30 bits of a zero vector into 1, which means that 30 dimensions are randomly selected from the 60-dimensional features, and a random chromosome sequence meeting the conditions is shown in Table 1.

Random generation of satisfying chromosome sequences. After n rounds of iteration, we can obtain an initial population M_0 composed of n chromosomes, each chromosome is unique in number and characteristics.

3.6 Population Chromosome Fitness Selection

By selecting d-dimensional features, the original Sonar dataset is turned into a 208* d-dimensional matrix to determine the fitness value $f(m)$ of a particular chromosome, which is derived from their previous closeness and similarity.

Based on the selection probability $p(f(m))$, chromosomes are sampled from the population. These samples undergo specific operations to proliferate and generate the next generation population M_{t+1}.

In fact, the sum of chromosome fitness in the population is 1, so a range from 0 to 1 can be listed as the sampling range. The fitness values of each chromosome in the population are added one by one to obtain some interval from 0 to 1, as shown in Table 2.

Table 2. Distribution of chromosome fitness.

Sampling range	Chromosome fitness value
[0, 0.14)	0.14
[0.14, 0.63)	0.49
[0.63, 0.69)	0.06
[0.69, 1]	0.31

As shown in the Table 2, it is divided into four chromosomes, and the corresponding fitness of each chromosome is 0.14, 0.49, 0.06 and 0.31, respectively, and the sum obtained is 1. And the four sections represent each of these four chromosomes. Using the random number algorithm to get a random number between 0 and 1, the chromosome represented by the random number belongs to the interval is taken out as a member of the new population. After this is repeated n times, we will get a new population of n, and the number of the population will not change compared to the previous generation, ensuring that the number is uniform.

3.7 Crossover

There are two types of crossover operations known in genetic algorithms that are most frequently used. The first is a single point crossing, that is, the crossing operation occurs at a crossing point. The second is a two-point crossover, that is, the operation of pairing and swapping takes place at two intersections.

How to get the chromosomes for crossing. We made some improvements to the study, randomly divided the initial population into two parts, and ensured that the size of the two parts was the same, and then scrambled the two parts respectively as the parent population of cross-chromosome extraction. The chromosomes used for cross-chromosome extraction were extracted from the two parent populations, so that a random matching process could be completed. However, the chromosome characteristic dimension after crossing needs to be kept unchanged, so the following methods are adopted for operation.

First, on each parent chromosome, k fragments are randomly selected and how many of them are genes (1) are recorded. Then, on these matched parent chromosomes, search for 1 and k fragments, and if found, continue the process. If we don't find anything, we keep searching, and eventually, we can thoroughly traverse the entire parent population.

3.8 Mutation

It should be noted that when performing mutation operations, it is necessary to ensure that the new individuals obtained after mutation still meet the constraints and requirements of the problem, otherwise illegal individuals may be generated, resulting in algorithm failure. To ensure that the number of features in the chromosome remains stable after mutation, we can take the following steps.

According to the given mutation probability, the whole population can be classified. If the classification result is correct, an individual will be randomly selected in the whole population, and a gene within it will undergo a reverse modification. If the value of this gene is changed, another 0 will be selected again, and then modified again to achieve a similar result. By comprehensively examining the entire species, we can determine the independence of each chromosome, thereby avoiding genetic differences.

3.9 Iteration

Through the iterative breeding process, a new generation of population M_{t+1} can be obtained based on the previous generation of population M_t. Repeat the process described above. In addition, the genetic algorithm can also obtain more results through continuous iteration. When the iteration times of the algorithm exceed t, the algorithm will be temporarily shut down, so that the screening results of D-dimensional features can be realized.

3.10 Improved Fitness Function

To utilize a decision tree in making decisions within a genetic algorithm, it is necessary to incorporate a fitness function. In the genetic algorithm process, the direction and strategy of evolution are decided by individual fitness. Once the fitness function is introduced, an individual's fitness value can be evaluated using the decision tree.

To accomplish this, we first train a decision tree model with strong generalization capabilities. Then, within the context of the genetic algorithm, each individual should be represented as a feature vector that is fed into the decision tree. The fitness is then determined based on the tree's assessment. Integrating a decision tree in this manner boosts search efficiency and addresses optimization challenges commonly encountered in genetic algorithms, thereby ensuring high precision.

Since the steps in implementing a genetic algorithm are fairly standard, they will not be delved into here. The distinction lies in the adjustment of the fitness function, which enhances the genetic algorithm. The decision tree serves as a classifier and undergoes six-fold cross-validation. If all fitness values are 0, then indeed, all are 0.

4 Experiments

4.1 Dataset and Experimental Settings

In our experiments, the two datasets are selected: Iris dataset and Sonar dataset. Iris dataset contains 150 data samples involving three different varieties of Iris. Each sample has 4 characteristic variables, namely calyx length, calyx width, petal length and petal width, with a total of 12 measurements. Sonar dataset is a binary classification used to categorize sonar signals as either metal or rock. It includes 208 samples with 60 real features each, indicating the direction of the transmitted signal and recorded echoes. Samples are labeled M for metal and R for rock.

The experiments are specifically tailored to match the distinct characteristics of the two datasets used. Given that the Iris dataset possesses only four dimensions, the experimental design involved selecting from 1 to 4 dimensions to test the convergence of the genetic algorithm. This setup allows for the observation of how the population's fitness value evolved with an increasing number of iterations. On the other hand, when the experiments are conducted on the Sonar dataset, the following parameters are established: the maximum number of generations is set to 100, the population size is determined to be 50, the crossover probability is fixed at 0.25, and the mutation probability is set to 0.01. This structured approach ensures that the experimental conditions are optimized for each dataset, aiming to extract meaningful results from the evolutionary algorithm's performance.

4.2 Analysis of Improved Genetic Algorithm

Verify the genetic algorithm on the Iris dataset. The fitness of the optimal individual, post-population selection with varying dimensions, is shown in Fig. 2, with dimension on the horizontal axis and fitness on the vertical axis.

Verify the genetic algorithm on the Sonar dataset with 100 iterations, a population size of 50, crossover probability of 0.25, and mutation probability of 0.01. The fitness values improve with each iteration. In 30 dimensions, the optimal fitness value is 0.064629, which is higher than the traditional genetic algorithm. The result is shown in Fig. 3, with iterations on the horizontal axis and fitness on the vertical axis.

The improved genetic algorithm has better global search ability and more easily finds the global optimum.

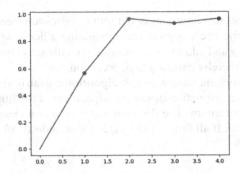

Fig. 2. Optimal individual fitness of the population.

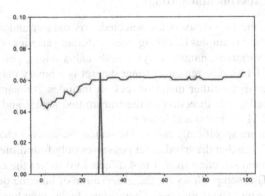

Fig. 3. Iterations (Fitness).

4.3 Analysis of Enhanced Fitness

Set the number of iterations to 100, the population size to 100, the crossover probability to 0.25, and the mutation probability to 0.01, is shown in Fig. 4, with iterations on the horizontal axis and optimum fitness on the vertical axis.

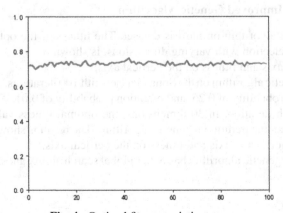

Fig. 4. Optimal fitness variation curve.

The best fitness of the first 10 iterations is shown in Table 3.

Table 3. Best fitness of the first 10 iterations.

Iterations	Optimum fitness
1	0.7211484593837536
2	0.7163865546218487
3	0.6973389355742298
4	0.707703081232493
5	0.7170868347338937
6	0.7120448179271709
7	0.7268907563025211
8	0.7173669467787116
9	0.7358543417366947
10	0.7362745098039216

The best fitness of the last 10 iterations is shown in Table 4.

Table 4. Best fitness of the last 10 iterations.

Iterations	Optimum fitness
99	0.726610644257703
98	0.730952380952381
97	0.7261904761904763
96	0.73109243697479
95	0.730952380952381
94	0.7308123249299721
93	0.7267507002801121
92	0.7261904761904762
91	0.7358543417366947
90	0.7119047619047619

Set the number of iterations to 100, the population size to 50, the crossover probability to 0.25, and the mutation probability to 0.01, as shown in Fig. 5, with iterations on the horizontal axis and optimum fitness on the vertical axis.

Set the number of iterations to 100, the population size to 50, the crossover probability to 0.25, and the mutation probability to 0.005, as shown in Fig. 6, with iterations on the horizontal axis and optimum fitness on the vertical axis.

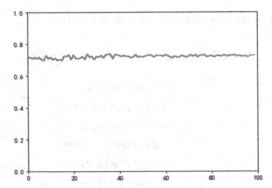

Fig. 5. Optimal fitness variation curve after varying population size.

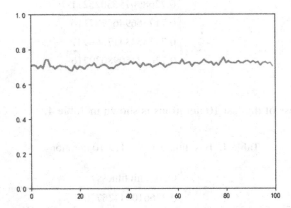

Fig. 6. Optimal fitness variation curve after varying mutation probability.

The genetic algorithm using $J_d(x)$ as fitness improvement has higher efficiency in feature selection than the traditional method. Introducing the decision tree into the fitness function enhances search efficiency and accuracy in the improved genetic algorithm.

5 Conclusions

In this paper, we propose a novel Feature Selection Method Based on Improved Genetic Algorithm (FS-IGA). Firstly, the fitness function is improved by introducing the separability criterion based on the distance between classes. Secondly, the decision tree is introduced as the classifier in the genetic algorithm to optimize the performance of the algorithm. It conducts specific experiments on two datasets to explore enhancements in genetic algorithms by improving fitness functions. The findings confirm the superior performance of the improved algorithm on both datasets, significantly enhancing feature selection and optimal fitness. These results provide a new perspective on improving genetic algorithms for feature selection.

Acknowledgement. This work was supported by the Social Science Planning Fund Program of Liaoning Province of China under Grant No.L23BJY018, the Science and Technology Program Major Project of Liaoning Province of China under Grant No. 2022JH1/10400009, the Natural Science Foundation of Liaoning Province of China under Grant No.2022-MS-171, 2022-KF-13-06, the National Natural Science Foundation of China under Grant No. 62072220.

References

1. Wettschereck, D., Aha, D.W., Mohri, T.: A review and empirical evaluation of feature weighting methods for a class of lazy learning algorithms. Artif. Intell. Rev. 273–317 (1997)
2. Holland, J.H.: Adaptation in natural and artificial systems: an introductory analysis with applications to biology, control, and artificial intelligence. University of Michigan Press (1975)
3. Chandrashekar, G., Sahin, F.: A survey on feature selection methods. Comput. Electr. Eng. **40**(1), 16–18 (2014)
4. Liu, H., Setiono, R.: A probabilistic approach to feature selection - a filter solution. In: International Conference on Machine Learning (1996)
5. Langley, P.: Selection of relevant features in machine learning. Comput. Sci. 245–271 (1994)
6. Kohavi, R., John, G.H.: Wrappers for feature subset selection. Artif. Intell. **97**(1–2), 273–324 (1997)
7. Almuallim, H., Dietterich, T.G.: Learning Boolean concepts in the presence of many irrelevant features. Artif. Intell. **69**(1–2), 279–305 (1994)
8. Estevez, P.A., Tesmer, M., Perez, C.A., et al.: Normalized mutual information feature selection. IEEE Trans. Neural Networks **20**(2), 189–201 (2009)
9. Brunato, M., Battiti, R.: X-MIFS: exact mutual information for feature selection. In: 2016 International Joint Conference on Neural Networks (IJCNN), pp. 3469–3476 (2016)
10. Michalak, K., Kwasnicka, H.: Correlation-based feature selection strategy in classification problems. Int. J. Appl. Math. Comput. Sci. **16**(4), 503–511 (2006)
11. Yang, J.B., Shen, K.Q., Ong, C.J., et al.: Feature selection for MLP neural network: the use of random permutation of probabilistic outputs. IEEE Trans. Neural Networks **20**(12), 1911–1922 (2009)
12. Cao, T.T., Zhang, M.J., Andreae, P., et al.: A wrapper feature selection approach to classification with missing data. Lecture Notes in Computer Science (2016)
13. Jim'enez-Cordero, A., Morales, J.M., Pineda, S.: A novel embedded min-max approach for feature selection in nonlinear Support Vector Machine classification. Eur. J. Oper. Res. **293**, 24–55 (2020)
14. Maldonado, S., López, J.: Dealing with high-dimensional class-imbalanced datasets: embedded feature selection for SVM classification. Appl. Soft Comput. **67**, 94–105 (2018)
15. Li, C., Huang, L., Shao, Y., Guo, T., Mao, Y.: Feature selection by Universum embedding. Pattern Recognit (2024)

Shearlet Transform Based Multiscale Fusion Network for Image Super Resolution

Wei Wei[1,2(✉)]

[1] Weihai Ocean Vocational College, Weihai, Shandong 264200, China
shuiyouhan2002@foxmail.com
[2] Weihai Big Data Intelligent Application Engineering Technology Research Center Shandong, Weihai 264200, China

Abstract. Recently, convolutional neural networks (CNNs) based methods have achieved great success in the field of image super-resolution. However, most methods produce over-smoothed outputs, and suffer from degradation for very low resolution images. We present an accurate CNN-based structure based on shearlet transform in this paper. Firstly, we propose a multi-scale fusion network (MSFN) to fully explore features from low resolution images. And then, we integrate shearlet transform to make MSFN further get the texture details for super-resolved images. The experiments demonstrate that the proposed network achieves superior super resolution results and outperforms the state-of-the-art.

Keywords: convolutional neural networks · multi-scale fusion network (MSFN) · super-resolution

1 Introduction

Super resolution (SR) refers to the technique of reconstructing a high resolution (HR) image from an observed low resolution (LR) input. It is a domain specific general image SR method.

As the powerful learning ability, convolutional neural networks (CNNs) based methods have demonstrated superiority over other learning paradigms [1]. However, most CNN-based methods tend to produce fuzzy and overly smooth output. SRCNN [2] is the first CNN-based super resolution method. With the development of CNN-based SR, it proves that deeper CNN can achieve better effects of super resolution, such as CDCSR [3] and RCAN [4]. Especially, the number of depth reaches 400 for RCAN. Besides, the effective deep super resolution methods are series of feature ex-traction blocks. The advantage of each block can have a major effect for the final super resolution performance. To this end, we present an effective multi-scale fusion block (MSFB) to explore features effectively.

As the texture and context information of each level of images can be retained in multi-scale geometric transform, super resolution reconstruction in multi-scale geometric transform has aroused more scholars' concern. Wavelet analysis is a classical transform method, which has been used for image super resolution. But, using the tensor product of two 1-dimensional wavelets to directly extend from 1-dimensional wavelet to 2-dimensional wavelet by is not any more the best way to represent an image with features along smooth curves. To avoid this deficiency, shearlet transform is introduced for super resolution reconstruction in this paper; it is a 2-dimensional image representation method with anisotropic, multi-directional and multi-scale features.

2 The Contributions

The contributions made by this paper are as follows:

1) We proposed MSFB, a new feature extraction block, to establish our MSFN architecture in a stacked manner. For each MSFB, the multi-scale features are effectively explored by convolution kernels of different sizes. By adopting global feature fusion and local residual learning methods, MSFN can learn hierarchical features as a whole and adaptively.
2) The existing super resolution methods based on CNN mostly focus on the spatial domain, resulting in too smooth reconstruction results; so, shearlet transform is employed to recover the local edge details and global topology of high-resolution images effectively.
3) Experimental results show that our MSFN with shearlet transform prediction is obviously superior to other outstanding methods according to PSNR/SSIM and image texture detail and edge structure enhancement.

Proposed Method: Network structure Fig. 1 shows our proposed MSFN network which contains 3 parts: shallow feature extraction, feature extraction of multiple receptive fields and deconvolution. Therefore, the following problem can be solved:

$$\hat{\theta} = \arg\min_{\theta} \frac{1}{N} \sum_{i=1}^{N} L^{SR}(F_{\theta}(I_i^{LR}), I_i^{HR}) \tag{1}$$

where θ indicates the weights and bias convolutional layer. Stands for the loss function for minimizing the difference between LR input IiLR and target HR IiHR. After shallow feature module, feature extraction part consists of a series of stacked MSF blocks. We fuse the global feature G from all MSFBs:

$$F_{GF} = H_{GFF}([F_1, ..., F_D]) \tag{2}$$

where [F1,...,FD] is the concatenation of feature maps generated by MSFBs. HGFF is a composite function of 1×1 and 3×3 convolutional layers. Then, the global residual learning is used to get the feature maps:

$$F_{DF} = F_0 + F_{GF} \tag{3}$$

where F_0 denotes the shallow features. All other layers before global feature fusion are fully used for our MSFBs. Finally, we input them to 12×12 deconvolutional layers to get the output of SR.

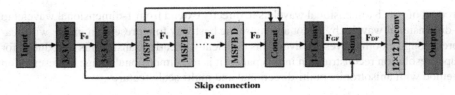

Fig. 1. The network structure of the proposed MSFN.

3 Proposed Method

MSFB Based on traditional framework [5], a new MSFB structure, is proposed in Fig. 2. In each MSFB, we built a three-by-pass network and different convolutional kernel is used.

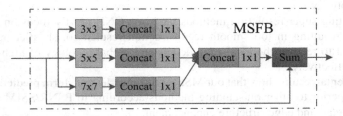

Fig. 2. The structure of the proposed MSFB.

Information between bypasses can be shared and image features at different scales could be detected. The operation can be defined as:

$$C_3 = \sigma\left(w^1_{3\times 3} * F_{d-1} + b^1\right) C_5 = \sigma\left(w^1_{5\times 5} * F_{d-1} + b^2\right)$$
$$C_7 = \sigma\left(w^1_{7\times 7} * F_{d-1} + b^3\right) H_3 = \sigma\left(w^2_{1\times 1} * [C_3, C_5] + b^4\right)$$
$$H_5 = \sigma\left(w^2_{1\times 1} * [C_3, C_7] + b^5\right) H_7 = \sigma\left(w^2_{1\times 1} * [C_5, C_7] + b^6\right) \quad (4)$$

$$F_d = w^3_{1\times 1} * [H_3, H_5, H_7] + b^7 + F_{d-1}$$

where w and b are weight and bias respectively.σ denotes the ReLU function. Fd-1 and Fd are the input and output of the d-th MSB respectively.

Proposed Method: Shearlet transform Success of wavelet in expressing digital signals [6] can be ascribed to the good sparse representation of one-dimensional signals far away from discontinuous points. But, multi-dimensional signals with singularity are not easy to be effectively processed by wavelet transform. Contrarily, many other signal representations, such as curvelets [7] and bandelets [8], can exploit the anisotropy of the surface along the edges. But, multi-resolution geometric representation cannot be provided by curvelets; and high computational cost is required by bandelets transform to search the best geometry. Then, Labate et al. [9] proposed a multi-scale, multi-resolution shearlet transform. Main advantage of this transform is that it can be performed by general multi-resolution analysis. High-frequency of shearlet transform are compared with that of wavelet transform (WT) in Fig. 3, it can be seen that Shearlet transform keeps more accurate curvature.

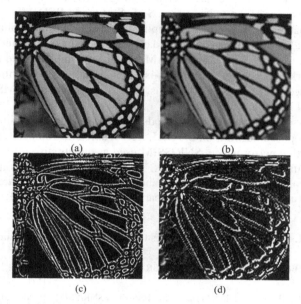

Fig. 3. *Comparison of shearlet transform and WT coefficients on the Butterfly*: (a) 256 × 256 high resolution Butterfly; (b) the low-frequency shearlet transform coefficients; (c) the fusion of high-frequency shearlet transform coefficients; (d) the fusion of DWT ("Haar" type) high-frequency coefficients.

In Fig. 4, Shearlet transform is used to make MSFN preserve more structural details than spatial domain. Detailed processes of shearlet transform could be seen in [8].

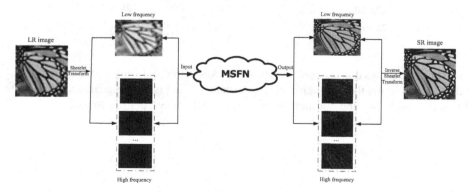

Fig. 4. ST domain coefficients prediction.

4 Experimental Result

Details Quality and quantity of the proposed method are estimated with PSNR and SSIM [10]. Recently, In Image recovering applications Timofte et al. [11] provided a DIV2K dataset. The 800 images are used a training network and 10 validation images are used in the process of the training. The Set5, Set14, and BSD100 datasets are used as testing. The patch size 48x48 of RGB input is trained from low resolution input. We randomly extract low-resolution patches and enhance them by turning and rotating them horizontally or vertically. Torch7 and ADAM optimizer is implemented to conduct our proposed network. The size of mini-batch is 32, and the rate of learning begins with 0.0004 and reduces half for every 50 epochs. Our MSFN contains 15 MSF blocks, and all convolutional layers have 32 filters.

Comparison with state-of-the-art methods in the following, we estimate the performance of the proposed network on the Set5, Set14, and BSD100 datasets. To estimate the super resolution performance, SNR/SSIM used as quantitative estimation standard to demonstrate the super resolution results. To make a fair comparison, the provided codes of the compared networks are used and the same training sets are used to train. The PSNR (dB) and SSIM values for comparison are shown in Table 1.

Table 1. Average PSNR/SSIM values for scaling factor $\times 2$, $\times 3$, and $\times 4$.

Datasets	Scale	Bicubic	SRCNN [2]	VDSR [12]	DWSR [13]	MemNet [14]	Ours
		PSNR/SSIM	PSNR/SSIM	PSNR/SSIM	PSNR/SSIM	PSNR/SSIM	PSNR/SSIM
Set5	×2	33.66/0.93	36.66/0.95	37.53/0.96	37.43/0.96	37.78/0.96	37.92/0.96
	×3	30.39/0.87	32.75/0.91	33.66/0.92	33.82/0.92	34.09/0.92	34.23/0.93
	×4	28.42/0.81	30.48/0.86	31.35/0.88	31.39/0.89	31.74/0.89	31.81/0.89
Set14	×2	30.24/0.87	32.45/0.91	33.03/0.91	33.07/0.91	33.28/0.91	33.39/0.91
	×3	27.55/0.77	29.30/0.82	29.77/0.83	29.83/0.83	30.00/0.84	30.11/0.84
	×4	26.00/0.70	27.49/0.75	28.01/0.77	28.04/0.77	28.26/0.77	28.32/0.77
BSD100	×2	29.56/0.84	31.36/0.899	31.90/0.90	31.80/0.89	32.08/0.90	32.19/0.90
	×3	27.21/0.77	28.41/0.77	28.82/0.80	–/–	28.96/0.80	29.04/0.80
	×4	25.96/0.67	26.90/0.71	27.29/0.72	27.25/0.72	27.40/0.73	27.51/0.73

The table shows that our proposed method takes better performance than other methods. Figure 5 shows the qualitative results on scale × 4. In Fig. 5(a) Fig. 5(b) and Fig. 5(c), From the results we find our method provides an effective super resolution.

Fig. 5. Comparison of Visual results.

5 Experimental Result

Performance Comparison on High-level Task To further verify the effectiveness of the proposed method, we will study the performance of different SR methods in image classification task. In this section, the classic ResNet50 network [15] is used as the evaluation model, and 1000 images are selected as the test images on the ImageNet verification set. The linear interpolation method is used to sample the test image, and the resolution is 56 × 56. Then, we use bicubic interpolation method and five super-resolution methods to sample the test image 4 times. Table 2 shows the image classification results of different super-resolution methods. Errors of top-1 and top-5 show that the proposed method achieves the optimal classification performance, which proves that the proposed method has greater application value in high-level tasks.

Table 2. Performance of image classification

Methods	Top-1 error	Top-5 error
Bicubic	0.506	0.266
SRCNN	0.484	0.291
VDSR	0.463	0.276
DWSR	0.402	0.203
MemNet	0.395	0.195
Ours	0.375	0.185
HR images	0.260	0.072

6 Experimental Result

Ablation Study the structure design and the effectiveness of shearlet transform will be verified. First, the relationship between the network depth and performance of the model is analysed to determine the final network structure. In the image super-resolution model based on CNN, increasing the network depth plays an important role in improving the performance, but it also brings the increase of parameters, which affects the application value of the model. Here, we will show the relationship between different depth and super-resolution performance, and select the reasonable structure design. Table 3 shows the PSNR results of reconstructed images of convolutional networks with different depth density and the parameters of corresponding models when the data set is 4 times of the scale. In Table 3, A represents the number of MSFB contained in the model; B represents the number of convolution layers in each MSFB, and each convolution layer contains 32 filters. The increasing depth can bring about continuous improvement of recovery performance, when reaching a certain depth, the improvement of performance starts to slow down, while the number of parameters continues to increase rapidly. In order to achieve a good balance between recovery performance and network parameters, the proposed method includes 7 MSFB, each MSFB contains 6 convolution layers, and the rest are designed with this network structure.

Table 3. Comparison of different network depths

PSNR(dB)/ Parameter(K)	A = 6	A = 7	A = 8
B = 4	27.06/555	27.14/679	27.19/812
B = 6	27.34/1063	27.43/1272	27.46/1490
B = 8	27.50/1793	27.52/2123	27.48/2463

Next, we will verify the effectiveness of shearlet transform. Based on the deep convolution network, shearlet transform is introduced to transform the prediction problem of spatial pixels into the prediction problem of shearlet subband coefficients, which significantly improves the super-resolution performance of the model, especially the recovery ability of texture details. Here, the 4X super-resolution on the Set5, Set14 and BSD100 datasets will be tested. In Table 4, the super-resolution results of different transform domains are given, and the model including 7 MSFB is selected as the benchmark, which is combined with wavelet transform and shearlet transform respectively. In different data sets, the performance of combining shearlet transform is better than the other two models, which proves the effectiveness of shearlet transform.

Table 4. Comparison of different transform domains

	Set 5	Set 14	BSD 100
Spatial domain(baseline)	31.92	28.39	27.43
Wavelet trasform	32.12	28.55	27.62
Shearlet transform	32.61	29.25	28.32

7 Conclusion

A shearlet transform - based method for CNN super-resolution is proposed. Our feature extraction block MSFN is composed of a series of stacked MSF blocks to explore image features and effectively improve super resolution performance. In addition, shearlet transformation is used in model structure to effectively preserve more texture details compared with spatial domain. The super-resolution performance is further improved. The subjective and objective results show that the network proposed in this paper has better performance than other excellent networks and improves the ability to recover low-resolution images.

References

1. Schulter, S., Leistner, C., Bischof, H.: Fast and accurate image upscaling with super-resolution forests. In: CVPR, Boston, USA, pp 3791–3799 (2015)
2. Dong, C., Loy, C.C., He, K.M., Tang, X.O.: Learning a deep convolutional network for image super-resolution. In: European Conference on Computer Vision (ECCV), pp. 184–199 (2014)
3. Wei, W., Feng, G.Q., Zhang, Q., Cui, D.L., Zhang, M., Chen, F.: Accurate single image super-resolution using cascading dense connections. Electron. Lett. **55**(13), 739–742 (2019)
4. Zhang, Y., Li, K., Li, K., Wang, L., Zhong, B., Fu, Y.: Image super-resolution using very deep residual channel attention networks. In: Proceedings of the European Conference on Computer Vision (ECCV), Munich, Germany, pp. 294–310 (2018)
5. Gedeon, T., Wong, K.M., Lee, M.: Neural Information Processing, Springer Science and Business Media LLC (2019)
6. Mallat, S.: A wavelet tour of signal processing: the sparse way. Academic press (2008)
7. Candès, E.J., Donoho, D.L.: New tight frames of curvelets and optimal representations of objects with piecewise C2 singularities. Commun. Pure Appl. Math. **57**(2), 219–266 (2004)
8. Pennec, E.L., Mallat, S.: Sparse geometric image representations with bandelets. IEEE Trans. Image Process. **14**(4), 423–438 (2005)
9. Labate, D., Lim, W.Q., Kutyniok, G., Weiss, G.: Sparse multidimensional representation using shearlets. In: Wavelets XI (2005)
10. Wang, Z., Bovik, A.C., Sheikh, H.R., Si-moncelli, E.: Image quality assessment: from error visibility to structural similarity. IEEE Trans. Image Proces. **13**(4), 600–612 (2004)
11. Agustsson, E., Timofte, R.: Ntire 2017 challenge on single image super-resolution: Dataset and study. In: Proceedings of the IEEE Conference on Computer Vision and Pattern Recognition (CVPR) Workshops, pp. 126–135 (2017)
12. J.Kim, J., Lee, J.K., Lee, K.M.: Accurate image super-resolution using very deep convolutional networks. In: Proceedings of the IEEE Conference on Computer Vision and Pattern Recognition (CVPR), pp. 1646–1654 (2016)
13. Guo, T.T., Mousavi, H.S., Vu, T.H., Monga, V.: Deep wavelet prediction for image super-resolution. In: The IEEE Conference on Computer Vision and Pattern Recognition Workshops (CVPRW), pp. 104–113 (2017)
14. Tai, Y., Yang, J., Liu, X.M., Xu, C.Y.: Memnet: a persistent memory network for image restoration. In: Proceedings of the IEEE Conference on Computer Vision and Pattern Recognition (CVPR), pp. 4539–4547 (2017)
15. He, K., Zhang, X., Ren, S., Sun, J.: Deep residual learning for image recognition. In: Proceedings of the IEEE Conference on Computer Vision and Pattern Recognition (CVPR), pp. 770–778 (2016)

Application of Segmental-Based Transfinite Mapping Method for Quadrilateral Mesh Generation of Complex Domain

Dao-Ju Qin(✉) and Pei-Pei Shang

Weihai Ocean Vocational College, Weihai 264300, Shandong, China
shangpeipei@whovc.edu.cn

Abstract. Algebraic interpolation method is widely used in almost all CAE tools because of its high computing efficiency and well meshing uniformity. Firstly, the advantages and disadvantages of transfinite mapping method are given, according to the disadvantages, a segmental-based transfinite mapping method is prompted, which is of great value for engineering analysis. Secondly, the computing methods of node in Parametric Domain and Physical Domain are described, Finally the reliability and quality of the prompted method are confirmed by comparing with other CAE toos. This method has been applied in thermo-structural coupling analysis for axis-symmetrical components of aero-engine, Results show that the prompted method is effective, reliable and fully meets the needs of FEM.

Keywords: Quadrilateral Mesh Generation · Segmental · Transfinite Mapping · FEM (finite Element Method)

1 Introduction

Finite element analysis is usually divided into four steps: mesh generation, establishing a finite element model, matrix solving, and visualization post-processing. As the starting point of the analysis work, the quality of the mesh affects the reliability, convergence, and computational efficiency of the subsequent analysis process, so the mesh generation is one of the core technologies in finite element analysis.

In the past 20 years, many scholars such as Hu Enqiu [1], Owen [2], Bern [3], Guan Zhenqun [4], and so on have discussed the finite element mesh generation method in an overview. From their studies, it can be seen that two types of two-dimensional regions are discretized using quadrilateral meshes and triangular meshes. The triangular meshing method is relatively mature, the quadrilateral mesh generation is closer to the discrete domain, and it is better than the triangular cells in both mesh quality and computational accuracy. Therefore, two-dimensional problems generally use quadrilateral cells, and the quadrilateral mesh generation algorithm has become the focus of research and application. In 1991, Blacker [5] and Stephenson proposed the Paving algorithm first arranges the quadrilateral cells along the boundary, and then develops deeper inside the region until the whole region is filled up, and the key of the algorithm is the cell gap and the intersection of the cells, so it is more difficult and more efficient. The key of this algorithm is the cell gap and cell intersection problems, so it is difficult and inefficient, in addition, it allows a small number of triangles to be generated. In 2009, the quadrilateral mesh dissection method based on the scattering model proposed by Chen Tao et al. [6] is a further enhancement of this algorithm; the triangular transformation method proposed by Lo [7] et al. first adopts the mature triangular mesh algorithm to discretize the target region and then transforms it into the quadrilateral cells through the merging or decomposition operation, which has a better advantage relative to the other methods. Compared with other methods, this method has better arbitrariness and versatility, but the quality is poor, and the discretized results must be topologically optimized and mesh smoothed in order to obtain a high-gradient, high-quality quadrilateral mesh suitable for finite-element analysis, and the improved algorithm proposed by Min Weidong [8] and others improve the algorithm's efficiency, but the problem of the mesh quality is still unresolved; In 2007, Ma Xinwu and Zhao Guoqun [9] et al. proposed a regional decomposition method combining density control and geometric decomposition, which can adaptively generate a quadrilateral mesh according to the magnitude of curvature of the geometric boundary of the object, the size of the mesh density transition requirements, and the number of cells to be divided. However, this method is first generated into boundary cells, and then the internal region is sub-domain dissected by bisection, the sub-domain dissected results are decomposed into quadrilaterals by pattern matching and other operations, so the algorithm is less efficient; in 1983, Yerry and Shephard [10] applied the quadtree spatial decomposition method to mesh generation and quickly gained attention, and many people have proposed improved algorithms on the basis of it and achieved favorable results. This is an efficient regional discretization method, but the method does not easy to control the density of the grid, and its basic idea is to "approximate the boundary", the effect of the complex boundary approximation is not very satisfactory, and the quality of the boundary grid is inferior. Therefore, this method is seldom used in Chinese research. Since it was proposed by Gordon [11] and Zienkiewicz [12] in the early 1870s, the mapping algorithm has been everlasting, and has been enhanced by researchers to give vigorous vitality. This method can simulate the boundary of the region very well, produce small geometric errors, and have high computational efficiency, uniform mesh distribution, and neat arrangement of the characteristics, so almost all the finite element pre-processing software use this method [13]. However, it is difficult to control the shape and density of the mesh, especially for complex boundaries, and it is difficult to construct the mapping function.

According to the above analysis, this paper focuses on the mapping method for quadrilateral meshing by analyzing the existing finite element calculation software, summarizing the advantages and disadvantages of the current mapping algorithms, and proposing an improved mapping algorithm based on the segmental transfinite mapping.

2 Conventional Transfinite Mapping Method

The basic principle of the mapping method is to map the physical domain to be dissected into the parameter space to form a regular parametric domain through the appropriate mapping function [10, 11]; the mesh of the regular parametric domain is dissected;

the mesh of the parametric domain is inversely mapped back to the physical space, so as to obtain the finite element mesh of the physical domain. The mapping method can be divided into three categories: Conformal Mapping Method, partial differential equation-based method (P.D.E- based Method), and Algebraic Interpolation Method. The algebraic interpolation method is a widely used mesh generation method, which describes the mapping relationship between the parameter space and the physical space through algebraic interpolation, and one of the most important categories is Transfinite Interpolation, which can adapt to the special shape of regional boundaries and control the shape and density of the generated cells, and it has become a standard method for algebraic interpolation mesh generation. It can not only adapt to the special region boundary shape, but also can control the shape and density of the generated cells, and has become a standard method of algebraic interpolation, the equal-parameter mapping proposed by Zienkiewicz is a special form of Transfinite Interpolation.

2.1 Basic Principle

In the region formed by the end of any quadrilateral, the approximation of the physical plane by the computational plane is realized by establishing the mapping function between the physical coordinate system (x, y) and the computational plane (u, v), which is called the surjective mapping method (as shown in Fig. 1). Compared with other mesh generation methods, the transfinite mapping method has the advantages of high accuracy of boundary approximation and an unlimited number of discrete points. If the region has one or more irregular boundaries (i.e., there is no precise mathematical description of the equations), we can use a larger number of control points in the boundary to fit the curve equations, so that the advantages of the surjective mapping method can be fully utilized.

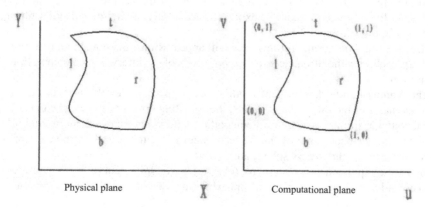

Fig. 1. Identification of generalized edges in the transfinite mapping method.

Define the up, down, left, and right relations of (x,y) to (u,v) on the four edges. The relationship between the upper and lower edges of the physical domain and the parameter domain u is xt (u), yt (u), xb (u), yb (u), and the relationship between the left and right edges of the physical domain and the parameter domain v is xl (v), yl (v), xr (v), yr (v), in which t and b stand for the top and bottom edges, respectively, and l and r stand for the left and right edges, respectively. The mapping relationship between the physical region and the computational plane shown in Eq. (1) can be obtained by using Boolean and interpolation operators [14].

$$\begin{cases} x(u,v) = (1-v)x_b(u) + vx_t(u) + (1-u)x_l(v) + ux_r(v) \\ \quad -[uvx_t(1) + u(1-v)x_b(1) + v(1-u)x_t(0) + (1-u)(1-v)x_b(0)] \\ y(u,v) = (1-v)y_b(u) + vy_t(u) + (1-u)y_l(v) + uy_r(v) \\ \quad -[uvy_t(1) + u(1-v)x_b(1) + v(1-u)y_t(0) + (1-u)(1-v)y_b(0)] \end{cases} \quad (1)$$

The variables (u, v) are the coordinates of the points in the parameter domain and (x, y) are the coordinates of the points in the physical domain.

The parameter domain coordinates (u, v) are easy to obtain. From Eq. (1), it can be seen that the basic idea of transfinite interpolation is to find the coordinates of each physical point inside based on the equations of the four edges of the region.

2.2 Disadvantages of the Algorithm

The Transfinite Mapping Method in the usual sense usually adopts the Vandermonde determinant method, Lagrange interpolation method, and Newton interpolation method, etc. The mapping equations are constructed through the key points of the geometric edges in the physical domain, and the order of the equations is the number of key points, which causes the following difficulties:

1) The accuracy of the mapping function is too low because the number of key points of geometric edges is too small. For example, if a geometric edge consists of two straight line segments, there are three key points of the geometric edge, and the two straight line segments cannot be expressed accurately by the second-order fitting, as shown in Figs. 2-a and 2-b;
2) Higher-order polynomial fitting curve will appear Runge phenomenon, resulting in a sharp decline in the fitting accuracy at both ends of the interval, as shown in Figs. 2-c and 2-d.
3) The Vandermonde determinant method needs to solve for the n-th power of the coordinates of the interpolated points, the rounding error of numerical calculation is unavoidable in the software implementation, and when there are more interpolated points (the critical value of the 32-bit system is 9), there will be the case of non-convergence of the matrix solver, and so on.
4) When the input is a curve segment, (elliptic) arc or spline curve, it is also necessary to consider the shape error when approximating the curve with a line segment.

The boundary of the region in engineering applications is extremely complex, a single region may have dozens of hundreds of geometric edges, these geometric edges are of different lengths and types, when applying the mapping algorithm, the first thing to do is to divide these edges into four super edges in accordance with the head-to-tail relationship, and then to establish the equations of the four super edges (through the solution of the super equations to get the approximation equations of the super edges). Therefore, there is usually a big error between the approximate equations of the super edges and the actual geometric model, which affects the accuracy of the mapping algorithm, and this is also the reason why the mapping algorithm is difficult to use directly on a wide scale.

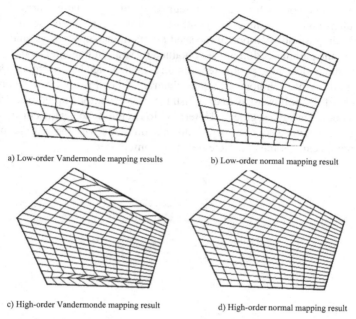

Fig. 2. Low-order and high-order polynomial fitting anomalies in the regular mapping algorithm.

3 Segmental Transfinite Mapping Algorithms

The starting point of the Boolean and interpolation operators is to preserve the interpolation properties while guaranteeing computational accuracy, which is the root of the transfinite mapping algorithm. If the algorithm can solve the above limitations after some kind of deformation of the above algorithms, it will inject new vitality into the transfinite mapping algorithm.

3.1 Algorithm Flow

In scientific computing, interpolation and fitting are inextricably linked, so we can first fit the geometric edges segmentally, and then apply the surjective mapping method to obtain xt (u), yt (u), xb (u), yb (u), and xl (v), yl (v), xr (v), yr (v), and then apply Eq. 1 to obtain the coordinates of the grid points in the physical domain. The execution flow of the segmental transfinite mapping algorithm is given here, followed by the parametric domain grid point calculation method in Sect. 3.2, and the segmental transfinite mapping function formation method as well as the physical domain grid point calculation method are given in Sect. 3.3.

As shown in Fig. 3, firstly, four super edges are specified for the geometric region by the designer according to the geometric properties of the region, then the dimension field information is formed by synthesizing the meshing dimension requirements [15], seed nodes are arranged on each super edge, segmented fitting equations are established for the four super edges, and the corresponding seed points in the parameter domain are solved according to the positions of the seed points in the super edges, and then the internal nodes of the parameter domain are obtained by straight line intersections. Based on this, the internal nodes of the parameter domain are calculated by intersecting straight lines, and the number of each node and quadrilateral grid cell is obtained according to the arrangement of the points, so as to establish the index relationship between the grid cells and grid nodes. Through segmented interpolation mapping to get the corresponding physical domain coordinates of the internal nodes, combined with the original physical domain seed points can get the complete physical domain grid data.

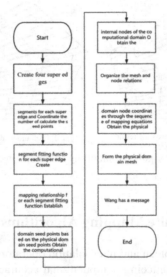

Fig. 3. Processing flow of mesh delineation based on hyperbolic mapping.

In this paper, we focus on the mapping function generation method and the generation method of physical domain nodes of the transfinite mapping method under complex boundary conditions, and the seed point acquisition and mesh optimization shown in Fig. 3 can be found in the previous research results [15].

3.2 Parametric Domain Mesh Node Acquisition

According to the principle of the mapping algorithm, the mesh nodes in the physical domain are mapped according to the nodes in the parameter domain, so it is necessary to obtain the points in the parameter domain first. In order to keep the dimension information of the meshing cells, the corresponding discrete points of the parameter domain are first solved according to the boundary discrete points of the physical domain, and the internal nodes in the parameter domain are solved according to these discrete points in the way of straight-line intersections, and the coordinates of the physical domain are solved through the mapping relationship between the physical domain and the parameter domain. The way of calculating the nodes of the parameter domain is given below:

1) Calculate the length of each geometric edge of the region, and find the length L of each super edge.
2) Calculate the number of splitting shares of each super-edge according to the mesh size constraints, and coordinate the number of splitting shares of odd and even super-edges upwards. For example, if the first super edge is divided into 6 parts and the third super edge is divided into 8 parts, then the odd super edge is divided into 8 parts;
3) Allocate the number of splitting shares based on the ratio of each geometric edge to the total length of the super edge. If a geometric edge explicitly specifies the number of splitting shares, the value will remain unchanged.
4) Generate seed nodes on each geometric edge according to the coordinated number of splitting shares, and organize the seed node Pi on each super edge [15].
5) Based on the ratio li/Ll of the curvilinear distance li of the seed node Ni relative to the starting point of the super-edge (i.e., the distance traveled along the super-edge from the starting point to this seed point) relative to the total length L of the super-edge, solve for the coordinates of the corresponding boundary seed nodes in the parameter domain:

Left boundary $\left(0, \frac{l_i}{L_1}\right)$ Right boundary $\left(1, \frac{l_i}{L_1}\right)$.
Top Boundary $\left(\frac{l_i}{L_1}, 1\right)$ Bottom boundary $\left(\frac{l_i}{L_1}, 0\right)$

6) Solve the internal seed nodes according to the boundary seed nodes in the parameter domain, connect the nodes on the bottom edge (b) and the top edge (t) to get the longitude lines, and connect the nodes on the left (l) edge and the right (r) edge to get the latitude lines, and the intersection point of the longitude and latitude lines is the internal node.

3.3 Formation and Use of Segment Mapping Function

The region where the geometric curvature varies greatly during finite element analysis is often the place where the load varies drastically. Considering the computational efficiency and mapping accuracy, the segmental linear fitting method can be used to establish the function after the boundary is discretized, and the super edge is expressed by the linear equation sequence to avoid the problem of using the conventional polynomial method for the transfinite mapping. The process of forming the segmental mapping function for each super-edge is given below.

1) Extract nodes from the seed points obtained in the parameter domain mesh generation to form the control points of the mapping equations: if the geometric edges are straight line segments, only the first and the last points need to be used; for other types of complex geometric edges, all the seed points are used in sequence;

2) Find the sequence of mapping equations for each super edge: the mapping equation of each interval is a first-order function, which has certain valid sub-intervals [start, end) in the parameter domain (u, v), the start and end correspond to the front and rear endpoints of the seed point straight line segment respectively. These subintervals are sequentially connected and their union set is the total interval [0,1], and the intersection set is empty. Therefore, the mapping equation corresponding to the super-edge is:

$$\begin{cases} x_b(u) = \{A_{xbi}u + B_{xbi}, u_{start_{bi}} \leq u \leq u_{end_{bi}}\} \\ y_b(u) = \{A_{ybi}u + B_{ybi}, u_{start_{bi}} \leq u \leq u_{end_{bi}}\} \\ x_t(u) = \{A_{xti}u + B_{xti}, u_{start_{ti}} \leq u \leq u_{end_{ti}}\} \\ y_t(u) = \{A_{yti}u + B_{yti}, u_{start_{ti}} \leq u \leq u_{end_{ti}}\} \\ x_l(v) = \{A_{xli}v + B_{xli}, v_{start_{li}} \leq v \leq v_{end_{li}}\} \\ y_l(v) = \{A_{yli}v + B_{yli}, v_{start_{li}} \leq v \leq v_{end_{li}}\} \\ x_r(v) = \{A_{xri}v + B_{xri}, v_{start_{ri}} \leq v \leq v_{end_{ri}}\} \\ y_r(v) = \{A_{yri}v + B_{yri}, v_{start\_ri} \leq v \leq v_{end\_ri}\} \end{cases} \qquad (2)$$

3) Use of segmented mapping equations (solving for physical domain coordinates): For each internal node in the parameter domain, Eq. (1) is applied to solve for the corresponding node in the physical domain. The subterms xb (u), yb (u), xt (u), yt (u), xl (v), yl (v), xr (v), yr (v) used in Eq. are the segmented linear equations in Eq. (2). When these two equations are applied to solve for the physical domain coordinates Pi corresponding to pi (ui, vi), it is necessary to apply the appropriate segmented equations in Eq. (2) according to the intervals to which u, v belongs to obtain the values of each subterm. In order to improve the calculation speed, the corresponding sub-interval can be obtained by bisecting and halving, and the linear equation of the interval can be applied directly to obtain the values of the sub-terms. The boundary nodes of the physical domain can directly use the original seed points.

4 Algorithm Verification

Based on the above principles, this paper adopts the.net platform to develop a segmental transfinite mapping mesh delineation system and is applied to an axisymmetric part solid-heat coupling analysis system. In order to verify the correctness of the method, a hexagonal area with a side length of 68 mm and a circular area with a radius of 70 mm are used as examples, and the Ansys system is used for comparison. Figure 4(a) shows the result of equal-size meshing for the hexagonal region, where the four super vertices of upper-left, lower-left, lower-right, and upper-right are defined to form four super edges, and the mesh sizes of the six vertices are all 15 mm, and Fig. 4(b) shows the result of the same mesh division for the same case in the Ansys system. Figure 4(c) shows the result of equal-size meshing for the circular region, defining the four super vertices of top, bottom, left, and right to form four super edges, the mesh size of the top and bottom two vertices is 4 mm, the mesh size of the left and right two vertices is 12 mm, so as to obtain the result of gradient-size meshing for the circular region; Fig. 4(d) is the result of mesh delineation for the same case in the Ansys system.

From the two examples in Fig. 4, it can be seen that the method in this paper can well establish the internal nodes according to the geometrical characteristics of the region

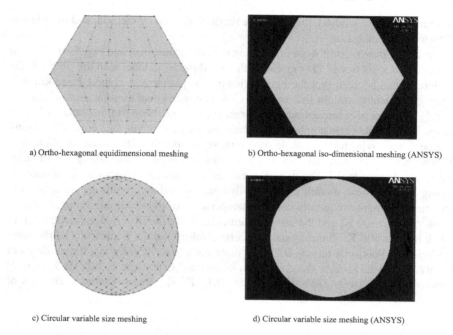

Fig. 4. Comparison of the results of this algorithm and Ansys mapping method for discretizing a simple region.

boundary, and form a good-quality region mesh. Specifically, in the case of equal-size meshing, the results obtained by this method and the Ansys system are exactly the same. For variable-size meshing, the number of boundary divisions obtained by this method is 1 smaller than that obtained by the Ansys system, so the meshing results are not exactly the same, but the distribution trends of mesh nodes and cells are similar. By comparing the two examples of variable size meshing, it can be seen that the method in this paper can get a better transition effect of mesh size, and at the same time, the number of mesh nodes and mesh cells is relatively small.

5 Case Study

The research team designs and develops the finite element platform from the perspective of multi-field coupling analysis, and thoroughly researches many aspects involved in finite element analysis, such as geometric modeling, mesh partitioning, finite element model building and solving, and scientific computation visualization, etc., and forms the "Aero-engine Axisymmetric Parts Thermoset Coupling Analysis System" with all independent intellectual property rights. The system is able to carry out finite element analysis with three/six-node triangular cells and four-node quadrilateral cells as models, in which the available triangular mesh algorithms include the advancing wavefront method, Delaunay method, and mapping method, and the quadrilateral mesh algorithms include the mapping method, the quad-tree method, the laying method, and the triangle transformation method.

Figure 5(a) shows a prototype of a certain type of turbine disk and Fig. 5(b) shows its meridional plane sectioned 2D graphic with each dimension labeled in the figure. From the figure, it can be seen that the boundaries of this planar region have three kinds of geometric elements: straight line segments, B-specimens, and circular arcs. Due to its relatively complex geometric modeling, direct meshing of the meridian plane will result in anomalous meshes (grid points and grid cells are located outside the geometric region), so it is necessary to perform a simple geometric segmentation of the meridian plane first, and then perform a segmental transfinite mapping meshing of the subregions directly, respectively. Figure 5(c) is the quadrilateral mesh obtained by segmental superlattice mapping, for this mesh model, the thermodynamic boundary conditions are applied to form a finite element model: the constant temperature of the upper edge (upper edge of the disk rim) is 1400 K, and the constant temperature of the lower edge (lower edge of the disk hub) is 600 K. The material is a certain alloy, and the parameters of the alloy are as follows: Poisson's ratio is 0.27, the mass density is 7850 kg/m^3, the modulus of elasticity is 2.06 E11N/m^2, the coefficient of thermal expansion is 1.384E-5, thermal conductivity 81 W/(m.K), specific heat 800 J/(kg.K).The thermodynamic analysis of

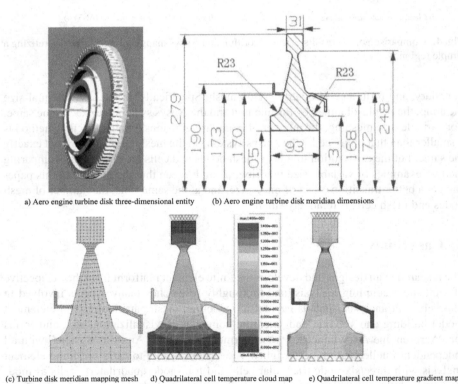

a) Aero engine turbine disk three-dimensional entity   (b) Aero engine turbine disk meridian dimensions

(c) Turbine disk meridian mapping mesh   d) Quadrilateral cell temperature cloud map   e) Quadrilateral cell temperature gradient map

Fig. 5. Aero-engine turbine solid model, meshing and finite element analysis results.

this finite element model gives the temperature field distribution shown in Fig. 5(d), and the corresponding temperature gradient field is shown in Fig. 5(e).

From the figure, it can be seen that when the high- and low-temperature boundary conditions are applied, the radial surface of the turbine disk shows a gentle transition state with a gradual decrease in temperature from the disk rim to the spoke plate and then to the disk hub. Due to the existence of asymmetric mounting edges on the left and right sides, which affect the temperature distribution on the symmetric turbine disk body, the neck of the disk body is the place with the largest temperature gradient, and the temperature gradient on the mounting edges is smaller. According to these analyses, the mapping algorithm proposed in this paper can provide a good-quality quadrilateral mesh, and the finite element analysis based on this mesh can well simulate the real temperature distribution and temperature gradient distribution of the temperature field.

For the same problem, the researchers in the "Aero-engine axisymmetric parts thermoset coupling analysis system" in the application of three-node triangular and six-node triangular cells were analyzed, and similar results were obtained. In order to ensure the correctness of the calculation results, the researchers used UG NX Nastran 6.0 and Ansys 10.0 to validate the case respectively, which also achieved consistent temperature distribution results and temperature gradient results.

6 Conclusion

The conventional transfinite mapping method is difficult to be widely used in engineering due to the Runge problem and so on. In order to give full play to the advantages of the transfinite mapping method, this paper proposes a method for constructing a sequence of segmental transfinite mapping equations based on the inextricable correlation between interpolation and fitting, and gives a method for calculating mesh nodes in the parameter domain and the physical domain based on this method. Compared with other quadrilateral meshing methods such as Paving, this method has the features of high boundary approximation accuracy, unrestricted number of discrete points, strong ability of complex boundary processing, fast solving speed, and stable and reliable algorithm, which extends the application scope of the transfinite mapping meshing.

Combined with the current needs of China's scientific and technological development, the research team to which the author belongs has set up a thermoset coupling analysis platform from a modular point of view, and has conducted in-depth research on geometrical modeling, mesh partitioning, finite element modeling and solving, as well as scientific computational visualization techniques necessary for finite element analysis, and has obtained a series of results with independent intellectual property rights, which have provided a very good way of thinking for the domestic practitioners of finite element research.

References

1. Enqiu, H., Xinfang, Z., Wen, X., et al.: A review of mesh generation methods for finite element computation. J. Comput.-Aided Design Comput. Graph. **9**(4), 378–383 (1997). (in Chinese)

2. Owen, S.J.: A survey of unstructured mesh generation technology. In: Proceedings of the 7th International Meshing Roundtable, Dearborn, pp. 239–267 (1998)
3. Bern, M., Plassmann, P.: Mesh generation[C], pp. 291–332. Elsevier Science, Handbook of computational geometry, North-Holland (2000)
4. Zhenqun, G., Chao, S., Yuanxian, G., et al.: Recent advances of research on finite element mesh generation methods J. Comput.-Aided Design Comput. Graph. **15**(1), 1–14 (2003). (in Chinese)
5. Blacker, T.D., Stephenson, M.B.: Paving: a new approach to automated quadrilateral mesh generation. Int. J. Numer. Meth. Eng. **32**, 811–847 (1991)
6. Tao, C., Guangyao, L., Xu, H.: Generation of quadrilateral mesh from discretized models. J. Mech. Eng. **45**(5), 121–127 (2009). (in Chinese)
7. Lo, S.H.: Generating quadrilateral elements on plane and over curved surfaces. Comput. Struct. **31**(3), 421–426 (1989)
8. Weidong, M., Zesheng, T.: Generating Quadrilateral Elements from a triangular mesh. J. Comput.-Aid. Design Comput. Graph. **8**(1), 1–6 (1996). (in Chinese)
9. Xinwu, M., Guoqun, Z., Fang, W.: The automatic generation of quadrilateral mesh-I: the method of domain division. J. Plast. Eng. **14**(4), 105–109 (2007). (in Chinese)
10. Yerry, M.A., Shephard, M.S.: A modified quadtree approach to finite element mesh generation. IEEE Comput. Graph. Appl. **3**(1), 39–46 (1983)
11. Gordon, W.J., Hall, C.A.: Construction of curvilinear coordinate systems and applications to mesh generation. Int. J. for Numer. Methods Engineering **7**, 461–477 (1973)
12. Zienkiewicz, O.C., Philips, D.V.: An automatic mesh generation scheme for plane and curved surfaces by isoparametric coordinates. Int. J. Numer. Methods Eng. **3**, 519–528 (1971)
13. Subramanian, G., et al.: An algorithm for two-and three-dimensional automatic structured mesh generation. Comput. Struct. **61**(3), 471–477 (1996)
14. Tao, W.: Numerical heat transfer (Second Edition). Xi'an: Xi'an Jiaotong University Publishing Company (2001). (in Chinese)
15. Wang, C., Cui, D., Xu, Y., et al.: Research and application of finite element triangle mesh generation in planar area. Comput. Integr. Manuf. Syst. **17**(2), 256–260 (2011). (in Chinese)

Augmenting Knowledge Tracing: Personalized Modeling by Considering Forgetting Behavior in Learning Process

Hongxin Yang, Yuefeng Du, Tingting Liu, and Linlin Ding(✉)

Liaoning University, Shenyang 110036, Liaoning, China
dinglinlin@lnu.edu.cn

Abstract. With the development of internet online education system, knowledge tracking (KT) is becoming more and more widely used. This technology can accurately model students' learning process, so as to provide students with a personalized knowledge push. However, most previous KT methods don't take into account the learning ability (LA) and forgetting behavior (FB) of student when assessing the state of knowledge. In fact, each student has their own LA and FB, which plays an important role in the KT process. However, at present, the personalized LA and FB of different students are not given in advance, which makes it more challenging to predict the learning status of each student. To address students' challenges in modeling KT, we design augmenting knowledge tracing (AKT), which first uses concept-wised percent correct (CPC) to describe students' overall mastery of knowledge, and builds an individualized forgetting rate (IFR) to describe the degree of forgetting during student learning process. The relationship between student history learning situation and time is considered, moreover, the knowledge mastery and FB are measured from the interaction with the topic during student learning process. The final experimental results show that the performance of proposed model is better than that of traditional methods.

Keywords: knowledge tracking · augmenting knowledge tracing · concept-wised percent correct · individualized forgetting rate

1 Introduction

Knowledge tracking (KT) is the key of constructing adaptive education system, which has attracted more and more researchers' attention in recent years. This technology can model students' historical interactions with question types to obtain students' current knowledge mastery, so as to predict how a student will perform on future interactions. The more accurate the knowledge tracking is, the more efficient the personalized teaching is.

Most existing solutions to KT problems can be divided into traditional Bayesian Knowledge Tracing (BKT), Deep Knowledge Tracing (DKT) and Dynamic Key-Value Memory Networks (DKVMN) [1–3]. BKT is a classic, widely used student learning model that defines two knowledge parameters and two performance parameters for all students. However, BKT model doesn't take into account differential information about individual students and topics. DKT introduces deep learning into KT for the first time, which uses the learning sequence as the input of long and short-term memory networks (LSTMs). This method can express the knowledge state of students with hidden states and reflect long-term knowledge relationships [4, 5]. DKVMN can accurately indicate whether students are good at a particular concept, DKVMN is not only superior to the latest DKT, but also can trace the student's understanding of each concept over time.

Predicting a student's knowledge state precisely is a difficult task because the knowledge mastery and forgetting behavior differ from student to student. To better illustrate them, a concrete example is given in the Fig. 1, where 3 students answered 7 exercises related to 3 different question types. As shown in the Fig. 1, student 2 can quickly grasp the new question type after fewer errors, indicating that student 2 learns faster than student 1 and student 3. At the same time, student 3 tends to make mistakes after doing the right questions (see Q1 and Q2), indicating that he is more likely to forget the key concepts than student 1 and student 2. Unfortunately, the knowledge mastery and forgetting behavior are not given in advance, which makes it very challenging to measure them.

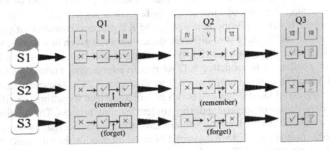

Fig. 1. An example of the learning process

To address the challenges in personalized modeling of student, we proposed an augmenting knowledge tracing (AKT) model to measure the learning ability and forgetting behavior of student. This model first uses concept-wised percent correct (CPC) to consider the overall mastery of the students' knowledge points, and then uses individualized forgetting rate (IFR) to consider the degree of forgetting of the students. After a comprehensive experiment, this method can obtain better knowledge tracking results, and the method has better performance compared to other existing models.

2 Augmenting Knowledge Tracing

In our study, a new knowledge tracking model is proposed. This model first uses Knowledge Mastery Rate (KMP) to consider the overall mastery level of student knowledge points, and then uses Individualized Forgetting Rate (IFR) to measure the behavior of students being forgotten. Considering that the historical learning interaction of students is a time series, a gated linear unit is used to simulate the hidden state. Finally, a fully connected network is used to predict the knowledge state of students at each moment. The overall flow is shown in Fig. 2.

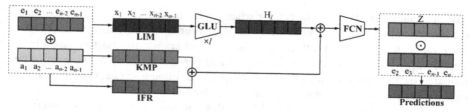

Fig. 2. AKT process

2.1 Basic Embedding

In our study, the embedding matrix x_n of student learning interaction is represented as **LIM**.

$$x_n = e_n \oplus a_n \qquad (1)$$

where e_n is the embedding of knowledge points corresponding to the question, and a_n is the correct answer of the student.

The correct percentage matrix reflects the student's comprehensive grasp of all knowledge concept which is represented as **KMP**. Measuring a student's mastery of global knowledge, the matrix is made up of the percentage of students who are correct about each knowledge concept. Calculate this matrix to get the student grading rate:

$$\mathbf{KMP}(m) = \frac{\sum_{n=1}^{t} \Pi(a_n^m = 1)}{t} \qquad (2)$$

where t is the number of answered m-th type questions, $\Pi(*)$ is the indicating function. **KMP** represents the accuracy of each student's answer to each type of question, which can preliminarily and accurately reflect the student's mastery of knowledge.

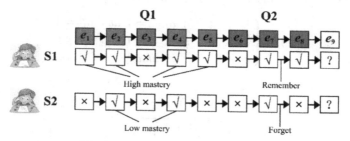

Fig. 3. Knowledge mastery of student.

2.2 Forgetting Behavior

Prior knowledge is hidden in students' history learning interactions. Firstly, some studies have shown that students may get similar scores in similar exercises. Secondly, students have a certain probability of forgetting when answering questions, that is, the longer the questions they have answered last time, the easier it is to forget, and the students' score rate can be seen as a reflection of their overall knowledge mastery (see Fig. 3). Therefore, from two perspectives: history-related performance and the number of historical answers, combined with the student's personal forgetting rate, comprehensively measure the student's personalized prior knowledge.

Integrating the questions answered by students in history with the time interval at this moment can better take into account the rate of forgetting of students. Through different weight assignments, the historical answers are linked to the time interval, that is, the longer the question is assigned a relatively small weight, the closer the question is assigned a higher weight, and the student's learning level at this moment is weighted and matrix calculated.

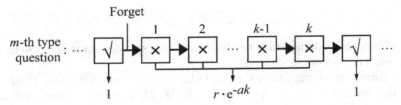

Fig. 4. Individualized Forgetting Rate model

In our study, it is assumed that when students consistently answer the questions correctly, their memory retention is 100%. When answering the wrong question for the first time, it can be considered that the student begins to forget, and this process should continue to decrease until the student answers the question correctly again (see Fig. 4). Considering that the sample size for each type of question is not fixed, the model uses Wilson confidence interval to compensate for the accuracy issue of small samples. The individualized forgetting rate (IFR) can be expressed as

$$\omega(k) = \begin{cases} 1 & \text{if } a_{k-1}^m = 1 \text{ and } a_k^m = 1 \\ r & \text{if } a_{k-1}^m = 1 \text{ and } a_k^m = 0 \\ r \cdot e^{-ak} & \text{if } a_{k-1}^m = 0 \end{cases} \quad (3)$$

where

$$r = \frac{\hat{p} + \frac{1}{2t}z_{1-\frac{\alpha}{2}}^2}{1 + \frac{1}{t}z_{1-\frac{\alpha}{2}}^2} \quad (4)$$

where $a_i^m \in \{0, 1\}$ is the accuracy of the answer to the i-th question of the m-th knowledge point. $\hat{p}$ indicates the correct proportion of the student answering the m-th type of question. t is the total number of answers to the m-th type of question. z represents the statistic corresponding to a certain confidence level. In our study, 1.96 is taken as the value of z at a 95% confidence level. The IFR can be modeled as:

$$IFR(m) = \sum_{i=1}^{t} \omega_m(i) x_i^m \quad (5)$$

2.3 Knowledge Tracking

Considering that the interaction of students in history learning is a time series, this article uses Gated Linear Units (GLUs) to simulate hidden states:

$$H_l = (W * LIM + b) \otimes \sigma(V * LIM + c) \tag{6}$$

where $W \in R^{2d \times d}$, $V \in R^{2d \times d}$, $b \in R^d$ and $c \in R^d$. $\sigma(*)$ is sigmoid function. The knowledge state of element product between matrices is generated by three parts: **LIM**, **KMP** and **IFR**. In our study, **KMP** and **IFR** are connected as inputs to a fully connected network (FCN) to obtain the knowledge state of students.

$$Z = \max(0, (H_l \oplus \text{KMP} \oplus IFR) * W + b) \tag{7}$$

To evaluate student performance, we use student knowledge status and new practice embeddings for prediction.

$$\begin{cases} y_{n+1} = z_n \cdot e_{n+1} \\ p_{n+1} = \sigma(y_{n+1}) \end{cases} \tag{8}$$

where z_n is current student knowledge status, and e_{n+1} is the embedding of new exercises. In order to train the new model, this paper adopts the cross entropy logarithmic loss function as the objective function.

$$L = -\sum_{t=1}^{N} (a_n \log p_n + (1 - a_n) \log(1 - p_n)) \tag{9}$$

3 Experiments

3.1 Data Set

In order to evaluate the performance of the model, four real public datasets and one synthetic dataset is used in the comprehensive experiments. Table 1 shows the statistics for all datasets.

Table 1. The statistics for all datasets.

Data set	Students	Questions	Records
ASSIST2009	4151	110	325637
ASSIST2015	19840	100	683801
ASSISTChall	1709	102	942816
Statics 2011	333	1223	189297
Synthetic-5	4000	50	200000

ASSIST2009: An online tutoring system collected from ASSISTMENS created in 2004. This data is collected from the question side, and students need to practice to master the question type.

ASSIST2015: It covers the records of 2015.

ASSISTChall: Used in the 2017 Assisted Data Mining Competition. The researchers collected it from a longitudinal study that tracked students' use of assistants.

Statics 2011: is obtained from a college-level engineering statics course.

Synthetic-5: Published in the DKT paper, which simulates virtual students learning virtual concepts. It is worth noting that the simulated virtual learning process does not take into account the individualized learning rate of students.

3.2 Method Comparison

To verify the effectiveness of the new model, we compared it using methods such as CKT-ONE, CKT-CPC, CKT-ILR, DKT, and CKT-IFR [6].

CKT-ONE: CKT that contain only one convolutional layer.

CKT-CPC: Only prior knowledge from THE CPC is considered.

CKT-ILR: Only personalized learning rates are simulated.

DKT: Use recurrent neural networks to assess students' knowledge status.

CKT-IFR: Only prior knowledge from IFR is considered.

3.3 Performance Prediction

To evaluate the performance of new method, we conducted extensive experiments. To provide robust evaluation results, Area Under Curve (AUC), Accuracy (ACC), and the square of Pearson correlation (r^2) were evaluated (Table 2).

Table 2. Performance of AUC

	CKT-ONE	CKT-CPC	CKT-ILR	DKT	CKT-IFR	CKT	AKT
ASSIST2009	0.825	0.82	0.815	0.82	0.815	0.825	0.827
ASSIST2015	0.734	0.732	0.729	0.734	0.728	0.735	0.736
ASSISTChall	0.71	0.71	0.708	0.7	0.71	0.72	0.715
Statics 2011	0.823	0.829	0.826	0.824	0.824	0.83	0.831
Synthetic-5	0.827	0.826	0.805	0.825	0.828	0.826	0.826

By observing the performance of AUC, it can be seen that the model we propose can better simulate the personalized learning rate and prior knowledge of different students in most data experiments, and obtain better prediction results. In addition, our model has achieved a higher improvement on the Statics 2011 and ASSIST2015 datasets (Table 3).

Through the ACC indicator, it can be seen that the accuracy of the model we propose is higher than that of other models in most datasets, 0.2 percentage points higher than the

Table 3. Performance of ACC

	CKT-ONE	CKT-CPC	CKT-ILR	DKT	CKT-IFR	CKT	AKT
ASSIST2009	0.775	0.773	0.77	0.77	0.769	0.775	0.777
ASSIST2015	0.754	0.752	0.752	0.754	0.749	0.755	0.754
ASSISTChall	0.685	0.684	0.682	0.681	0.678	0.685	0.69
Statics 2011	0.813	0.815	0.813	0.813	0.81	0.815	0.815
Synthetic-5	0.757	0.755	0.735	0.755	0.755	0.756	0.755

accuracy of the CKT model by 77.7% on the ASSIST2009 dataset, and 0.5 percentage points higher than that of the CKT model in the ASSISTChall dataset (Table 4).

Table 4. Performance of r^2

	CKT-ONE	CKT-CPC	CKT-ILR	DKT	CKT-IFR	CKT	AKT
ASSIST2009	0.32	0.315	0.31	0.315	0.316	0.32	0.322
ASSIST2015	0.137	0.135	0.132	0.135	0.131	0.139	0.138
ASSISTChall	0.13	0.129	0.128	0.125	0.125	0.126	0.13
Statics 2011	0.25	0.265	0.26	0.258	0.255	0.264	0.266
Synthetic-5	0.31	0.305	0.27	0.305	0.316	0.314	0.316

The square of Pearson correlation(r^2) shows that the models we design have a high effect on each data set and can better predict the learning status and ability of students.

In this experiment, we evaluate the effectiveness of AKT by predicting the student's performance in each student interaction, as shown in the figure, from the figure, it is easy to see that HCKT and some other algorithms perform on several real data sets, and we can conclude that HCKT can obtain better prediction results. AKT achieved higher predictive results on the ASSISTchall dataset. This suggests that AKT can make better predictions about learning information with longer sequences than shorter sequences. In addition, CKT, which only models personalized learning rates, and CKT-IFR, a method that only considers prior knowledge from IFR, show more weaknesses in predicting student performance.

4 Conclusions

In this paper, we present Augmenting Knowledge Tracing (AKT) to model personalized learning information for students in KT tasks. The model uses Concept-wised Percent Correct (CPC) to describe the overall knowledge mastery of the students, and the Individualized Forgetting Rate (IFR) is designed to describe the forgetting characteristics of the students, and the personalized prior knowledge for the students is obtained. Hierarchical convolution is then used to extract personalized learning rates based on continuous learning interactions. Finally, through a large number of experiments, the results show that AKT can better use students' historical learning records for knowledge tracking and obtain better prediction results than other existing prediction methods. In this study, the forgetting characteristics of students are considered, but the model is a linear description, and people's grasp of knowledge points is non-linear, obeying the Ebbinghaus Forgetting Curve model, so we will integrate the model with KT in our follow-up research to better achieve recommendation.

Acknowledgments. This study was funded by the National Natural Science Foundation of China (No. 62072220); Natural Science Foundation of Liaoning Province (2022-KF-13-06); Open Project of National Engineering Research Centre of Advanced Network Technologies (No. ANT2023003); Social Science Planning Fund Program of Liaoning Province of China (No. L23BJY018).

References

1. Albert, C., John, A.: Knowledge tracing: modeling the acquisition of procedural knowledge. UMUAI **4**(4), 253–278 (1994)
2. Yann, D., Angela, F., et al.: Language modeling with gated convolutional networks. ICML **70**, 933–941 (2017)
3. Kaiming, H., et al.: Deep residual learning for image recognition. CVPR, 1–12 (2016)
4. Liu Q., Huang Z., Yin Y., et al.: EKT: exercise-aware knowledge tracing for student performance prediction. IEEE Trans. Knowl. Data Eng. 452–463 (2019)
5. Gan, W., Sun, Y., Peng, X., et al.: Modeling learner's dynamic knowledge construction procedure and cognitive item difficulty for knowledge tracing. Appl. Intell. **3**(50), 3894–3912 (2020)
6. Pei H., Wei B., Chang C., et al.: Geom-GCN: geometric graph convolutional networks. ICLR, 1–10 (2020)

A Time-Aware Sequential Recommendation Based on Attention Mechanism

Tingting Liu[1], Tianrui Li[1,2], Baoyan Song[1], Yuefeng Du[1], Hongxin Yang[1], and Linlin Ding[1(✉)]

[1] Liaoning University, No. 66 Chongshan Middle Road, Huanggu District, Shenyang, Liaoning, China
dinglinlin@lnu.edu.cn
[2] BYD Auto Industry Co., Ltd., No. 3001, 3007 Hengping Road, Pingshan New District, Shenzhen, Guangdong, China

Abstract. Sequential recommendation is a technique used to predict a user's next purchased items by modeling their historical behaviors. However, traditional sequential recommender methods typically capture only sequential patterns and without taking other temporal information into account. As a result, their performance may be limited. In this paper, a time-aware sequential recommendation model based on attention mechanism (named TASR) is proposed. To enhance the modeling of user preferences and improve recommendation performance, our model captures temporal features of the sequence and combines them with item representations. Specifically, our approach utilizes an aspect-aware convolution to extract user and item representations from the embedding matrix. These representations are then merged with the temporal features during the temporal dynamic modeling process. Finally, multi-layer attention is used to derive users' short-term preferences and the ratings of items are predicted using our TASR model. Extensive experiments on three datasets demonstrate the superiority of TASR against several state-of-art baselines.

Keywords: Recommender system · Attention mechanism · Deep learning · Temporal features

1 Introduction

The development of the Internet has led to a massive increase in network information. Internet users now face information overload, and the information available on the Internet has a variety of complex characteristics. Users interact with applications in various scenarios, resulting in a large amount of user behavior data. User characteristics extracted from their behavior can be used by recommender systems to recommend items of interest to users. Collaboration filtering-based recommendations [1–3], content-based recommendations [4, 5], point-of-interest recommendations [6, 7], graph-based recommendations [8, 9], and hybrid recommendations [10] are examples of existing recommendation methods.

A user's behavior usually occurs in a sequence, and the interactions in the sequence are interdependent. Sequential recommendation [11, 12] is a method that captures shifts in users' interests by modeling their behavioral sequences. Traditional sequential recommendation models rely on sequential pattern mining [13, 14] and Markov chain models [15, 16]. However, with the emergence of deep learning, newer models have been developed that can capture sequential information and improve the performance of recommendations [17–20]. Nonetheless, traditional sequential recommendations only take into account the order of items in a sequence, and they disregard other temporal information such as the time interval of items. As a result, their recommendation performance is limited. In reality, users' interests dynamically change over time, and even within the same sequence, different time intervals between items can yield different recommendation results. Shorter interval items usually have a more significant impact on candidate items. Despite its widespread use in e-commerce scenarios, traditional sequential recommendation's lack of consideration for temporal information limits its effectiveness. As shown in Fig. 1, user A accomplished the last interaction within the last week, whereas user B completed the last one two months ago. Considering the impact of time intervals, the model recommends different items for two users regardless of whether their historical interaction sequences are identical. As a result, in order to effectively use multiple temporal features implied in the sequence, we propose a time-aware sequential recommendation model (TASR), which is optimized on the basis of RNS [21] and takes into account the influence of two types of temporal information, namely, timestamps and time intervals, on user preference modeling, to complete users' long- and short-term interests. The proposed TASR model first converts the review document into word embeddings and uses CNN to extract the features. The model then combines timestamps and time intervals with item features to capture changes in the user's preferences over time. Multi-layer attention is then utilized to evaluate the weight of various items and derive an embedding representation of the user's interests. Finally, the model predicts the user's rating towards target items and recommends the top-k items to the user.

2 Method

This section introduces the time-aware sequential recommendation model based on attention mechanism (TASR), the framework of which is shown in Fig. 1. The model is made up of aspect-ware convolution, temporal dynamic modeling, and multi-layer attention. The review documents of users and items, as well as the corresponding historical purchase sequences, are firstly converted into word embedding, and the feature inside is extracted using aspect-ware convolution as the long-term user preferences and the embedding representations of items. Following that, a temporal dynamic modeling comprised of a temporal feature fusion layer and a GRU layer is applied to integrate temporal features into item representation. The temporal feature fusion layer can integrate the timestamp and time interval correlating to the items into the items' representation, and the GRU layer can capture the explicit connection between the sequence patterns in the user's behavior sequence and the items. Multi-layer attention module allows the model to focus on the items that are most relevant to the target. Eventually, the model predicts the user's rating towards candidate items based on the user's long-term and short-term preferences.

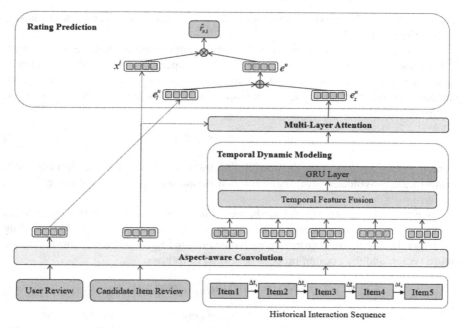

Fig. 1. The network architecture of our proposed TASR.

2.1 Problem Formulation

Given the user and item set U and I. Each user $u \in U$ is associated with a sequence of items $S^u = \left(S_1^u, S_2^u, ..., S_{|S^u|}^u\right)$ and timestamps $T^u = \left(T_1^u, T_2^u, ..., T_{|T^u|}^u\right)$, where $S_i^u \in I$ denotes the items purchased by the user at the time step i, and T_i^u denotes the time stamp corresponding to items in the sequence. Reviews of users and items are respectively combined to form the user review document D^u and the item review document D^i. The task of the recommendation is to predict the user's rating s_{uj} for the candidate item based on the user's most recent purchase sequence and the associated review documents.

2.2 Aspect-Aware Convolution

This section will describe how to process the user's review documents using aspect-aware convolution. First, each word of the user's review document D^u is mapped to corresponding d-dimensional vector. As a result, the user document D^u is transformed into an embedding matrix $M^u \in R^{|v| \times d}$, where $|v|$ is the number of words, and d is the embedding dimension corresponding to the words.

The associated word vector is not unique because the meaning of the same word in different scenes and contexts can vary significantly. Therefore, an embedding matrix focusing on specific aspects of words $W_a \in R^{d \times d}$ is introduced in the embedding layer. A three-dimensional tensor M^u is used to represent the user review matrix:

$$E_a^u = W_a \cdot M^u \tag{1}$$

Here, $E^u \in R^{|v| \times d \times m}$ is the embedding of a particular aspect of the user review document, and m is the number of aspects.

Convolutional neural networks (CNN) are being used to extract long-term user interest from the review documents D^u. Convolutional operations are used to efficiently extract user preferences and contextual features from the user's review. Using the embedding of the user review E^u as the input to the CNN, the features are extracted as follows:

$$z_k = f\left(W_k \odot E^u + b_k\right) \quad (2)$$

where $1 \leq k \leq n$, $\odot$ denotes the convolution operation, W_k is the weight of the k-th feature in the convolution layer, b_k is the corresponding bias and $f()$ is the activation function.

Since not all features are useful for the recommendation, the max pooling operation is used to process the feature vectors obtained from each convolution kernel. Max pooling can extract the maximum vector from each convolution result, effectively shortening the feature vector, and the pooling process is defined as follows:

$$e_l^u = [\max(z_1), \max(z_2), ..., \max(z_n)] \quad (3)$$

Note that n is the total number of convolution kernels. Finally, the output vector e_l^u is the ultimate result in this layer, representing users' long-term preferences. Because the same operation is performed on both user review documents and item review documents in this module, only the processing of user review documents D^u is described. The same operation is performed on the candidate item review and the purchase sequence, yielding the feature vectors x^i and $X^u = \left(x_1^u, x_2^u, ..., x_L^u\right)$, respectively.

2.3 Temporal Dynamic Modeling

The research [21] combines review information and sequential pattern for the recommendation. However, in e-commerce scenarios, users' preferences typically change over time, and users' recent purchases can also influence their purchase behavior. Users' interest is significantly influenced by temporal information. As a result, in the temporal feature fusion layer, temporal information is fused with item representation so that the timestamp and time interval can be considered during the subsequent calculation. The model details of the temporal dynamic modeling and multi-layer attention are shown in Fig. 2.

First, using Eq. (4), calculate the time interval Δt_i between two user purchases behaviors:

$$\Delta t_i = \lfloor \log(t_i - t_{i-1} + 1) \rfloor \quad (4)$$

where t_i is the timestamp for the user's most recent purchase and t_{i+1} is the timestamp for the user's next purchase. Given a user u and his most recent purchase of L items $x_1^u, x_2^u, ..., x_L^u$, the item embedding is fused with two types of time information: timestamp and time interval. A logarithmic function is used to rescale the time interval, and the bottom function converts the scaled time interval to a positive integer. Finally, the item representation with fused time features is obtained as follows:

$$x_i^u = x_i^u * (\Delta t_i + t_i + 1) \tag{5}$$

Here, x_i^u is the embedding vector of the i-th item in the sequence and Δt_i is the time interval between two interactions.

After incorporating temporal information into the item representation, users' short-term preferences are attained using a gated recurrent unit (GRU) [22] and a multi-layer attention mechanism [23]. GRU is one of the most popular variants of LSTM, which is frequently used when dealing with sequential information because its structure is simpler than LSTM and can achieve comparable results to LSTM. GRU not only effectively extracts the sequential patterns contained in the user's sequence, but it also evaluates the influence of previous behavior on current actions. The GRU layer's node update process is as follows, using the user's historical sequence $X^u = \left(x_1^u, x_2^u, ..., x_L^u\right)$ as input:

$$r_t = \sigma\left(W_r[h_{t-1}, x_t^u] + b_r\right) \tag{6}$$

$$z_t = \sigma\left(W_Z[h_{t-1}, x_t^u] + b_z\right) \tag{7}$$

$$\tilde{h}_t = \tanh\left(\left(W_{\tilde{h}}[r_t * h_{t-1}, x_t^u]\right)\right) \tag{8}$$

$$h_t = (1 - z_t) * h_{t-1} + z_t * \tilde{h}_t \tag{9}$$

where x_t^u is the t-th item in the sequence, $W_r, W_z, W_{\tilde{h}}$ is the weight matrix of the GRU, $*$ denotes the Hadamard product, and σ denotes the Sigmoid function. Finally, the output of GRU is expressed as $H^u = (h_1, h_2, ..., h_L)$.

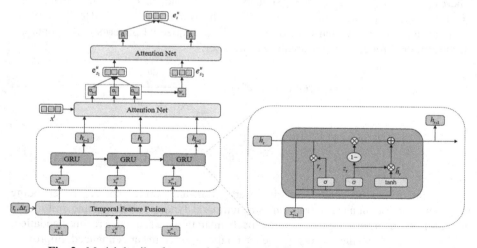

Fig. 2. Model details of temporal dynamic modeling and multi-layer attention.

2.4 Multi-layer Attention

Different historical purchase items bring different impacts on the user's current decision, and the attention mechanism can calculate the importance of various items in the sequence for the candidate items. Inspired by the research [21], this paper uses multi-layer attention. First, the attention weights between each item representation $h_i (1 \leq i \leq L)$ and candidate item representation c_i in the sequence are calculated as follows:

$$\alpha_m = \textit{soft} \max(x^i \cdot h_m) = \frac{\exp(x^i \cdot h_m)}{\sum_{k=1}^{L} \exp(x^i \cdot h_k)} \tag{10}$$

The user's short-term interest in the candidate item is expressed as a weighted sum of the items in the following sequence:

$$e_{s_1}^u = \sum_{m=1}^{L} \alpha_m \cdot h_m \tag{11}$$

where $e_{s_1}^u$ is the user's sequential pattern at the union-level, which is encoded jointly by the L items in the sequence. However, sometimes there are cases where only one of the historical items has a large influence on the candidate item, in which case the item with the highest attention weight is chosen as follows:

$$w_m = \arg\max_m (\alpha_m) \tag{12}$$

$$e_{s_2}^u = h_{w_m} \tag{13}$$

where $e_{s_2}^u$ is the individual-level user sequential pattern, indicated by the item that has the greatest influence on the candidate item. The multi-layer attention is used to differentiate the importance of individual-level and union-level sequential patterns and to obtain a short-term preference e_u^s representation of the user:

$$\beta_n = \textit{soft} \max\left(x^i \cdot e_{s_n}^u\right) = \frac{\exp(x^i \cdot e_{s_n}^u)}{\sum_{i=1}^{L} \exp(x^i \cdot e_{s_i}^u)} \tag{14}$$

$$e_u^s = \sum_{n=1}^{2} \beta_n \cdot e_{s_n}^u \tag{15}$$

In conclusion, this module obtains users' short-term preferences e_u^s by encoding their behavioral sequence information using temporal dynamic modeling and multi-layer attention. By incorporating the attention mechanism into the sequence recommendation, the features that have the greatest influence on the target items are enhanced, improving the recommendation model's performance.

2.5 Prediction and Model Optimization

Following the above modules, the user's interest is effectively represented. The user's long-term and short-term interests are linearly fused to obtain the user's embedding, which is then used to predict the next item e^u:

$$e_u = \gamma \cdot e_u^s + e_u^l \tag{16}$$

where the parameters γ dictate the importance of short-term interest in the rating prediction. Finally, the representation of the user and items are combined to produce the final prediction results:

$$s_{uj} = \sigma\left(e^{u} x^i\right) \tag{17}$$

Here, s_{uj} is the probability of a user purchasing a candidate item and σ is the sigmoid function. Furthermore, in this paper, the loss function employs a binary cross-entropy loss with an L2 regularization. A training instance is created for each user by extracting consecutive L items and the next item from the user sequence S_u as the target item.

$$\mathcal{L} = \sum_u \sum_{t_j \in C^u} \left(-\log(s_{uj}) + \sum_{i \in N(j)} -\log(1 - s_{ui}) \right) + \lambda \|\Theta\|^2 \tag{18}$$

In the equation, C_u is the set of all training instances of user u, $N(j)$ is a randomly selected x negative samples, λ is a regularization parameter, and Θ represents all parameters of the model. Adam optimizer is used to train the model.

3 Experiment

In this paper, experiments will be conducted around the following three questions.

3.1 Experimental Settings

Datasets. To validate the model's effectiveness, an Amazon public dataset with 24 subsets is used as the experimental dataset, three of which are chosen in this paper. They are Instant Video, Pet Supplies, Tools and Home Improvement (referred to as Tools and Home in the following). The dataset's main information includes user ID, item ID, the ratings of items, timestamps, and review documents.

Since the model in this paper is an enhancement of the literature [21], it is preprocessed using the same data preprocessing steps. For all datasets, ratings are regarded as purchases, items are sorted by timestamp, and users who purchase fewer than 10 items in the dataset are removed. Table 1 displays statistical data about the current state of the dataset.

Evaluation Metrics. To evaluate recommendation performance, we divide the dataset into two parts: the training set and the testing set. The first 70% of the dataset serves as training, and the remaining 30% is used for testing. The experiment's evaluation metrics are Precision@N, Recall@N, NDCG@N, and HR@N.

Table 1. Statistics of the three datasets.

Dataset	# users	# items	# interactions	Sparsity
Instant Video	1372	7957	23181	99.79
Pet Supplies	7417	33798	117385	99.95
Tools and Home	10076	66710	169245	99.97

Baseline Methods. The same comparison models as RNS are used to evaluate the performance of the TASR model proposed in this paper. The following methods were used in the experiment: POP, RUM(I) [24], RUM(F) [24], FPMC [25], GRU4Rec [26], Caser [27], SASRec [28], and RNS [21].

3.2 Results and Discussion

Table 2 compares the experimental results of all methods on three datasets to the state-of-the-art recommendation method proposed. The baseline method's results are from the literature [3], with the best results bolded and the second best underlined. The following conclusions are drawn from the experimental results:

Table 2. Performance comparison of different methods across the datasets. The best and second best are highlighted in boldface and underlined respectively.

Baseline	Instant Video				Pet Supplies				Tools and Home Improvement			
Measures@5	Precision	Recall	NDCG	HR	Precision	Recall	NDCG	HR	Precision	Recall	NDCG	HR
Pop	0.0783	0.0876	0.0873	0.2648	0.0615	0.0622	0.0750	0.2376	0.0542	0.0575	0.0587	0.2105
FPMC	0.1067	0.1214	0.1187	0.3900	0.0985	0.1009	0.1075	0.3761	0.0785	0.0828	0.0911	0.3047
GRU4Rec	0.1036	0.1166	0.1137	0.3804	0.0943	0.0940	0.1015	0.3646	0.0712	0.0747	0.0821	0.2915
RUM(I)	0.1160	0.1258	0.1245	0.3921	0.1013	0.1005	0.1086	0.3790	0.0790	0.0853	0.0936	0.3166
RUM(F)	0.1142	0.1274	0.1274	0.3928	0.1041	0.1047	0.1146	0.3973	0.0803	0.0855	0.0925	0.3199
Caser	0.1152	0.1321	0.1456	0.4060	0.1038	0.1136	0.1252	0.3975	0.0819	0.0873	0.1029	0.3378
SASRec	0.1183	0.1295	0.1436	0.4054	0.1026	0.1105	0.1241	0.3922	0.0820	0.0875	0.1036	0.3384
RNS	<u>0.1329</u>	<u>0.1531</u>	<u>0.1648</u>	<u>0.4446</u>	<u>0.1146</u>	<u>0.1252</u>	<u>0.1388</u>	<u>0.4362</u>	<u>0.0894</u>	<u>0.0943</u>	<u>0.1120</u>	<u>0.3614</u>
TASR	**0.1359**	**0.1565**	**0.1690**	**0.4526**	**0.1147**	**0.1348**	**0.1479**	**0.4480**	**0.0897**	**0.1022**	**0.1193**	**0.374**
Improvement	2.26%	2.22%	2.55%	1.80%	0.09%	7.67%	6.56%	2.71%	0.34%	8.38%	6.52%	3.49%

(1) The performance of the non-personalized recommendation method POP is much less effective in all metrics. It is obvious that personalized recommendation methods can better reflect user preferences compared to non-personalized recommendation methods.
(2) The recommendation performance of the RNN-based recommendation method GRU4Rec and the traditional recommendation method FPMC without neural networks is slightly better than POP, but not as good as the sequential recommendation

methods Caser, HGN, SASRec, and RNS. This is because sequential recommendation makes full use of the temporal information contained in the user's sequence, which leads to more accurate prediction scores.

(3) Among all baseline methods, RNS has the best experimental results. The method combines sequential recommendation with the review-based method, which results in a significant performance improvement over previous methods.

(4) The TASR model proposed in this paper achieves the best results on all metrics of the three datasets, which optimizes the RNS model, and incorporates the temporal information into the item representation. When compared to the previous models, TASR improves recommendation accuracy much more, and the model in this paper improves accuracy and recall by up to 2.26% and 7.67%, respectively, when contrasted to the RNS model.

3.3 Parameter Sensitivity

In order to discuss the effect of parameter settings on model validity, this section conducts experiments with different parameter values on three data subsets and analyzes the effect of sequence length L and parameter size on the experimental results.

(1) Effect of sequence length L: The left image in Fig. 3 depicts the effect of L values ranging from 1 to 5 on the recommended performance while all other parameters remain constant. By analyzing the experimental results, it is found that the performance of the TASR model increases with the increase of L in most cases in all three datasets, and the best results are achieved when $L = 5$. This is due to the fact that, when $L < 5$, the model can use less information to learn user preferences, and as L increases, the model can learn more and more information from the historical sequence.

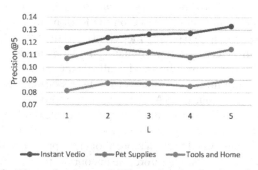

Fig. 3. The performance of different L values on three datasets.

(2) Effect of parameters γ: The right image of Fig. 4 depicts the effect of changing the value γ on the recommendation performance while keeping all other parameters constant. This parameter determines whether the user's long-term or short-term preferences are more important in the recommendation. The results show that the model achieves worse performance when $\gamma = 0$ or $\gamma = 1$, because only the user's long-term preferences

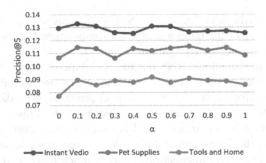

Fig. 4. The performance of different γ values on three datasets.

or only the user's short-term preferences are considered in the recommendation; when $\gamma = 0.1$, the model achieves the best recommendation performance; when $\gamma > 0.1$, the recommendation performance changes continuously with increasing, but the performance is always lower than the model performance when $\gamma = 0.1$. This indicates that in the sequential recommendation, both the long-term and short-term preferences of users are combined to improve the user's decision.

4 Conclusion

In this work, we proposed a time-aware sequential recommendation model based on attention mechanism (TASR). Our model is composed of three components: aspect-aware convolution, temporal dynamic modeling, and multi-layer attention. This architecture allows our model to capture the dynamics of user preferences over time, taking into account temporal features such as time intervals. Our proposed TASR model offers several advantages over prior methods. Firstly, our model incorporates multiple temporal information into item embeddings based on the RNS model, resulting in improved recommendation results compared to only modeling the sequential pattern of items. Secondly, our approach introduces a temporal dynamic modeling module to fuse timestamps and time intervals with item embeddings. This allows the model to dynamically select the representation that are more important for the user's current decision. The results of the experiments showed that the TASR model can improve its performance of the model. Future research could explore the inclusion of contextual features in historical sequences, such as time intervals, geographic location, and so on. Incorporating such features could facilitate our model to learn more about user preferences and provide more accurate recommendation results. Therefore, we encourage further investigation into the use of contextual features to improve the performance of recommendation systems.

Acknowledgement. This study was funded by the National Natural Science Foundation of China (No. 62072220); Natural Science Foundation of Liaoning Province (2022-KF-13–06); Liaoning Provincial Department of Education Youth Project (LJKQZ2021023); Ministry of Education Key Laboratory Open Fund Project (93K172022K06); University-level Teaching Reform Project (d272454002).

References

1. Xiao, T., Shen, H.: Neural variational matrix factorization for collaborative filtering in recommendation systems. Appl. Intell. **49**, 3558–3569 (2019)
2. Zhao, C., Shi, X., Shang, M., Fang,Y.: A clustering-based collaborative filtering recommendation algorithm via deep learning user side information. In: Web Information Systems Engineering-WISE 2020: 21st International Conference, Amsterdam, The Netherlands, October 20–24, 2020, Proceedings, Part II 21, pp. 331–342. Springer (2020)
3. Yuan, Z., Yu, T., Zhang, J.: A social tagging based collaborative filtering recommendation algorithm for digital library. In: Digital Libraries: For Cultural Heritage, Knowledge Dissemination, and Future Creation: 13th International Conference on Asia-Pacific Digital Libraries, ICADL 2011, Beijing, China, October 24–27, 2011. Proceedings 13, pp. 192–201. Springer (2011)
4. Brusilovsky, P., Kobsa, A., Nejdl, W. (eds.): The Adaptive Web. LNCS, vol. 4321. Springer, Heidelberg (2007). https://doi.org/10.1007/978-3-540-72079-9
5. Markapudi, B., Chaduvula, K., Indira, D., Sai Somayajulu, M.V.: Content-based video recommendation system (cbvrs): a novel approach to predict videos using multilayer feed forward neural network and monte carlo sampling method. Multimed. Tools Appl. **82**(5), 6965–6991 (2023)
6. Shi, H., Chen, L., Xu, Z., Lyu, D.: Personalized location recommendation using mobile phone usage information. Appl. Intell. **49**, 3694–3707 (2019)
7. Laroussi, C., Ayachi, R.: A deep meta-level spatio-categorical poi recommender system. Int. J. Data Sci. Anal. 1–15 (2023)
8. Yin, F., Ji, M., Wang, Y., Yao, Z., Feng, X., Li, S.: Enhanced graph recommendation with heterogeneous auxiliary information. Complex Intell. Syst. **8**(3), 2311–2324 (2022)
9. Do, P., Pham, P.: Heterogeneous graph convolutional network pre-training as side information for improving recommendation. Neural Comput. Appl. **34**(18), 15945–15961 (2022)
10. Guan, Y., Wei, Q., Chen, G.: Deep learning based personalized recommendation with multi-view information integration. Decis. Support Syst. **118**, 58–69 (2019)
11. Li, Y., Ding, Y., Chen, B., Xin, X., Wang, Y., Shi, Y., Tang, R., Wang, D.: Extracting attentive social temporal excitation for sequential recommendation (2021). arXiv preprint arXiv:2109.13539
12. Sun, F., Liu, J., Wu, J., Pei, C., Lin, X., Ou, W., Jiang, P.: Bert4rec: Sequential recommendation with bidirectional encoder representations from transformer. In: Proceedings of the 28th ACM International Conference on Information and Knowledge Management, pp. 1441–1450 (2019)
13. Yap, G.-E., Li, X.-L., Yu, P.S.: Effective next-items recommendation via personalized sequential pattern mining. In: Database Systems for Advanced Applications: 17th International Conference, DASFAA 2012, Busan, South Korea, April 15–19, 2012, Proceedings, Part II 17, pp. 48–64. Springer (2012)
14. Bin, C., Gu, T., Sun, Y., Chang, L.: A personalized poi route recommendation system based on heterogeneous tourism data and sequential pattern mining. Multimed. Tools Appl. **78**, 35135–35156 (2019)
15. Garcin, F., Dimitrakakis, C., Faltings, B.: Personalized news recommendation with context trees. In: Proceedings of the 7th ACM Conference on Recommender Systems, pp. 105–112 (2013)
16. Nasir, M., Ezeife, C.: Semantic enhanced Markov model for sequential e-commerce product recommendation. Int. J. Data Sci. Anal. 1–25 (2022)
17. Hochreiter, S., Schmidhuber, J.: Long short-term memory. Neural Comput. **9**(8), 1735–1780 (1997)

18. Quadrana, M., Karatzoglou, A., Hidasi, B., Cremonesi, P.: Personalizing session-based recommendations with hierarchical recurrent neural networks. In: Proceedings of the Eleventh ACM Conference on Recommender Systems, pp. 130–137 (2017)
19. Wang, D., Deng, S., Xu, G.: Sequence-based context-aware music recommendation. Inf. Retr. J. **21**, 230–252 (2018)
20. Li, B., Liu, K., Gu, J., Jiang, W.: Review of the researches on convolutional neural networks. Comput. Era **4**, 12–17 (2021)
21. Li, C., Niu, X., Luo, X., Chen, Z., Quan, C.: A review-driven neural model for sequential recommendation (2019). arXiv preprint arXiv:1907.00590
22. Cho, K., et al.: Learning phrase representations using RNN encoder-decoder for statistical machine translation (2014). arXiv preprint arXiv:1406.1078
23. Vaswani, A., et al.: Attention is all you need. Advances in neural information processing systems 30 (2017)
24. Rendle, S., Freudenthaler, C., Schmidt-Thieme, L.: Factorizing personalized Markov chains for next-basket recommendation. In: Proceedings of the 19th International Conference on World Wide Web, pp. 811–820 (2010)
25. Chen, X., Xu, H., Zhang, Y., Tang, J., Cao, Y., Qin, Z., Zha, H.: Sequential recommendation with user memory networks. In: Proceedings of the Eleventh ACM International Conference on Web Search and Data Mining, pp. 108–116 (2018)
26. Hidasi, B., Karatzoglou, A., Baltrunas, L., Tikk, D.: Session-based recommendations with recurrent neural networks (2015). arXiv preprint arXiv:1511.06939
27. Tang, J., Wang, K.: Personalized top-n sequential recommendation via convolutional sequence embedding. In: Proceedings of the Eleventh ACM International Conference on Web Search and Data Mining, pp. 565–573 (2018)
28. Kang, W.-C., McAuley, J.: Self-attentive sequential recommendation. In: 2018 IEEE International Conference on Data Mining (ICDM), pp. 197–206 (2018). IEEE

Integrating User Sentiment and Behavior for Explainable Recommendation

Dong Li[1], Zhicong Liu[1], Qingyu Zhang[2], Yue Kou[2], Tingting Liu[1(✉)], and Haoran Qu[1]

[1] Liaoning University, Shenyang 110036, China
dongli@lnu.edu.cn
[2] Northeastern University, Shenyang 110819, China

Abstract. With the advancement of internet technology, responsibly recommending the most suitable items to users has emerged as a pivotal challenge for online e-commerce platforms. However, existing works fail to combine users' sentiments with behavioral characteristics to provide effective and explainable recommendations. In this paper, we propose a novel Explainable Recommendation method, termed USB-ER, which integrates User Sentiment and Behavior, thereby enhancing the accuracy and explainability of the recommendation. Specifically, we first propose a sentiment classification model based on user reviews. Different from traditional classification models, our model utilizes a multi-task learning framework to extract fine-grained sentiments at both the rating and review levels, subsequently fusing them to ensure a precise quantification of the Positive Review Rate (PRR). Then we propose a PRR-based explainable recommendation model that integrates collaborative filtering with items' PRR values, thereby enhancing the recommendation quality and enabling more personalized explanations. We conduct extensive experiments on real-world dataset. The results demonstrate the effectiveness of our proposed method.

Keywords: Explainable Recommendation · User Sentiment · User Behavior · Collaborative Filtering · Sentiment Classification

1 Introduction

Users require effective methods to identify and obtain valuable personalized information quickly. In the early stages of recommendation system development, researchers primarily focused on offering basic recommendation services based on users' historical behaviors and preferences. This included collaborative filtering algorithms that utilized user-item matrices, recommending items by calculating similarity. However, these methods are challenged by the cold start problem and data sparsity, limiting their practical effectiveness. With the advancement of big data and machine learning technologies, there were significant breakthroughs in recommendation system research. Researchers started to apply state-of-the-art techniques like deep learning, extracting more profound features from user behavior enhancing the accuracy and efficiency of recommendations.

To cater to higher demands, recommendation systems evolved to become more personalized, capable of providing more accurate recommendations tailored to individual interests, preferences, and behavioral patterns [1]. With improvements in recommendation accuracy, the importance of explainability grew. Explainable recommendations involve rationales alongside suggested items, helping users understand why they are recommended specific content. Such explanations improve greater trust and satisfaction with the recommendation system [2].

Despite the notable advancements, real-world application of recommendation systems still grapples with challenges. Firstly, meeting higher service demands and providing accurate explanations remains a hurdle. Secondly, capturing dynamic shifts in user preferences and updating recommendations in real-time is crucial. Effectively integrating such changes into recommendation models is a critical challenge for explainable recommendation systems. More specifically, we make the following contributions.

(1) We propose a novel Explainable Recommendation method, termed USB-ER, which integrates User Sentiment and Behavior to enhance the accuracy and explainability of the recommendation.
(2) For the review and rating of items, we first propose a sentiment classification model based on user reviews, that is based on TextCNN, BERT and LSTM models. Unlike traditional sentiment classification models, our model uses the multi-task learning framework to mine and integrate the fine-grained sentiments of the user's rating and review level, which effectively guarantees the accuracy of the quantified praise rate.
(3) For the recommendation generation and explainability generation task, we propose a PRR (Positive Review Rate)-based explainable recommendation model. This model combines the principles of collaborative filtering with item's positive review rates. This model first generates an initial list of recommendations based on user similarities; then refines and filters this list based on item positive feedback rates to produce the final recommendations with explanations.
(4) We conduct experiments to verify the the performance of our proposed model.

2 Related Works

2.1 Recommendation Methods

Collaborative Filtering Recommendation. Collaborative filtering-based recommendation [3, 4] systems are widely used personalized recommendation technologies that predict content or items a user may be interested in by analyzing and comparing user behavior, preferences, or rating data. The core idea of collaborative filtering is "birds of a feather flock together", meaning that similar users may be interested in similar items, or users may also be interested in other items like those they have liked in the past. Collaborative filtering recommendation systems are mainly divided into two types, which include user-based collaborative filtering and item-based collaborative filtering.

Content-Based Recommendation. Content-based recommendation [5, 6] is a technique in recommendation systems that generates recommendations by analyzing users' historical behaviors, preferences, and metadata associated with items, such as movies, music, books, news articles, etc. The key of this approach lies in understanding and matching the similarity between user interests and item content.

Rule-Based Recommendation. Rule-based recommendation [7] is a method that relies on predefined rules or logic for generating recommendations. This approach often depends on expert knowledge or historical data to formulate recommendation rules, which are then used to match user needs with item attributes, resulting in personalized recommendation.

2.2 Explainable Recommendation

In recent years, the focus of recommender systems has expanded beyond mere algorithmic accuracy to include the provision of explanations to enhance user acceptance and satisfaction [8]. Explanations aim to help users understand the rationale behind recommended items and facilitate informed decision-making.

Similarity-Based Explanation. One widely adopted type of explanation is based on similarity, leveraging either item-to-item or user-to-user comparisons [9–11]. Herlocker et al. [12] introduced an approach using collaborative filtering results to generate explanations, which significantly improved user acceptance. This approach is exemplified by popular services like Amazon's "Users who bought... Also bought..." feature, as well as similar mechanisms employed by Netflix, Spotify, and other platforms.

Content-Based Explanation. Another approach includes content-based information such as attributes and reviews within explanations [13, 14]. For instance, Vig et al. [15] designed a method to calculate tags for items and use them as explanations. Xian et al. [16] proposed a three-step framework to provide recommendations along with important attributes as explanations. These content-based approaches have shown promise in increasing user satisfaction.

KG-Based Explanation. The knowledge graph (KG) is incorporated to generate knowledge-related content explanations [17, 18]. For example, Ma et al. [19] propose a multi-task learning framework to utilize rules in knowledge graph to recommender systems. Zhao et al. [20] take the temporal information into account and design a time-aware path reasoning method for explainable recommendations. Geng et al. [21] propose a path language modeling recommendation framework to tackle the recall bias in knowledge path reasoning. It is shown that comprehensive information in knowledge graphs is quite useful for explainable recommendations.

3 Sentiment Classification Model Based on Review

3.1 Problem Statement

Sentiment analysis technology has become a focal point of research in the field of natural language processing in recent years. However, there remain some challenges and issues in practical applications. Firstly, existing sentiment analysis models often struggle to accurately identify users' true sentiments when dealing with complex and diverse commentary data. For example, some comments may contain complex linguistic phenomena such as sarcasm or metaphor, which are difficult for traditional sentiment analysis models to capture. Secondly, the generalization capability of sentiment analysis models is limited. Due to significant differences between commentary data across different domains and languages, existing sentiment analysis models often struggle to adapt to these differences, leading to performance degradation in different domains or languages, which restricts the broad application of sentiment analysis technology. Finally, the real-time performance and scalability of sentiment analysis models are also current issues.

To address these issues, we first proposed a sentiment classification model based on user reviews. By training the existing three sentiment analysis models based on reviews, we conducted in-depth analysis and processing of the review data. By comparing the training results of the three models, we attempted to identify a model that can more accurately recognize the sentimental tendencies of users.

3.2 Model Overview

Sentiment analysis is an important branch of text classification, and its core is also classification. The overall model framework is divided into two main parts, which are upstream task and downstream task. The main responsibility of the upstream task is to transform the text into a vector representation that provides input to the downstream task without losing text semantics. The downstream task focuses on using these vector representations to classify sentiments and identify sentiments through neural network models. The model architecture of the sentiment analysis task is shown (see Fig. 1). The BERT [22] model, TextCNN [23] model, and LSTM [24] model are selected in this paper to achieve the classification task.

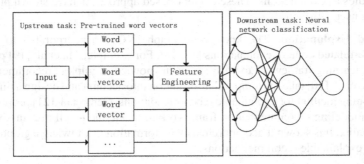

Fig. 1. Architecture of sentiment classification model based on review.

The advantage of the sentiment classification model based on TextCNN is that it can efficiently capture local features and patterns in the text, and quickly extract keywords and phrases through the convolutional layer to achieve accurate recognition of sentimental tendencies. The advantage of the BERT-based sentiment classification model is that it can provide deep semantic representation through bidirectional context understanding, combined with pre-training and fine-tuning strategies, to achieve high-precision and flexible sentiment analysis. LSTM-based sentiment classification models provide powerful tools in natural language processing through their ability to effectively deal with long-term dependencies, mitigate gradient problems, and understand context, especially for capturing and analyzing delicate sentimental expressions in text.

4 Explainable Recommendation Model Based on PRR

4.1 Problem Statement

Designing a recommendation model that can fully utilize positive review information while providing explainable recommendation results has become an urgent issue to address. In response to this problem, this study aims to propose an explainable recommendation model integrating user sentiment and behavior. The model will combine users' historical purchase records, browsing behaviors, and the positive review rates of items, and model and predict user interests and item quality through collaborative filtering algorithms, recommending items that not only meet users' interests but also have a high positive review rate. At the same time, the model will use explainability techniques to explain and clarify the recommendation results, allowing users to understand the reasons behind the recommendations, thereby improving the acceptance and satisfaction of the recommendations [25].

4.2 Model Overview

In this paper, we propose integrating User Sentiment and Behavior for Explainable Recommendation method (USB-ER), which is as shown Fig. 2. Based on sentiment analysis of item reviews, we can determine the positive and negative reviews of items, and then calculate the positive review rate for each item. Subsequently, using a user-based collaborative filtering algorithm, we can identify other users with similar interests to the target user and generate personalized item recommendations for the target user based on the preferences of these similar users.

During the recommendation process, the recommendation reasons are provided in conjunction with the item's positive review rate to enhance the persuasiveness and accuracy of the recommendations. The overall flow of the recommendation algorithm is shown (see Fig. 3).

4.3 Recommendation Strategy

We adopt a recommendation strategy that combines collaborative filtering algorithms with positive review rates (PRR). By using a user-based collaborative filtering algorithm, we identify the interaction relationships between users and items, thereby obtaining a preliminary item recommendation list. Through the sentiment analysis module based on review, the review is classified sentimentally. Based on the predicted results, the PRR of the items can be calculated, and the recommendation list obtained through collaborative filtering can be screened. The items with higher positive review rates are recommended to the target user, and a recommendation explanation combined with the PRR is provided.

User-Item Model. By collecting user information, such as ratings, purchase records, etc. A sample of user-item rating matrix is shown in Table 1.

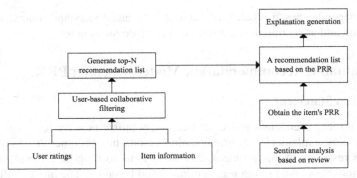

Fig. 2. The overall of USB-ER.

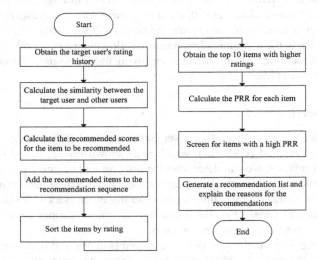

Fig. 3. The process of the recommendation algorithm.

Table 1. Rating matrix.

	User 1	User 2	User 3	User 4
Item 1	5		2	4
Item 2		4	4	5
Item 3	4		5	
Item 4	3			3

Nearest Neighbor Search. The purpose of nearest neighbor search is to identify other users who share the most similar interests with the target user, known as "nearest neighbors". Based on the rating matrix, the similarity between the target user and all other users is calculated, which is based on the items they have both rated. According to the results of the similarity calculation, a certain number of the most similar users are selected as neighbors for the target user. There are three methods for calculating similarity.

Cosine Similarity. In user-based collaborative filtering, users are represented as multidimensional vectors, where the dimensions correspond to the number of items in the system. Each specific value in the vector reflects the user's rating or preference for the corresponding item. To measure the similarity between users, we can use the cosine similarity metric. Cosine similarity assesses their similarity by calculating the cosine value of the angle between two user vectors, which can effectively capture the similarity of user preferences, thus providing a valuable reference for the recommendation system. Equation (1) illustrates this definition of cosine similarity.

$$sim(i,j) = \frac{i \cdot j}{||i|| \cdot ||j||} \quad (1)$$

Pearson Correlation. In collaborative filtering algorithms, pearson [26] correlation is often used to calculate the similarity between users or items. Equation (2) illustrates this definition of Pearson Correlation.

$$sim(i,j) = \frac{\sum_{p \in P}(R_{j,p} - \overline{R_i})(R_{j,p-\overline{R_j}})}{\sqrt{\sum_{p \in P}(R_{i,p} - \overline{R_i})^2}\sqrt{\sum_{p \in P}(R_{j,p} - \overline{R_j})^2}} \quad (2)$$

In the Eq. (2), $R_{i,p}$ represents the rating of item P by user i, $\overline{R_i}$ represents the average rating of all items by user i, and P represents the set of all items.

Adjusted Cosine Similarity. Adjusted cosine similarity is an adjustment based on cosine similarity, mainly used to handle scenarios such as rating data, by normalizing the mean to eliminate the influence of rating magnitude on similarity calculation. Equation (3) illustrates this definition of Adjusted Cosine Similarity.

$$sim(i,j) = \frac{\sum_{p \in I_{ij}}(R_{j,p} - \overline{R_i})(R_{j,p} - \overline{R_j})}{\sqrt{\sum_{p \in I_i}(R_{i,p} - \overline{R_i})^2}\sqrt{\sum_{p \in I_j}(R_{j,p} - \overline{R_j})^2}} \quad (3)$$

In the Eq. (3), $R_{i,p}$ represents the rating of item p by user i, $\overline{R_i}$ represents the average rating of all items by user i, and I_{ij} represents the set of items that both user i and user j have rated.

Recommendation Generation. Based on the predicted ratings, a recommendation list is generated for the target user, sorted from high to low predicted scores, and items with higher scores are included in the recommendation list. Equation (4) illustrates this concept.

$$P_{i,p} = \overline{R_i} + \frac{\sum_{j \in N} sim(i,j) * (R_{i,p} - \overline{R_j})}{\sum_{j \in N}(|sim(i,j)|)} \quad (4)$$

In the Eq. (4), $sim(i,j)$ represents the similarity between user i and user j, $P_{i,p}$ represents the rating of item p by the nearest neighbor user j, $\overline{R_i}$ and $\overline{R_j}$ represent the average ratings of users i and j, respectively.

For the explainable recommendation model based on the positive review rate, we calculate the positive review rate for each item based on the results obtained from the text classification of reviews by the sentiment analysis model. Equation (5) illustrates this concept.

$$K_i = \frac{\sum_1^n P_i}{\sum_1^n (P_i + N_i)} \tag{5}$$

In the Eq. (5), P_i represents the number of positive reviews for each item, N_i represents the number of negative reviews for each item, and K_i represents the positive review rate for each item.

For the item recommendation list obtained from the user-based collaborative filtering algorithm, the list is re-sorted according to the PRR by comparing the rates of the recommended items. The items with higher positive review rates are ultimately recommended to the user, accompanied by an explanation of the recommendation based on the positive review rate.

5 Experiments

5.1 Dataset

We use the JD.com sentiment analysis dataset as the foundation for the experiment, which is widely used to evaluate the performance of sentiment analysis algorithms, as shown in Table 2.

Table 2. Dataset.

Dataset	Information	Size
Training set	User ID, item ID, review content, rating	40000
Test set	User ID, item ID, review content	10000
Item information sheet	Item ID, item name	5822

In the sentiment analysis task, we adopted a simplified binary classification, categorizing ratings of 1 to 3 as negative sentiment (labeled as 0), and ratings of 4 to 5 as positive sentiment (labeled as 1), to facilitate model learning and performance evaluation.

5.2 Sentiment Classification Experiment

In this paper, the BERT model, TextCNN model, and LSTM model are selected to realize the classification task, and the experimental results of the three models are compared to select the best model for sentiment analysis.

The hyperparameters of model training are shown in Table 3, and during training, we monitored the accuracy and loss on both the training and test sets to ensure that the models were not overfitting.

Table 3. Hyperparameters.

Hyperparameters	TextCNN	LSTM	BERT
Word vector dimension	128	128	128
Learning rate	0.001	0.001	0.001
Batch size	64	64	16
Activation function	ReLU	ReLU	ReLU
Loss function	Cross-entropy loss function	Cross-entropy loss function	Cross-entropy loss function
Dropout rate	0.5	0.5	0.1
Number of epochs	8	5	5

The TextCNN model demonstrated precision of 0.99 in identifying positive sentiments (Label 1) yet exhibited a relatively lower recall rate of only 0.34 for negative sentiments (Label 0). Conversely, the LSTM model showcased the highest precision of 0.92 in handling negative sentiments, although its recall rates were not as impressive as those of the TextCNN for both labels. The BERT model, on the other hand, displayed a commendable balance across both labels, achieving an F1 score of 0.96 for Label 1, indicating a favorable equilibrium between precision and recall (see Table 4).

Table 4. Performance evaluation of sentiment classification model.

Sentiment Classification Model	Label	Precision	Recall	F1	Support
TextCNN	0	0.77	0.34	0.44	259
	1	0.99	0.93	0.86	741
LSTM	0	0.92	0.34	0.44	150
	1	0.99	0.80	0.75	957
BERT	0	0.80	0.56	0.54	785
	1	0.75	0.87	0.96	566

Analyzing the accuracy and loss rate changes for the TextCNN, LSTM, and BERT models (see Fig. 4, Fig. 5, Fig. 6), all three models exhibit high accuracy rates in sentiment classification, with the TextCNN model showing a particularly impressive accuracy peak. The LSTM model maintains a high accuracy level, although there is a suggestion of a gap between training and validation accuracy that warrants further investigation. The BERT model demonstrates a strong learning trajectory, evidenced by a sharp decline in loss rate, indicating its adaptability and robustness in learning from the training data.

The ROC curves of the three models on JD dataset are shown (see Fig. 7). The BERT model has the highest AUC score of 0.917, suggesting superior performance in

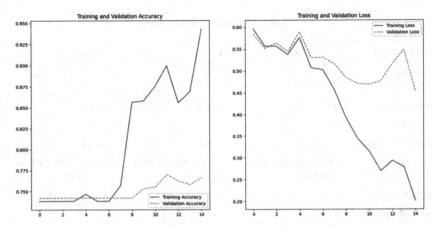

Fig. 4. Accuracy and loss rate changes for TextCNN.

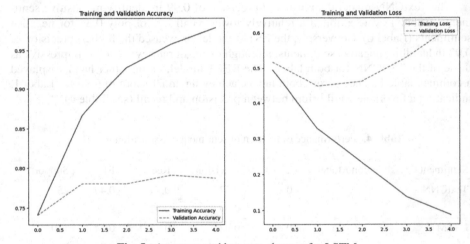

Fig. 5. Accuracy and loss rate changes for LSTM.

distinguishing between sentiment classes compared to the LSTM and TextCNN models, which have AUC scores of 0.727 and 0.777, respectively. All three models achieve the best results when using cross-entropy as the loss function and Adam as the optimizer.

5.3 Explainable Recommendation Experiment

For the explainable recommendation model based on PRR, we calculate the PRR for each item based on the results obtained from the text classification of reviews by the sentiment analysis model.

In the JD.com item sentiment analysis dataset, the relevant information and PRR of the recommended items are calculated (see Table 5). When the system identifies the recommended items and their corresponding PRR, it sorts the recommended items

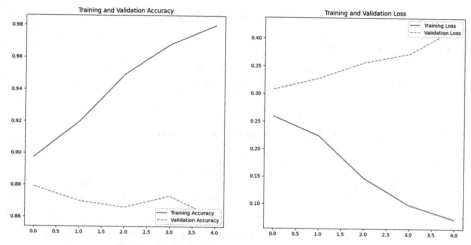

Fig. 6. Accuracy and loss rate changes for BERT.

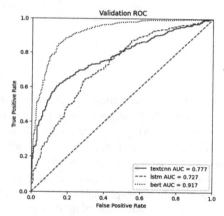

Fig. 7. ROC curve graph.

according to the PRR and recommends the top 10 items along with the reasons for the recommendations (see Table 6).

Table 5. Item positive review rate (PRR).

Item ID	Label	Name of item	Category	Count
ITEM_80196	1.0	Apple MacBook Pro MD101CH/A…	66, 332, 423	11
…	…	…	…	…

(*continued*)

Table 5. (continued)

Item ID	Label	Name of item	Category	Count
ITEM_50908	0.0	Galanz microwave ovenG70F20C…	313, 1136, 852	1
ITEM_517551	0.0	ikodoo MacBook Air 13.3 Case…	145, 980, 139	1

Table 6. Generate recommendation and explanation.

Recommended item ID	Explanation
Apple MacBook Pro MD101CH/A…	This computer is highly favored by student users and has received a 100% positive review rate
…	…
Galanz Microwave OvenG70F20C…	This microwave oven is highly favored by household users and has received a 100% positive review rate
ikodoo MacBook Air 13.3 Case…	This protective case is highly favored by Apple computer users and has received a 100% positive review rate

6 Conclusions

In this paper, we propose integrating User Sentiment and Behavior for Explainable Recommendation (USB-ER) method, which integrates user sentiment and behavior, thereby improving the accuracy and explainability of the recommendation system. In detail, we first propose the sentiment classification model based on user reviews. Fine-grained sentiment is extracted at both the rating and review levels using a multi-task learning framework and then fused to ensure accurate quantification of positive comment rate (PRR). Next, we propose a PRR-based explainable recommendation model that combines collaborative filtering with the PRR value of an item to improve recommendation quality and provide more reasonable explanations. We are in the real-world dataset on a wide range of experiments. The results demonstrate the effectiveness of our proposed method.

Acknowledgment. This work was supported by the Social Science Planning Fund Program of Liaoning Province of China under Grant No. L23BJY018.

References

1. Zhou, Z.W., Cao, D., Xu, Y.F., Liu, B.: Overview of research on recommendation systems. J. Hebei Univ. Sci. Technol. **41**(1), 76–87 (2020)
2. Zhang, Y., Chen, X.: Explainable recommendation: a survey and new perspectives. Found. Trends® Inf. Retrieval **14**(1), 1–101 (2020)

3. Guo, B., Xu, S., Liu, D., Niu, L., Tan, F., Zhang, Y.: Collaborative filtering recommendation model with user similarity filling. In: 3rd IEEE Information Technology and Mechatronics Engineering Conference (ITOEC), pp. 1151–1154 (2017)
4. Alatrash, R., Priyadarshini, R., Ezaldeen, H.: Collaborative filtering integrated fine-grained sentiment for hybrid recommender system. J. Supercomput. **80**(4), 4760–4807 (2024)
5. Shivaram, K., Liu, P., Shapiro, M., Bilgic, M., Culotta, A.: Reducing cross-topic political homogenization in content-based news recommendation. In: the 16th ACM Conference on Recommender Systems, pp. 220–228 (2022)
6. Zanon, A.L., Souza, L., Pressato, D., Manzato, M.G.: A user study with aspect-based sentiment analysis for similarity of items in content-based recommendations. Expert Syst. **39**(8) (2022)
7. Yu, Y., Si, X.S., Hu, C.H., Zhang, J.X.: A review of recurrent neural networks: LSTM cells and network architectures. Neural Comput. **31**(7), 1235–1270 (2019)
8. Gedikli, F., Jannach, D., Ge, M.: How should I explain? A comparison of different explanation types for recommender systems. Int. J. Hum. Comput. Stud. **72**(4), 367–382 (2014)
9. Berkovsky, S., Taib, R., Conway, D.: How to recommend? User trust factors in movie recommender systems. In: the 22nd International Conference on Intelligent User Interfaces, pp. 287–300 (2017)
10. Damak, K., Khenissi, S., Nasraoui, O.: Debiased explainable pairwise ranking from implicit feedback. In: Proceedings of the 15th ACM Conference on Recommender Systems, pp. 321–331 (2021)
11. Tanwar, M., Khatri, S.K., Pendse, R.: A Framework for feature selection using natural language processing for user profile learning for recommendations of healthcare-related content. Int. J. Bus. Anal. (IJBAN) **9**(3), 1–17 (2022)
12. Herlocker, J.L., Konstan, J.A., Riedl, J.: Explaining collaborative filtering recommendations. In: The 2000 ACM Conference on Computer Supported Cooperative Work, pp. 241–250 (2000)
13. Liu, Y., Miyazaki, J.: Knowledge-aware attentional neural network for review-based movie recommendation with explanations. Neural Comput. Appl. **35**(3), 2717–2735 (2022)
14. Musto, C., Lops, P., Gemmis, M., Semeraro, G.: Justifying recommendations through aspect-based sentiment analysis of users reviews. In: the 27th ACM Conference on User Modeling, Adaptation and Personalization, pp. 4–12 (2019)
15. Vig, J., Sen, S., Riedl, J.: Tagsplanations: explaining recommendations using tags. In: The 14th International Conference on Intelligent User Interfaces, pp. 47–56 (2009)
16. Xian, Y., et al.: EX3: explainable attribute-aware item-set recommendations. In: the 15th ACM Conference on Recommender Systems (RecSys), pp. 484–494 (2021)
17. Lyu, Z., Wu, Y., Lai, J., Yang, M., Li, C., Zhou, W.: Knowledge enhanced graph neural networks for explainable recommendation. IEEE Trans. Knowl. Data Eng. **35**(5), 4954–4968 (2022)
18. Ma, T., Huang, L., Lu, Q., Hu, S.: KR-GCN: knowledge-aware reasoning with graph convolution network for explainable recommendation. ACM Trans. Inf. Syst. **41**(1), 1–27 (2022)
19. Ma, W., et al.: Jointly learning explainable rules for recommendation with knowledge graph. In: The World Wide Web Conference, pp. 1210–1221 (2019)
20. Zhao, Y., et al.: Time-aware path reasoning on knowledge graph for recommendation. ACM Trans. Inf. Syst. **41**(2), 1–26 (2022)
21. Geng, S., Fu, Z., Tan, J., Ge, Y., de Melo, G., Zhang, Y.F.: Path language modeling over knowledge graphs for explainable recommendation. In: Proceedings of the ACM Web Conference 2022, pp. 946–955. (2022)
22. Kim, Y.: Convolutional neural networks for sentence classification. In: The 2014 Conference on Empirical Methods in Natural Language Processing (EMNLP), pp. 1746–1751 (2014)

23. Devlin, J., Chang, M. W., Lee, K., et al.: BERT: pre-training of deep bidirectional transformers for language understanding. In: Conference of the North American Chapter of the Association for Computational Linguistics: Human Language Technologies, vol. 1, pp. 4171–4186 (2018)
24. Hochreiter, S., Schmidhuber, J.: Long short-term memory. Neural Comput. **9**(8), 1735–1780 (1997)
25. Chen, Z., Wang, X., Xie, X., et al.: Co-attentive multi-task learning for explainable recommendation. In: The 28th International Joint Conference on Artificial Intelligence (IJCAI), pp. 2137–2143 (2019)
26. Armstrong, R.A.: Should Pearson's correlation coefficient be avoided? Ophthalmic Physiol. Opt. **39**(5), 316–327 (2019)

MADM Workshop

SACC: Secure-Cooperative Adaptive Cruise Control for Unmanned Vehicles

Wen Ran, Changlong Li(✉), and Edwin H.-M. Sha

Department of Computer Science, East China Normal University, Shanghai, China
72275900023@stu.ecnu.edu.cn, {clli,edwinsha}@cs.ecnu.edu.cn

Abstract. With the development of autonomous driving, the data security of the collaboration between vehicles plays a more important role. The research of auto drive systems is still in its infancy, this paper shows that the security mechanisms either depend on hardware or introduce high latency. This paper proposes a novel data protection strategy for efficient coordination across vehicle systems. Our design is based on an observation that the security demand in different phases varies. Specifically, for the external system, this paper proposes an information security scheme with a fine-grained attribute selection mechanism, which has the flexibility of encryption attribute selection. For the internal system, this paper proposes a dual-channel RSA encryption scheme that makes the control information transmission more robust and saves encryption delay. Experimental results illustrate that both security and low latency can be ensured with the proposed SACC.

Keywords: Data Security · Unmanned Vehicles · CACC · Attribute-based Encryption

1 Introduction

As the advancement of new energy vehicles and autonomous driving technologies, collaborative adaptive cruise control (CACC) system has garnered extensive scholarly attention in recent years. CACC manages an autonomous driving fleet through information exchange between vehicles [2]. Unfortunately, cross-vehicle communication can lead to information leakage and be vulnerable to attacks.

Existing studies ensure security based on customized hardware [1,6,7]. For example, Farias, Paulo V. G., et al. proposed a stability control scheme, which successfully improved the safety performance of coordinated vehicle platoons [2]. For internal data, RSA based on onboard hardware to solve communication failures and information loss issues. In addition, a multitude of investigations have been undertaken on the confidentiality of network connectivity information in the Vehicular Ad-Hoc Network (VANET) [3–5].

This work was partially supported by the Dreams Foundation of Jianghuai Advance Technology Center (No. 2023-ZM01Z011) and the National Natural Science Foundation of China (No. 62302169).

Nonetheless, these approaches cannot ensure low latency control as their encryption strategies are inflexible and inefficient.

In summary, real-time communication between unmanned vehicles is highly demanded for both latency and security, while the two metrics are always a trade off.

This paper explores an alternative approach improving the response speed and data security comprehensively and proposes a **S**ecure-cooperative **A**daptive **C**ruise **C**ontrol system for unmanned vehicles. In the design, a new data application classification scheme has been established corresponding to the proposed security enhancement and low latency encryption architecture, and a data category strategy has been proposed to better adapt to different control and encryption requirements for different data. In the external loop of SACC, CP-ABE encryption enhances security. RSA encryption is used in the internal loop for low latency and security. Furthermore, based on the proposed security and low-latency encryption architecture and different data classification schemes, a new response mechanism has been proposed to survive network attacks. The network connectivity attacks are divided into *external* and *internal* attacks. To defend against external attacks, SACC adopts stronger encryption strategies to keep data confidential, while internal attacks use a cloud data comparator to identify the attacking vehicles that forge data within the system. Furthermore, this paper proposes a dual-channel data transmission scheme, which enhances stability and security comprehensively. Experimental results showed that the new architecture can achieve low latency and high efficiency, meet the timely requirements of SACC system control, and ensure private data security. The contributions of this paper are summarized as follows:

- This paper proposes a novel data protection architecture, SACC. This is the first work that has realized efficient and flexible cross-vehicle encryption through software.
- This paper introduces a classification method for SACC. It categorizes data into authentication and control types based on the distinct application characteristics within the system. It assigns different security and real-time levels to establish a novel data application classification scheme.
- This paper introduces a dual-channel data communication scheme. The onboard control data is transmitted through the V2V network within the platoon, while the other is relayed to other vehicles within the platoon through RSU. Combined with the integration of the vehicle road cloud, two independent communication channels are constructed to effectively mitigate the issues of unstable communication or data loss when a network is under attack.
- This is the first work that co-designs the CP-ABE encryption strategies with vehicle systems. Experimental results demonstrate that the new architecture has achieved low latency and high efficiency, satisfy the stringent real-time requirements of SACC system control, and ensure private data security.

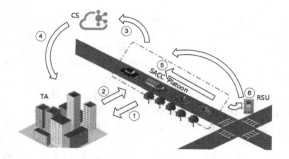

Fig. 1. The schematic diagram of a SACC architecture.

2 Design

2.1 Architecture Overview

This paper introduces SACC, an efficient and flexible encryption architecture for unmanned vehicles. As shown in Fig. 1, both data encryption and attack identification features are supported to protect the data under different security levels.

Based on the characteristics and security classification of the above data, the data are categorized into exclusive privacy authentication data and exclusive real-time control data. For the exclusive privacy authentication data, CP-ABE encryption is used to achieve anonymity, fine-grained access, and flexible data control. For the exclusive real-time control data, RSA encryption with lower real-time performance and better latency is used to better adapt to the synchronization control of vehicles within the platoon, thereby enhancing robustness.

Incorporating vehicle-road-cloud integration considerations, the paper further proposes a data communication methodology designed to enhance the stability of the system's communication architecture.

Roles in SACC. The encryption architecture of the system, which combines vehicle road cloud integration, consists of the following parts:

- **TA:** The trusted center of the system, which authenticates the identity of SACC vehicles, issues public and private keys, and prevents and monitors external network attacks.
- **Cloud Server (CS):** The cloud server stores the data and can perform calculations and encryption on big data.
- **RSU:** The component unit of the SACC, which is also a communication node of the system, facilitates the calculation and communication of in-vehicle data, as well as the verification of external authentication data.
- **Vehicle:** The component unit of the SACC, which is also a communication node of the system, achieves communication with the external system, RSU, and V2V data communication, and adjusts its status according to control parameters.

Workflow. According to different practical application scenarios, the architecture delineates the security scenario into two distinct stages. Firstly, as shown in the process ①, before applying to enter the vehicle platoon, the safety level of the scene currently is required to be high. The vehicle needs to first complete the security authentication of the vehicles entering the queue, which is completed through the identity confirmation registry issued by the trusted center. The vehicle is registered on the trusted center through CP-ABE, and the trusted center performs elliptic asymmetric operations through the cloud to confirm the registration information of the vehicle and hide the actual information of the vehicle. Temporary vehicle identification is provided and added to the vehicle registry. After obtaining information such as joining time, location, and temporary identity, the vehicle will align with the convoy and join to complete identity authentication with relatively high-security requirements, as shown in process ②. At this stage, using CP-ABE encryption with a higher level of security can prevent external network attacks and improve the defense level of attacks.

After joining the platoon, the communication of the platoon can be achieved through end-to-end data verification and monitoring. The data transmitted by vehicles in the platoon is control data with relatively low-security level requirements mentioned earlier, including speed, acceleration, distance, etc., These data change promptly and require high randomness and delay. For vehicles within the convoy, the similarity of data at the same time is also consistent, so the security level requirements for transmission within the convoy are relatively lower. What needs to be prevented for such transmission is only network attacks emitted by vehicles in the platoon with low probability. For this type of network attack, as shown in process ③, the control data sent by each vehicle can be transmitted to the cloud within a set cycle. A data comparator can be set up in the cloud server to quickly identify vehicles that have reported incorrect data, and the registration and authentication permissions of the vehicles can be deleted from the trusted center from the trusted center's SACC registry, hereby nullifying their identity authentication and expelling them from the vehicle queue, as shown in process ④. The prevention process is designed to counter internal attacks on vehicles within the platoon.

A dual-channel communication architecture has been designed to enhance stability to address the issue of temporary communication loss caused by attacks on the vehicle's exhaust system or low external signal coverage. Firstly, as shown in the process ⑤, real-time communication within the platoon is maintained, and normal control data is transmitted between vehicles through the internal vehicle networking channel, ensuring the synchronization of vehicle control information and reducing communication loss caused by external communication signals being attacked or interfered with. In addition, as shown in process ⑥, combined with RSU, another communication channel between vehicles was designed, which utilizes external communication devices and has stronger communication capabilities. We can transfer some data to RSU storage and then transmit it to other vehicles that need control data through RSU. The dual-channel design enhances the stability of the communication system, effectively solving the problem of fleet control instability caused by data attacks or signal loss in any channel, and enhancing the safety of vehicle platoon operation.

2.2 External Security Architecture

The external information and communication security process includes five parts: system initialization, device registration and authentication (vehicle, road, cloud) and information anonymity, external information encryption, external message forwarding and decryption, and information decryption outsourcing. In the initialization phase, TA (Trusted authority) inputs the safety parameter λ, and outputs PK (public key) and MSK (master private key), The derivation formula is Setup(λ) $\rightarrow$ PK, MSK. The generated public key and master private key will be used for the subsequent generation of user private keys.

In the device registration phase, TA registers the devices separately using authentication information provided by the vehicle, road, and cloud. After registering with real name information, TA anonymizes devices post-authentication to secure privacy in communications. At the same time, TA publishes a registration form, and the devices listed in the form list are trusted devices within the scope.

In the external information encryption stage, the trusted authority TA first needs to generate a private key SK. TA inputs public key PK, master private key MSK, attribute set S, and output user key SK. The derivation formula is KeyGen (PK, MSK, S) $\rightarrow$ SK. After SK is generated, TA shall use SK for data encryption. TA inputs system public key PK, access structure A, plaintext M, outputs ciphertext CT. The derivation formula is Ex (PK, A, M) $\rightarrow$ CT.

In the external information decryption stage, the vehicle receives and decrypts the ciphertext CT. Vehicle inputs public key PK, ciphertext CT, user private key SK, and the number of attributes associated with the ciphertext by itself S, if S $\in$ A, outputs message M, otherwise, decryption fails. The derivation formula is Dx (PK, CT, SK) $\rightarrow$ M.

In the outsourcing stage of information decryption, the vehicle uses a public key to encrypt the pseudonym of i-th vehicle and send it to road side unit RSU or cloud server CS. RSU or CS decrypts it based on attribute S, verifies the certified form list of vehicle and RSU pseudonyms, and if it is valid, establishes a trust relationship with the vehicle. After external decryption is finished, roadside unit RSU or cloud server CS will send plaintext M to the i-th vehicle.

It fully considers the security, fine-grained, and flexible characteristics of the CP-ABE algorithm in the authentication, encryption, and decryption process of external information, ensuring the security and reliability of the closed-loop architecture for external information transmission.

2.3 Internal Security Architecture

In the system initialization phase, the trusted authority TA system inputs safety parameters λ, Outputs system public key PK and private key SK. Setup(λ) $\rightarrow$

PK, SK, TA distributes public keys and corresponding private keys to authenticated vehicles within SACC.

In the information encryption stage, vehicles in the fleet receive public and corresponding private keys, and each encrypts the control information of the vehicles with RSA, and transmit it to the fleet through the internal local area network. At the same time, the externally encrypted CP-ABE public key is used to encrypt the same information and is transmitted to the external cloud server, where the control information of the fleet is monitored by comparing the controller in the cloud.

In the information decryption stage, the encrypted information transmitted by the vehicle is transmitted through the internal network. Other vehicles in the fleet receive the information and use their private key to decrypt it. After verification, plaintext is obtained, and the information is used as internal input for controlling by themselves and SACC.

In the attack monitoring stage, the cloud server receives encrypted information. It is decrypted by CP-ABE ciphertext. If the decryption is successful, the information is used as input for comparative information monitoring. If abnormal vehicle information is detected, it is reported to the trusted authority (TA), and the registry is canceled to initiate subsequent security defense and processing of the SACC fleet.

The system takes into account the fixed and periodic characteristics of internal information transmission for the authentication and encryption/decryption process of internal system information, meanwhile, it adapts to the requirements of closed-loop transmission of vehicle internal safety information.

3 Evaluation and Security Analysis

3.1 Performance Analysis

The verification test of this article was conducted on a 2GHz Intel Celeron CPU, 12 GB RAM, and x64 Ubuntu Linux operating system. The test code is based on the bilinear mapping, attribute pairing encryption CP-ABE algorithm, and the RSA algorithm based on the Euler function and modular inversion formula.

For the external security loop of SACC, the test content was set to a fixed format size text M. This article detects the decryption time of received attribute ciphertext and analyzes and evaluates the detection results. For the RSU or CS end attribute encryption and decryption process, the verification method is similar to the vehicle end information encryption and decryption process, and the analysis and evaluation process is also the same. No additional testing and evaluation was conducted in this article.

For the internal security loop of SACC, this article tested the RSA encryption and decryption process of the internal closed-loop and analyzed and evaluated the performance of the internal closed-loop encryption and decryption based on the test results. For the CP-ABE encryption and decryption process of information dissemination, as the principle and process are similar to the external closed-loop usage method, no additional testing and analysis were conducted.

The CP-ABE and RSA encryption and decryption test results are shown in Table 1.

Table 1. The encryption and decryption test results

CP-ABE external closed-loop	Total attr.	8	8	10	10	30	30	30
	User attr.	2	3	2	5	2	8	15
	En. (ms)	52.5	53.7	114.5	115.8	311.5	314.8	331.7
	De. (ms)	108.1	109.2	116	118.7	156.2	227.5	428.3
RSA internal closed-loop	Size (bit)	16	32	64	128	256	512	1024
	Time (ms)	3	8	61	144	176	305	6545

Test Result Analysis. Corresponding to the external closed-loop system, according to the test evaluation Table 1, CP-ABE encryption is applied to the external security closed-loop information using different attribute numbers, and the encryption process time is tested. From the test table, we can see that the total number of attributes in the given attribute set, as well as the number of attributes that the decryption end user satisfies the confidential architecture, have a significant impact on the encryption and decryption time of the entire external closed-loop system. When the number of user-matching attributes is fixed and the total number of attributes remains constant, the encryption time of the system significantly increases after the total number of attributes exceeds 10, and at the same time, the decryption time also significantly increases. This indicates that when using the CP-ABE encryption method for system encryption and decryption, the selection of the total number of attributes in the bilinear mapping and attribute pairing encryption algorithm has a significant impact on the encryption time after more than 10. At present, in the application of SACC system engineering, the impact of this result should be considered, and the total number of attributes should be reasonably selected within the range of encryption quantity and encryption cycle requirements, to achieve ideal practical application effects. The schematic diagram of the test results for this method is shown in Fig. 2(a).

On the other hand, if the total number of attributes is fixed and users are matched with changes in the number of attributes, the test results show that an increase in the number of attributes has little effect on the encryption time of the external closed-loop system. But when the number of user attributes is greater than 10, it has a significant impact on the decryption time of the system. This indicates that in the design of the SACC system if the encryption party needs to define finer-grained attributes and decrypt the encrypted text when passing it to the queue, the required decryption time will significantly increase with the number of attributes matched by the queue user greater than 10. This also needs to be optimized and selected according to application requirements

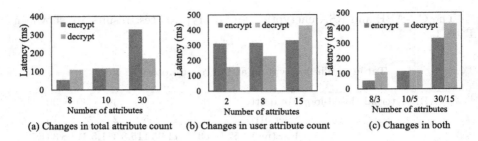

Fig. 2. CP-ABE Encryption and Decryption Time Diagram

in SACC system design. The schematic diagram of the test results for this test is shown in Fig. 2(b).

In addition, if the total number of attributes and the number of matching attributes for row users increase simultaneously, the schematic diagram of the test results is shown in Fig. 2(c). Select test items with a total number of verified attributes and a matching attribute count of 8/3, 10/5, and 30/8, respectively, for testing. The test results show that as the number of attributes increases, the trend of the encryption and decryption time of the external closed-loop system will be more significant. This indicates that the larger the total number of selected attributes and the number of matching attributes for the scheduling user, the more significant the impact on the delay performance of the SACC external closed-loop system.

For the internal closed-loop system, RSA encryption is used, and the amount of encrypted text information has a significant impact on the encryption time of the internal closed-loop system. The test results of RSA encryption are shown in Table 1. From the results, it is seen that as the text size increases, the encryption time will significantly increase. Since the SACC internal closed-loop system is a trustworthy, high real-time, and low latency system that has been verified through external closed-loop authentication, it is recommended to fix the text size and format according to real-time requirements when designing this part of the internal closed-loop system. This is easy to achieve in the system for text requirements of internal information. The information of the internal closed-loop system mainly transmits real-time data and control information of the vehicle, including self-driving speed, acceleration, constant spacing and time spacing, workshop distance, etc. This type of information has a fixed format and does not vary in size. Therefore, this design and encryption method ensures the real-time and low latency of the internal closed-loop system in practical applications, as shown in Fig. 3, 16-bit sized text is 3 ms of total encrypted and decrypted time, and 32-bit sized text is 8 ms.

The experimental results show that different encryption architectures and algorithms are applied to external and internal loops with different security levels, and the test verification results will vary. The experimental results meet the security and real-time requirements of the SACC system with different encryption levels. In this experiment, CP-ABE and RSA encryption were chosen for

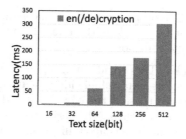

Fig. 3. Relationship between the en(/de)cryption latency and text size.

architecture design and experimental verification. In the next step of work, the security architecture and verification algorithm will be further adjusted and verified to better meet the security, reliability, and real-time requirements of the SACC system, based on the differentiated requirements of different security levels and real-time performance.

3.2 Security Analysis

Aligning with SACC application traits, this article designs a security architecture that addresses both internal and external scenarios, ensuring a comprehensive security analysis.

Data Confidentiality. SACC is divided into external and internal scenario applications. Externally, CP-ABE encryption facilitates information exchange, with bilinear mapping used to derive a public and master key. The system encrypts messages using the public key and an access structure, while a private key is generated from the master key and attribute set. This encryption process is completed through bilinear mapping calculation while generating random HASH values, used for verifying the correctness of ciphertext; Internally, the TA validates vehicle credentials for SACC, enabling certified vehicles to join the platoon via V2V networks for coordinated driving. The control information of vehicles has the characteristics of short cycle, limited content, and real-time. By using the modular operation of RSA to encrypt and decrypt information in a short time, attacks cannot complete the encryption and decryption of periodic content in a short time. The end-to-end encryption and decryption of RSA in the internal network ensures the security of the internal network.

Preventing Collusion. The vehicles' unique authentication information provided by TA after registration is bound to the vehicle's unique identity, effectively distinguishing the vehicle's identity. At the same time, The encryption process incorporates a random HASH value for verifying data integrity, ensuring that vehicles can only decrypt ciphertexts with matching attributes. Even if different vehicles collude for decryption, due to the binding of the access structure to the attribute mapping calculation, At the same time, there are random HASH values for verification. Even if colluding vehicles collude, the addition of attributes cannot meet the matching requirements of specific policy operations, and cannot pass the verification of random HASH values. Therefore, the transmission of external information is anti-collusion.

Prevent Replay Attacks. TA provides unique authentication of anonymous vehicle information, ensuring secure vehicle identification. At the same time, the HASH value verifies the correctness of the encrypted and decrypted information. The same information is rejected during the verification stage, and replay attacks are eliminated at the RSU information verification end. It has security under replay attacks.

Forward Security. When a vehicle enters the SACC fleet, the TA publishes a registry CertList, which is a certified form of SACC vehicle and RSU pseudonyms. The vehicles listed in the CertList are valid and can be used for internal and external information transmission within the fleet. The vehicle's information is encrypted using CP-ABE and uploaded to the cloud, where a comparative control unit assesses it against fleet-wide control data. The TA identifies and reports any information surpassing fleet thresholds for review and potential vehicle delisting from the CertList. This process is traceable, deterring internal security breaches and ensuring forward security.

Outsourced Decryption Security. The vehicle entrusts RSU or cloud with information decryption, encrypts the anonymous vehicle information and CertList list published by TA through public key encryption, and passes them to the roadside or cloud after CP-ABE encryption. The roadside and the cloud verify the vehicle information through CP-ABE decryption. If the vehicle information matches the registry, it is authenticated through decryption outsourcing. The outsourcing decryption randomly generates a HASH value to verify the correctness of decryption and ensure the security of external decryption.

4 Related Work

In terms of hardware handling of communication loss in CACC, Liu Y, Wang W. [1] proposes an SR-CACC (Safety Reinforced Cooperative Adaptive Cruise Control)dual branch control strategy, which solves the above problems by switching from the vehicle's hardware control to a sensor-based adaptive cruise control strategy. The article addresses communication loss, devising a control strategy and controller for seamless CACC to ACC mode transition, ensuring enhanced stability and maneuverability. CACC-DIFT (dynamic information flow topology) [6] is designed to mitigate the negative impact caused by communication faults. When communication faults occur, necessitating a control shift, the system transitions to a specific CACC mode rather than reverting to ACC, thereby enhancing control performance.

Tan, Ying Kiat. et al. [7] proposes a distributed CACC control framework with real-time prediction, which considers the delay of onboard sensors, communication, and actuators. A CACC controller based on real-time distributed MPC is also designed, which compensates for the delay through the real-time prediction distributed CACC control algorithm, improving the stability of the fleet under various traffic dynamic conditions.

In terms of vehicle information security encryption, Q. Nicolas et al. [3] proposes a privacy protection protocol for fleet settings using homomorphic encryption (HE), which achieves confidentiality operations on sensitive information such as fleet location and destination. Huang, H., Zhang, H., Zhao, Z. et al. [4] proposes an encrypted access control scheme based on variant attributes to achieve fine-grained access control of VANET data shared in the cloud and fog. Concurrently, a novel approach integrates roadside units into the blockchain, publishing device public signatures to authenticate sender identities, thus preventing signature tampering. L. Yilong et al. [5] proposes a lightweight CP-ABE scheme that supported direct attribute revocation in VANET. It uses a scalar multiplication on elliptic curves instead of bilinear pairing operations and employs a computational outsourcing techniques to reduce terminal decryption costs, meanwhile, it also improves the efficiency of the scheme. Taha, Mohammad Bany et al. [8] proposes a method of using vehicle clusters for CP-ABE encryption. Through a V2V network, CP-ABE encryption is allocated to various vehicles in the vehicle cluster using a designed algorithm. The CP-ABE micro-service constructed by the vehicle cluster is applied to ensure data confidentiality and reduces the computational cost of vehicle resources.

Based on this, we propose our information encryption architecture and security data classification scheme according to the actual application situation and characteristics and implement different encryption algorithms and processes for different application scenarios. Meanwhile, the integrated network of vehicle-road-cloud is used to enhance our encryption efficiency and communication reliability.

5 Conclusion

This article constructs a CP-ABE and RSA encryption system architecture and security processing mechanism suitable for the characteristics of SACC. This architecture classifies and encrypts data based on different security levels. Meanwhile, a dual channel RSA encryption scheme was proposed to enhance the robustness of control information transmission and saving encryption latency. In addition, a cloud data verifier has been built to prevent vehicles within the system from publishing incorrect data that could attack the system, enhancing the controllability and anti-attack capabilities of the secure platoon system. Finally, the computational performance of the CP-ABE and RSA encryption system was verified, and the system's architecture encryption time met the requirements of the SACC system.

References

1. Liu, Y., Wang, W.: A safety reinforced cooperative adaptive cruise control strategy accounting for dynamic vehicle-to-vehicle communication failure. Sensors **21**(18), 6158 (2021). https://doi.org/10.3390/s21186158

2. Farias, P.V.G., et al.: Mitigating message dissemination issues in safety applications for vehicular ad hoc networks. Concurr. Comput.: Pract. Exp. **35** (2020)
3. Quero, N., et al.: Towards privacy-preserving platooning services by means of homomorphic encryption. In: Proceedings Inaugural International Symposium on Vehicle Security & Privacy (2023)
4. Huang, H., Zhang, H., Zhao, Z., et al.: A traceable and verifiable CP-ABE scheme with blockchain in VANET. J. Supercomput. **79**, 16859–16883 (2023). https://doi.org/10.1007/s11227-023-05322-z
5. Liu, Y., et al.: A lightweight CP-ABE scheme with direct attribute revocation for vehicular ad hoc network. Entropy **25** (2023)
6. Gong, S., et al.: Cooperative adaptive cruise control for a platoon of connected and autonomous vehicles considering dynamic information flow topology. Transp. Res. Rec. **2673**, 185–198 (2018)
7. Tan, Y.K., Zhang, K.: Real-time distributed cooperative adaptive cruise control model considering time delays and actuator lag. Transp. Res. Rec. **2676**, 93–111 (2022)
8. Taha, M.B., et al.: A cluster of CP-ABE microservices for VANET. Procedia Comput. Sci. (2019)
9. Ozmen, M.O., et al.: Short: rethinking secure pairing in drone swarms. In: Proceedings Inaugural International Symposium on Vehicle Security & Privacy (2023)
10. Abidi, I., et al.: Privacy in urban sensing with instrumented fleets, using air pollution monitoring as a usecase. In: Proceedings 2022 Network and Distributed System Security Symposium (2022)
11. Dadras, S., et al.: Vehicular platooning in an adversarial environment. In: Proceedings of the 10th ACM Symposium on Information, Computer and Communications Security (2015)
12. Monteuuis, J.P., et al.: SARA: security automotive risk analysis method. In: Proceedings of the 4th ACM Workshop on Cyber-Physical System Security (2018)
13. Nazat, S., Abdallah, M.: Anomaly detection framework for securing next generation networks of platoons of autonomous vehicles in a vehicle-to-everything system. In: Proceedings of the 9th ACM Cyber-Physical System Security Workshop (2023)
14. Plappert, C., et al.: Secure role and rights management for automotive access and feature activation. In: Proceedings of the 2021 ACM Asia Conference on Computer and Communications Security (2021)
15. Bethencourt, J., et al.: Ciphertext-policy attribute-based encryption. In: 2007 IEEE Symposium on Security and Privacy (SP 2007), pp. 321–334 (2007)
16. Waters, B.: Ciphertext-policy attribute-based encryption: an expressive, efficient, and provably secure realization. IACR Cryptol. ePrint Arch. **2008**, 290 (2011)
17. Bishop, A., Waters, B.: Decentralizing attribute-based encryption. IACR Cryptol. ePrint Arch. **2010**, 351 (2011)
18. Ostrovsky, R.M., et al.: Attribute-based encryption with non-monotonic access structures. In: Conference on Computer and Communications Security (2007)
19. Wee, H.: Optimal broadcast encryption and CP-ABE from evasive lattice assumptions. IACR Cryptol. ePrint Arch. **2023**, 906 (2022)
20. Fan, K., et al.: A secure and verifiable outsourced access control scheme in fog-cloud computing. Sens. (Basel Switz.) **17** (2017)

A Systematic Mapping Study of LLM Applications in Mobile Device Research

Chong Chen, Bo Wang[✉], and Youfang Lin

School of Computer Science and Technology, Beijing Jiaotong University, Beijing, China
{22120350,wangbo_cs,yflin}@bjtu.edu.cn

Abstract. The extensive utilization of Large Language Models (LLMs) has significantly influenced academic research in mobile device-related fields, encompassing application testing, malware detection, voice control, and software development enhancement. Concurrently, the increasing demand for user access to LLMs on mobile devices for tasks like question answering has introduced new research directions, such as developing native LLMs by reducing parameter sizes. We aim to study the relationship between the evolution of LLMs and mobile device-related research, exploring their integration into traditional tasks and their adaptation to mobile platforms, from early transformer-based models to modern architectures like GPT-4. We have reviewed 55 recent papers, including 50 novel approaches, 1 benchmark, and 4 empirical studies, covering various aspects of LLM applications in mobile devices.

Keywords: Systematic Mapping · Large Language Models · Mobile Device · Mobile Testing

1 Introduction

Recently, Large Language Models (LLMs) have been widely applied to various academic fields, including research on mobile devices. On the one hand, LLMs can be used to conduct traditional tasks like mobile application development, testing, recommending, and maintenance to help humans free of mechanical labor. On the other hand, how to access LLMs conveniently on mobile devices has become a brand new question to researchers. Large numbers of research has been proposed for employing LLMs in mobile device-related tasks. It seems to be the time for systematically collecting and ordering those existing works for the future development of this area.

Because of the immense popularity of Large Language Models these years, researchers have proposed several empirical studies that conduct high-quality summarization for research using LLMs in mobile device areas. The survey presented by Cui et al. [3] extensively investigates the effectiveness of nine state-of-the-art LLMs in Android text input generation for UI pages. The study by Wang et al. [43] sorts out software testing approaches using LLMs, including mobile application testing. The study by Shi et al. [36] lays the groundwork for further developments in the field of human-GenAI interactions. The study by Hou et al. [10] focuses on LLMs for Software Engineering (LLM4SE), which also mentions LLMs for Android application management. Nevertheless, we find the coverage of these works is either slightly narrow for only a certain field in mobile devices research [3], or slightly wide for the usage of LLMs on a general topic that is not limited to mobile devices [3,10,36,43]. It is still necessary for us to conduct a study that systematically synthesizes research and approaches proposed for LLMs used in every academic field related to mobile devices.

Our mapping study includes a total of 55 related works, namely 50 for new methods, tools, or frameworks, 4 for empirical study, and 1 for benchmarks, from different popular academic conferences and journals in computer science.

Section 2 presents our research questions. Section 3 introduces the search keyword selection strategy we formulated to collect and specify the related studies. Section 4 displays the distribution of the selected studies in this paper from different aspects and provides the answers to research questions by analyzing the search results. Section 5 concludes the whole paper.

2 Research Questions

To fully understand how LLMs are employed in different academic fields for mobile device-related research, the study aims to answer the following research questions:

RQ1: What are the most common LLMs used in mobile device research?

RQ2: What are the main research areas in mobile devices that regularly use LLMs?

RQ3: What's the main purpose for researchers to involve LLMs in their work?

3 Search Strategy

Since the target of this study is to analyze the impact of LLMs brought to mobile application-related work, We select and filter those related studies by using keywords in the form of "$MB \wedge LLM$", where MB represents words related to mobile applications, including descriptions like "Mobile applications—Android applications—IOS applications—Mobile apps", and LLM represents keyword related to Large Language Models, including general description like "Large Language Model—LLM" and popular specific LLMs like "ChatGPT—GPT—BERT". We use the keyword combinations above to search in Google Scholar and artificially pick the ones that meet our requirements. For the selection of the range of years, we set the start point to 2018 when the first two widely used LLMs: GPT-1 and BERT were proposed.

4 Results

4.1 Search Results

Table 1 and Fig. 1 show the distribution of selected studies by publishing institutions and years respectively. From Fig. 1 we can see that the history of introducing Large Language Models to mobile device-related studies is not that long with only 3 years and a few related works in 2022. While the number of studies experienced a rapid growth from 3 in 2022 to 21 in 2023, and then to already 31 in merely the past 6 months in 2024. Besides, Table 1 reveals that related studies start to appear massively at top-level conferences of computer science (for example, 8 in ICSE) in 2024. All these messages indicate that researchers are paying greater attention to employing Large Language Models in the area of mobile device-related work.

Additionally, the academic field of mobile device-related research that uses LLMs to conduct experiments has also become much more abundant. At first, related studies centered on a few topics, including malware detection and code analysis, while until 2024, the topics have extended to various areas like software testing, security protection, task automation, etc. This phenomenon also indicates that it's necessary to perform a systematic summarization and arrangement for existing related works.

Table 1. Distribution of the researches by publication

Conference/Journal	Paper #
arXiv	25
ICSE	8
CHI	4
Springer	3
TSE	2
others	13

4.2 Answering to Research Questions

RQ1: What are the Most Common LLMs Used in Mobile Device Research? Table 2 shows the distribution of Language Language Models used in works selected in this study. It's worth noting that not every research included has a corresponding model. Some are empirical studies, some don't publish the model they use, and some framework approaches leverage numbers of LLMs for testing. We only count models that are claimed to be used in related works. It's shown that the GPT series models are the most popular ones for researchers. Correspondingly, studies based on open-source LLMs other than BERT are less sufficient.

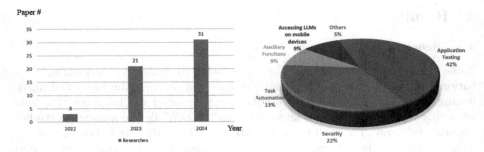

Fig. 1. Distribution of researches by year and academic field

For the techniques involved, 14 of the studies conducted few-shot in context learning, and 21 works used prompt tuning to train models. With the proposition of the idea of combining LLMs with multimodality, 2 studies that use the new GPT-4V model have been proposed. LLMs have been used by researchers for many different missions, including natural language understanding, code analysis, classification, and prediction, etc.

Table 2. Distribution of LLMs used in researches

Year	2022	2023	2024
BERT	1	3	3
GPT-2	1	0	0
GPT-3	0	3	7
GPT-3.5	0	2	3
GPT-4	0	5	8
GPT-4v	0	0	2
Open-Source LLMs	1	4	3

RQ2: What are the Main Research Areas in Mobile Devices that Regularly Use LLMs? Figure 1 displays the distribution of related studies by academic fields. Software testing occupies a large proportion because LLMs can work on both the test generation side by creating human-like interaction sets with software and the test maintenance side by replaying and fixing existing test scripts. For security protection, researchers mostly leverage LLM's code comprehension ability to obtain the purpose of certain applications and check if it's dangerous. Task automation allows users to interact with applications by simply giving a natural language instruction by asking LLMs to perform real operations instead based on their natural language understanding ability. For auxiliary functions, LLMs can help to develop or better use applications by giving advice with their commonsense knowledge.

Application Testing: Table 3 shows topics in mobile app testing that involve LLMs in research. The result indicates that LLMs have been most widely used in user interface testing with 14 related works (11 in component interaction and 3 in text input generation), following test script fixing (3), static analysis (2), and bug replay (2).

For UI testing, one popular approach is to transform the testing procedure described with natural language to executable test scripts or direct interaction sets. The study by Liu et al. [26], Zimmermann et al. [54,55], Wen et al. [48], Taeb et al. [39], Li et al. [21] and Kumar et al. [16] have used different method to achieve this. The same principle is to translate the app GUI state information and the available actions on the smartphone screen to natural language prompts and ask the LLM to choose actions.

Besides providing the detailed entire test procedure to LLMs, another approach is just giving some general requirements and asking them to form a test plan with rich commonsense knowledge. The work by Liu et al. [25] achieves this by formulating the mobile GUI testing problem as a Q&A task, asking LLM to chat with the mobile apps bypassing the GUI page information to LLM to elicit testing scripts. Yoon et al. [50,51] use GPT-4 to set relevant task goals automatically and subsequently try to achieve them by interacting with the app. The study by Wang et al. [46] achieves automatic user acceptance testing with the help of LLMs.

Another important task for UI testing is to generate various text inputs for components like a search box that asks users to type some message for transforming to the next state. Unlike buttons or choice boxes, the candidate input text is infinite but the output reachable state is limited. Thus, it's particularly important to cover the whole output state with as few inputs as possible. LLMs have been proven to be reasonably helpful for that. The work by Liu et al. [28] leverages the LLM to automatically generate unusual text inputs for mobile app crash detection by formulating the unusual inputs generation problem as a task of producing a set of test generators. The work by Liu et al. [24] boosts the performance of LLM by developing a prompt-based data construction and tuning method that automatically extracts the prompts and answers for model tuning. Besides, the study by Cui et al. [3] extensively investigates the effectiveness of nine state-of-the-art LLMs in Android text input generation for UI pages.

For bug replay, the study by Feng et al. [6] leverages few-shot learning and chain-of-thought reasoning to elicit human knowledge and logical reasoning from LLMs to accomplish the bug replay like a developer. Huang et al. [12] focus on automatically reproducing mobile application crashes directly from the stack trace by leveraging a pre-trained LLM to predict the exploration steps for triggering the crash.

For test analysis and repair, the study by Ali et al. [1] uses BERT to predict the severity of bug reports submitted by users for mobile app maintenance. The study by Pan et al. [2] repairs the broken UI test script caused by UI layout changes with LLMs. The study by Liu et al. [29] systematically investigates LLM-based approaches for detecting and repairing configuration compatibility bugs.

The study by Zhang et al. [52] uses BERT to capture contextual information and learns a matching model for test script migration.

LLMs can also be used to conduct program analyses for testing. The study by Mohajer et al. [33] and Wang et al. [42] performs static analysis on mobile applications with GPT-4, which can detect bugs, filter false positive warnings, and patch the detected bugs without human intervention. The study by Lemieux et al. [20] leverages Codex to help do Search-Based Software Testing on Android devices.

Table 3. Tasks for software testing and security

Testing	Paper #	Security	Paper #
UI testing	12	Malware Detection	5
Valid Input Text Generation	3	Vulnerability Detection	4
Bug Replay	2	Privacy Protection	2
Test Script Analysing/Fixing	3		
Static Code Analyse	2		
Others	2		

Security: Table 3 lists the contribution of LLMs to research related to mobile software security. The works concentrate on three aspects, namely malware detection (5), vulnerability detection (4), and user privacy protection (3).

Malware detection may be one of the earliest fields that use LLMs to conduct research in mobile device areas. Related works have been appearing since 2022 when GPT-4 has not even been published to the public. During the year 2022, The study by Lee et al. [18], Souani et al. [37] and Garcia et al. [8] have proposed their approach to distinguish Android malware by using GPT-2, BERT and CodeT5, respectively. The key point is to obtain the purpose of a certain application with the help of the code comprehension ability of LLMs. The study by Jones et al. [14] explores the efficacy of BERT in the domain of Android malware detection and compares its performance against CNNs and LSTMs, demonstrating that the former achieves better results. The study by Zhao et al. [53] detects Android malware and generates diagnostic reports by employing LLM to reason and generalize the functions and potential behaviors of certain software features like permission, API, URL, and uses-feature.

LLMs can also be used to uncover vulnerabilities in mobile devices spreading from hardware to software. The study by Ferrage et al. [7] and Ma et al. [30] demonstrate the ability of LLMs to detect network-related vulnerabilities on IoT devices. The study by Li et al. [23] combines the strengths of static analysis and LLMs for vulnerability detection without suffering their limitations. The study by Mathews et al. [32] focuses on building an AI-driven workflow to assist developers in identifying and rectifying vulnerabilities.

Another aspect is to leverage LLM for user privacy protection. The study by Jiang et al. [13] uses LLM for the detection of privacy leakage within WeChat Mini Programs. The study by Nguyen et al. [34] proposes a new method based on XLM-RoBERTa to detect inconsistencies between the permission extracted from the description application and privacy policy and the permission extracted from the application's source code and uncover unreasonable system permissions acquired by the application.

Task Automation: The role that LLMs play in task automation is somewhat like that in UI testing with detailed test descriptions in natural language. The common ground is to let LLMs comprehend a set of instructions written in natural language and transform them into specific actions. The study by Wang et al. [44,45] and Ma et al. [31] introduces an autonomous multi-modal mobile device agent with GPT-4V. The agent can autonomously plan and decompose the complex operation task, and navigate the mobile Apps through operations step by step, making the users free from interacting with UI components by just giving a simple demand. The study by Wen et al. [47] achieves task automation by combining the commonsense knowledge of LLMs and domain-specific knowledge of apps through automated dynamic analysis. The approach GptVoiceTasker provided by the study by Vu et al. [40] can handle voice form of instruction by excelling at intelligently deciphering user commands and executing relevant device interactions to streamline task completion. The study by Lee et al. [19] proposes a task automation tool that emulates the cognitive process of humans interacting with a mobile app-explore, select, derive, and recall. The study by Wang et al. [41] explores the probability of LLMs in enabling versatile conversational interactions with mobile UIs using GPT-3 and PaLM.

Auxiliary Functions: LLMs can bring convenience to both mobile software development and usage. The amount of related research is considerable.

From the aspect of developers, the study by Huang et al. [11] introduces a novel approach based on Large Language Models to automatically extract semantically meaningful macros from both random and user-curated mobile interaction traces. The study by Dlamini et al. [4] proposes an approach for detecting design patterns in the Android development domain to help write design documents. The study by Gu et al. [9] develops a new MUI element detection dataset and proposes an Adaptively Prompt Tuning (APT) module to take advantage of discriminating OCR information.

From the aspect of users, The study by Liu et al. [27] proposes a GPT-3 based tool that can extract hint text attribute in the text input component that cannot be identified by most screen reader tools. The study by Khaokaew et al. [15] employs Large Language Models (LLMs) and installed app similarity to perform application usage prediction.

Accessing LLMs on Mobile Devices. Different from the areas mentioned above, this is a newly appeared topic accompanied by the rapid growth of demand for users to use ChatGPT for question answering on mobile devices. A regular solution to this question is to deploy the model on cloud service. The queries for an LLM on a mobile device are processed in the cloud and the LLM output is sent back to the device. This is the standard workflow for the ChatGPT app and most other LLM-powered chat apps. But on some extreme occasions, this is infeasible. For example, when the device is in areas with limited connectivity or under strict monitoring and surveillance of Internet traffic, or when the locs of users contacting ChatGPT must be in strict confidence. The solution is to transplant the whole model to mobile devices by cutting off the parameter amount.

The study by Patil [35] in 2023 is probably the first attempt to migrate the whole model to mobile devices. The researchers succeeded in reducing the size of BERT from 438 MB to 181 MB with negligible accuracy degradation, which is a significant achievement. The work by Yin et al. [49] presents LLMS, which can decouple the memory management of app and LLM contexts. They managed to migrate an open-source Large Language Model, LLama2 to mobile devices with the help of LLMS. The study presented by Laskaridis et al. [17] builds a benchmark for LLMs across ML frameworks, devices, and ecosystems that can automate the interaction with instruction fine-tuned models and capture events and metrics of interest at a granular level, both in terms of performance as well as energy. The study by Fassold et al. [5] demonstrates theoretically how to port LLMs efficiently to mobile devices so that they run natively and at interactive speed on a mobile device. The study by Li et al. [22] proposes four optimization techniques to facilitate high-efficiency LLM deployment on device GPUs. The experiment result shows that compared with CPU-based FastLLM and GPU-based MLC-LLM, their approach attains over 10x speedup for the prefill speed and 2 3x speedup for the decoding speed on Chatglm2-6B.

RQ3: What's the Main Purpose for Researchers to Involve LLMs in Their Work? We have summed up 4 categories of missions that LLMs perform in mobile devices-related studies:

Instruction Understanding: LLM accepts a series of instructions about interacting with mobile applications described in natural language, parses the semantics and attempts to execute the provided tasks. Mostly happens in test script generation and task automation.

Code Understanding: LLM accepts the source code of an API or a whole application and analyzes the intention with its rich knowledge of code generation. Mostly happens in malware detection.

Predicting: LLM serves as a regular deep-learning model that learns features from history datasets and leverages the learned knowledge to do prediction or classification tasks.

Schemeing: LLM gives advice for software testing or development with its commonsense knowledge, usually in the form of a full plan by question answering.

5 Conclusion

LLMs are bringing significant changes to traditional fields and techniques for mobile device-related tasks. From the steep growth of the number of related studies since 2022, we could conclude that LLM's vast ability of code comprehension and natural language understanding implicitly matches the demands for mobile device-based tasks. Meanwhile, researchers have also been making efforts to promote the performance of LLMs in mobile device-related tasks. Sun et al. [38] proposed a new approach for better handling Andriod Bytecode on LLMs. A number of benchmarks and datasets have been built for tuning LLMs to better fit the tasks. The future of employing LLMs in mobile device-related research is outstanding.

In this paper, we present a survey on mobile device-related studies that employ LLMs in their approaches, classify these studies by academic fields, and discuss the search results with 3 research questions. We hope our research will guide and inspire future work on the usage of LLMs in the mobile device field.

References

1. Ali, A., Xia, Y., Umer, Q., Osman, M.: BERT based severity prediction of bug reports for the maintenance of mobile applications. J. Syst. Softw. **208**, 111898 (2024)
2. Cao, S., et al.: Comprehensive semantic repair of obsolete GUI test scripts for mobile applications. In: Proceedings of the IEEE/ACM 46th International Conference on Software Engineering, pp. 1–13 (2024)
3. Cui, C., Li, T., Wang, J., Chen, C., Towey, D., Huang, R.: Large language models for mobile GUI text input generation: an empirical study. arXiv preprint arXiv:2404.08948 (2024)
4. Dlamini, G., Ahmad, U., Kharkrang, L.R., Ivanov, V.: Detecting design patterns in android applications with codeBERT embeddings and CK metrics. In: International Conference on Analysis of Images, Social Networks and Texts, pp. 267–280. Springer (2023)
5. Fassold, H.: Porting large language models to mobile devices for question answering. arXiv preprint arXiv:2404.15851 (2024)
6. Feng, S., Chen, C.: Prompting is all you need: automated android bug replay with large language models. In: Proceedings of the 46th IEEE/ACM International Conference on Software Engineering, pp. 1–13 (2024)
7. Ferrag, M.A., et al.: Revolutionizing cyber threat detection with large language models: a privacy-preserving BERT-based lightweight model for IoT/IIoT devices. IEEE Access **12**, 23733–23750 (2024). https://doi.org/10.1109/ACCESS.2024.3363469
8. García-Soto, E., Martín, A., Huertas-Tato, J., Camacho, D.: Android malware detection through a pre-trained model for code understanding. In: International Conference on Ubiquitous Computing and Ambient Intelligence, pp. 1055–1060. Springer (2022)
9. Gu, Z., Xu, Z., Chen, H., Lan, J., Meng, C., Wang, W.: Mobile user interface element detection via adaptively prompt tuning. In: 2023 IEEE/CVF Conference on Computer Vision and Pattern Recognition (CVPR), pp. 11155–11164 (2023). https://api.semanticscholar.org/CorpusID:258741182

10. Hou, X., et al.: Large language models for software engineering: a systematic literature review. ArXiv abs/2308.10620 (2023). https://api.semanticscholar.org/CorpusID:261048648
11. Huang, F., Li, G., Li, T., Li, Y.: Automatic macro mining from interaction traces at scale. In: Proceedings of the CHI Conference on Human Factors in Computing Systems, pp. 1–16 (2024)
12. Huang, Y., et al.: CrashTranslator: automatically reproducing mobile application crashes directly from stack trace. In: Proceedings of the 46th IEEE/ACM International Conference on Software Engineering, pp. 1–13 (2024)
13. Jiang, L.: Utilizing large language models to detect privacy leaks in mini-app code. arXiv preprint arXiv:2402.07367 (2024)
14. Jones, R.K.: Beyond traditional learning: leveraging BERT for enhanced android malware detection. In: Multisector Insights in Healthcare, Social Sciences, Society, and Technology, pp. 208–228. IGI Global (2024)
15. Khaokaew, Y., Xue, H., Salim, F.D.: MAPLE: mobile app prediction leveraging large language model embeddings. Proc. ACM Interact. Mob. Wearable Ubiquit. Technol. **8**, 10:1–10:25 (2023). https://api.semanticscholar.org/CorpusID:262044117
16. Kumar, S., Yadav, M., et al.: GUI testing using random event-based test cases. In: 2023 12th International Conference on System Modeling & Advancement in Research Trends (SMART), pp. 512–520. IEEE (2023)
17. Laskaridis, S., Kateveas, K., Minto, L., Haddadi, H.: Melting point: mobile evaluation of language transformers. arXiv preprint arXiv:2403.12844 (2024)
18. Lee, S.C., Jang, Y., Park, C.H., Seo, Y.S.: Feature analysis for detecting mobile application review generated by AI-based language model. J. Inf. Process. Syst. **18**(5) (2022)
19. Lee, S., et al.: Explore, select, derive, and recall: augmenting LLM with human-like memory for mobile task automation. ArXiv abs/2312.03003 (2023). https://api.semanticscholar.org/CorpusID:265690585
20. Lemieux, C., Inala, J.P., Lahiri, S.K., Sen, S.: CODAMOSA: escaping coverage plateaus in test generation with pre-trained large language models. In: 2023 IEEE/ACM 45th International Conference on Software Engineering (ICSE), pp. 919–931. IEEE (2023)
21. Li, C., Xiong, Y., Li, Z., Yang, W., Pan, M.: Mobile test script generation from natural language descriptions. In: 2023 IEEE 23rd International Conference on Software Quality, Reliability, and Security (QRS) pp. 348–359 (2023). https://api.semanticscholar.org/CorpusID:266555632
22. Li, L., Qian, S., Lu, J., Yuan, L., Wang, R., Xie, Q.: Transformer-lite: high-efficiency deployment of large language models on mobile phone GPUs. arXiv preprint arXiv:2403.20041 (2024)
23. Li, Z., Dutta, S., Naik, M.: LLM-assisted static analysis for detecting security vulnerabilities. arXiv preprint arXiv:2405.17238 (2024)
24. Liu, Z., et al.: Fill in the blank: context-aware automated text input generation for mobile GUI testing. In: 2023 IEEE/ACM 45th International Conference on Software Engineering (ICSE), pp. 1355–1367. IEEE (2023)
25. Liu, Z., et al.: Chatting with GPT-3 for zero-shot human-like mobile automated GUI testing. ArXiv abs/2305.09434 (2023). https://api.semanticscholar.org/CorpusID:258714899
26. Liu, Z., et al.: Make LLM a testing expert: bringing human-like interaction to mobile GUI testing via functionality-aware decisions. In: Proceedings of the

IEEE/ACM 46th International Conference on Software Engineering, pp. 1–13 (2024)
27. Liu, Z., et al.: Unblind text inputs: predicting hint-text of text input in mobile apps via LLM. In: Proceedings of the CHI Conference on Human Factors in Computing Systems, pp. 1–20 (2024)
28. Liu, Z., et al.: Testing the limits: unusual text inputs generation for mobile app crash detection with large language model. In: Proceedings of the IEEE/ACM 46th International Conference on Software Engineering, pp. 1–12 (2024)
29. Liu, Z., et al.: LLM-CompDroid: repairing configuration compatibility bugs in android apps with pre-trained large language models. arXiv preprint arXiv:2402.15078 (2024)
30. Ma, W., Cui, B.: Fuzzing IoT devices via android app interfaces with large language model. In: International Conference on Emerging Internet, Data & Web Technologies, pp. 87–99. Springer (2024)
31. Ma, X., Zhang, Z., Zhao, H.: Comprehensive cognitive LLM agent for smartphone GUI automation. arXiv preprint arXiv:2402.11941 (2024)
32. Mathews, N.S., Brus, Y., Aafer, Y., Nagappan, M., McIntosh, S.: LLbezpeky: leveraging large language models for vulnerability detection. arXiv preprint arXiv:2401.01269 (2024)
33. Mohajer, M.M., et al.: SkipAnalyzer: an embodied agent for code analysis with large language models. arXiv preprint arXiv:2310.18532 (2023)
34. Nguyen, Q.N., Cam, N.T., Van Nguyen, K.: XLMR4MD: new Vietnamese dataset and framework for detecting the consistency of description and permission in android applications using large language models. Comput. Secur. **140**, 103814 (2024)
35. Patil, P., Rao, C., Meena, S.: Optimized BERT model for question answering system on mobile platform. In: International Conference on Speech and Language Technologies for Low-resource Languages, pp. 129–139. Springer (2023)
36. Shi, J., Jain, R., Doh, H., Suzuki, R., Ramani, K.: An HCI-centric survey and taxonomy of human-generative-AI interactions. arXiv preprint arXiv:2310.07127 (2023)
37. Souani, B., Khanfir, A., Bartel, A., Allix, K., Le Traon, Y.: Android malware detection using BERT. In: International Conference on Applied Cryptography and Network Security, pp. 575–591. Springer (2022)
38. Sun, T., et al.: DexBERT: effective, task-agnostic and fine-grained representation learning of android bytecode. IEEE Trans. Softw. Eng. (2023)
39. Taeb, M., Swearngin, A., Schoop, E., Cheng, R., Jiang, Y., Nichols, J.: AXNav: replaying accessibility tests from natural language. In: Proceedings of the CHI Conference on Human Factors in Computing Systems, pp. 1–16 (2024)
40. Vu, M.D., et al.: GPTVoiceTasker: LLM-powered virtual assistant for smartphone. arXiv preprint arXiv:2401.14268 (2024)
41. Wang, B., Li, G., Li, Y.: Enabling conversational interaction with mobile UI using large language models. In: Proceedings of the 2023 CHI Conference on Human Factors in Computing Systems (2022). https://api.semanticscholar.org/CorpusID: 252367445
42. Wang, C., Liu, J., Peng, X., Liu, Y., Lou, Y.: LLM-based resource-oriented intention inference for static resource leak detection (2023). https://api.semanticscholar.org/CorpusID:265050465
43. Wang, J., Huang, Y., Chen, C., Liu, Z., Wang, S., Wang, Q.: Software testing with large language models: survey, landscape, and vision. IEEE Trans. Softw. Eng. (2024)

44. Wang, J., et al.: Mobile-agent-v2: mobile device operation assistant with effective navigation via multi-agent collaboration. arXiv preprint arXiv:2406.01014 (2024)
45. Wang, J., et al.: Mobile-agent: autonomous multi-modal mobile device agent with visual perception. arXiv preprint arXiv:2401.16158 (2024)
46. Wang, Z., et al.: XUAT-copilot: multi-agent collaborative system for automated user acceptance testing with large language model. arXiv preprint arXiv:2401.02705 (2024)
47. Wen, H., et al.: AutoDroid: LLM-powered task automation in android. In: Proceedings of the 30th Annual International Conference on Mobile Computing and Networking, pp. 543–557 (2024)
48. Wen, H., Wang, H., Liu, J., Li, Y.: DroidBot-GPT: GPT-powered UI automation for android. arXiv preprint arXiv:2304.07061 (2023)
49. Yin, W., Xu, M., Li, Y., Liu, X.: LLM as a system service on mobile devices. arXiv preprint arXiv:2403.11805 (2024)
50. Yoon, J., Feldt, R., Yoo, S.: Intent-driven mobile GUI testing with autonomous large language model agents (2024)
51. Yoon, J., Feldt, R., Yoo, S.: Autonomous large language model agents enabling intent-driven mobile GUI testing. arXiv preprint arXiv:2311.08649 (2023)
52. Zhang, Y., et al.: Learning-based widget matching for migrating GUI test cases. In: Proceedings of the 46th IEEE/ACM International Conference on Software Engineering, pp. 1–13 (2024)
53. Zhao, W., Wu, J., Meng, Z.: AppPoet: large language model based android malware detection via multi-view prompt engineering. arXiv preprint arXiv:2404.18816 (2024)
54. Zimmermann, D., Koziolek, A.: Automating GUI-based software testing with GPT-3. In: 2023 IEEE International Conference on Software Testing, Verification and Validation Workshops (ICSTW), pp. 62–65. IEEE (2023)
55. Zimmermann, D., Koziolek, A.: GUI-based software testing: an automated approach using GPT-4 and selenium webdriver. In: 2023 38th IEEE/ACM International Conference on Automated Software Engineering Workshops (ASEW), pp. 171–174. IEEE (2023)

Application Framework for OpenHarmony Distributed Trusted Execution Environment

Yilong Wang[1,2], Yang Yu[1,2(✉)], and Dong Du[1,2]

[1] Institute of Parallel and Distributed Systems, Shanghai Jiao Tong University, Shanghai, China
yu_y@sjtu.edu.cn
[2] Engineering Research Center for Domain-Specific Operating Systems, Ministry of Education, Beijing, China

Abstract. OpenHarmony empowers numerous heterogeneous devices with the capability of intelligent connection and synergy. However, many of these devices lack TEE (Trusted Execution Environment) capabilities, making them incapable of handling various high-security scenarios. How to extend and share TEE capabilities in distributed environments is an urgent requirement and also introduces significant challenges. To address these, we propose DTee, the first general-purpose programming framework for OpenHarmony that simplifies the development and deployment of applications on a distributed TEE. DTee encapsulates the complexities of distributed TEE implementation and exposes user-friendly programming interfaces, enabling developers to build secure applications without deep expertise in TEE internals. DTee introduces a novel programming model that divides the development process into development and deployment stages, decoupling business logic from TEE-specific details and streamlining the development process. We also provide a suite of tools to significantly lower the barrier to entry for developers. Evaluations with real-world applications such as distributed face recognition and speech recognition demonstrate DTee's effectiveness in enhancing security, maintaining performance, and simplifying the development of distributed TEE applications.

1 Introduction

The rapid advancement of interconnected smart devices and the growing prevalence of distributed systems have raised significant concerns regarding the security and privacy of sensitive data [1,2,11,19]. OpenHarmony, a distributed operating system, is designed to operate seamlessly across multiple devices, offering a broad range of distributed capabilities. However, the integration of Trusted Execution Environments (TEEs) within such a distributed ecosystem poses several challenges. TEEs, like Intel SGX [5], ARM TrustZone [13], and RISC-V Penglai [6], are traditionally confined to single-device implementations, which limits their potential in a distributed context. Furthermore, not all devices are equipped with the necessary hardware support for TEEs, preventing these devices from performing tasks that are closely tied to security.

TEEs are critical for securing sensitive operations by providing isolated execution environments that protect code and data from unauthorized access and modification [15]. Despite their importance, extending TEE capabilities to support distributed systems introduces complexities related to programmability, secure communication, and interoperability among different TEE technologies.

To address these challenges, we propose DTee, the first general-purpose programming framework that simplifies the development and deployment of applications on a distributed TEE for OpenHarmony. The DTee framework encapsulates the complexities of distributed TEE implementation and exposes user-friendly programming interfaces, enabling developers to build secure applications without deep expertise in TEE internals.

The contributions of this paper are as follows:

Framework Design: We present the design of DTee, detailing its architecture and components. The framework introduces a novel programming model that divides the project development process into development and deployment stages. This model decouples business logic from TEE-specific details, streamlining the development process.

Implementation and Tooling: We implement a prototype of DTee and develop a suite of tools, including the dteegen tool, which supports project creation, building, and deployment using containerized environments. This tool significantly lowers the barrier to entry for developers.

Evaluation: We evaluate DTee through real-world applications, including distributed face recognition, distributed speech recognition, and distributed Lua script execution. These evaluations demonstrate the framework's effectiveness in enhancing security, maintaining performance, and simplifying the development of distributed TEE applications.

2 Background and Motivation

OpenHarmony and Distributed Operating Systems. OpenHarmony is a distributed operating system designed to seamlessly operate across multiple devices, ranging from smartphones and tablets to IoT devices and wearables. It aims to create a cohesive ecosystem where applications and services can function fluidly and cooperatively across diverse hardware platforms. This distributed nature of OpenHarmony enhances user experience by providing consistent and synchronized functionality regardless of the device in use.

The primary advantage of distributed operating systems like OpenHarmony lies in their ability to offer enhanced flexibility, scalability, and resource sharing. By leveraging the collective capabilities of various devices, these systems can achieve greater efficiency and provide more robust services. However, this distributed paradigm also introduces significant challenges, particularly concerning security and privacy. Ensuring that data remains secure and operations are trustworthy across multiple devices is critical for maintaining the integrity of the overall system.

Trusted Execution Environment. A Trusted Execution Environment (TEE) is a secure area within a processor that guarantees the confidentiality and integrity of the code and data loaded within it. TEEs are designed to protect sensitive operations from being compromised by external threats, including the operating system itself. Several notable TEE technologies include RISC-V Penglai, Intel SGX, ARM TrustZone, and AMD SEV.

The Significance of Supporting Distributed TEE with a General App Framework. The integration of TEEs into distributed systems like OpenHarmony is crucial for several reasons: (1) **Enhanced Security:** Distributed TEEs can ensure that sensitive data and operations are protected across all devices within the ecosystem. This is particularly important in scenarios where data is transmitted and processed by multiple nodes, each potentially vulnerable to different threats. (2) **Data Privacy:** By ensuring that data is processed within secure enclaves on each device, distributed TEEs can prevent unauthorized access and data breaches. This is essential for maintaining user trust and compliance with privacy regulations. (3) **Consistency and Integrity:** Distributed TEEs can help maintain the integrity of operations across devices, ensuring that the results of computations are trustworthy and consistent. This is vital for applications that require coordinated actions across multiple devices. (4) **Scalability:** A general app framework that supports distributed TEEs can scale security across a broad range of devices, from high-performance servers to resource-constrained IoT devices. This scalability is essential for supporting the diverse hardware landscape of modern distributed systems. (5) **Ease of Development:** A unified framework for developing distributed TEE applications simplifies the development process. Developers can focus on building secure applications without needing to manage the complexities of TEE interactions and secure communications manually.

The proposed DTee framework addresses these challenges by providing a comprehensive solution for developing and deploying applications on a distributed TEE. By encapsulating the complexities of TEE implementation and exposing user-friendly programming interfaces, DTee enables developers to build secure, efficient, and scalable applications for distributed systems.

3 Design

The DTee application framework encapsulates the distributed TEE implementation and exposes user programming interfaces to the upper layers. On this basis, DTee innovatively introduces a programming model that divides the project development process into development and deployment stages. During the development stage, distributed TEE is not introduced (developers do not even need to install the corresponding SDK), and only during deployment is the distributed

TEE introduced, simplifying the development process and unifying different TEE application development modes. Additionally, based on container technology, a tool called dteegen is developed, encapsulating the complete build environment, and supporting project creation, build, and deployment.

3.1 Architecture

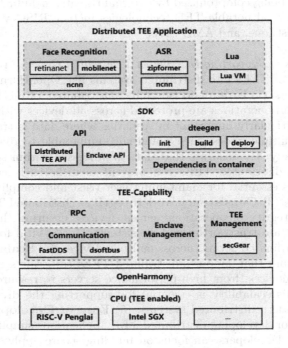

Fig. 1. Architecture of DTee

The architecture of DTee is shown in Fig. 1. At the bottom is the hardware layer, where devices have various architectures and use different types of TEEs, such as Penglai on RISC-V architecture machines and Intel SGX on x86-64 machines. Some devices may not include a TEE. The operating system layer runs an OpenHarmony operating system. Above it is the distributed TEE framework layer, supporting different communication methods, including OpenHarmony's distributed soft bus and FastDDS. Based on the communication foundation, the framework implements an RPC mechanism. Additionally, the distributed TEE framework needs to support enclave code file management and the free flow of enclaves between devices. TEE management is handled using the secGear framework to interface with different TEE implementations. In the upper layer, we implement an SDK to support application development, including APIs and the dteegen project development tool. The SDK has two types of APIs: one for distributed TEE APIs to enable application calls to distributed TEE capabilities and one for enclave APIs for programs within the TEE. The dteegen project development tool is based on Docker images, encapsulating the build environment. Dteegen supports project creation, building, deployment, and signing. At the top layer, this work implements three applications: distributed face recognition, distributed voice recognition, and distributed Lua script execution.

3.2 TEE-Capability

TEE-Capability interacts with TEE to enable distributed TEE capability sharing, abstracting this for upper layers. It consists of several modules:

Communication Module: Implements basic TEE-Capability functionality. Devices are categorized as light (no TEE) and rich (with TEE). The architecture uses a service-based abstraction, supporting communication frameworks like FastDDS and OpenHarmony's distributed soft bus.

TEE Perception Module: Detects the presence of TEE capabilities across devices and registers them. If local TEE is absent, it queries other nodes, enabling redirection of calls to actual TEE nodes.

TEE Invocation Module: Manages remote attestation and enclave execution, using frameworks like secGear to abstract differences between TEE platforms (e.g., Intel SGX, ARM TrustZone). The attestation process involves key exchange and establishing secure communication channels for data encryption.

Enclave Management Module: Handles the persistence and management of enclave files, updating versions as needed to minimize transmission overhead.

Enclave Transfer Module: Facilitates the transfer of enclave files across devices, ensuring efficient use of TEE capabilities in the network.

Multi-TEE Node Scheduling Module: Uses a proxy mechanism (Ribbon) to manage multiple TEE-enabled devices, forwarding client requests to the appropriate server based on specific policies.

This design ensures a robust, scalable, and efficient distributed TEE architecture, maximizing the utilization of TEE capabilities across diverse devices.

3.3 Programming Model

In the development of distributed TEE (Trusted Execution Environment) applications, developers write functions to run in the TEE and call these functions from the host side. During the debugging phase, a local TEE is required to execute these functions, and if using distributed TEE, the support of related distributed communication libraries is also necessary.

However, we observed that during the development phase, developers focus on implementing business logic without needing to use a TEE to run these functions or to support distributed execution through distributed TEE. In fact, during development, these functions can be executed as regular functions.

To decouple business logic from distributed TEE, this work proposes a new programming model characterized by dividing the project lifecycle into two stages: development and deployment, thereby minimizing the complexity during the development phase.

Development Phase. In the development phase, the project does not include the implementation of distributed TEE. Developers can focus on business logic and can call trusted functions as if they were local functions. In fact, during this phase, these functions are executed as local functions. To support the transition of the project to the deployment phase, developers need to follow certain conventions. Firstly, the project must be divided into two separate directories for untrusted and trusted code, named insecure and secure respectively. Developers write traditional business logic in the insecure directory and trusted functions in the secure directory. This separation allows the framework's codegen module to track cross-references between these directories, accurately analyze enclave calls (ecall), and external calls (ocall), and automatically generate the critical code needed for injecting distributed TEE capabilities. This design is based on the observation that traditional TEE development requires splitting the logic into trusted and untrusted parts. Therefore, by explicitly separating the project into two directories, the framework can use tools to automatically parse inter-directory calls and obtain more semantic information.

Deployment Phase. In the deployment phase, calls from the insecure directory to functions in the secure directory will have distributed TEE execution capabilities. If there is at least one node in the cluster with a TEE device, the function will always execute in the TEE. Specifically, if the local device has a TEE, the function will execute locally. Otherwise, if other nodes have a TEE, the function will execute on those nodes.

3.4 User Programming Interface

This work provides two types of user programming interfaces (APIs) to the upper layers. One type supports APIs related to initialization, invocation, and deregistration of the distributed TEE on the host side, namely the distributed TEE API. The other type involves APIs related to TEE capabilities on the enclave side, such as seal and unseal, known as the enclave API. The reason for providing an additional set of enclave APIs is that, according to the programming model proposed in the previous section, TEE invocations will not be introduced during the development phase. At this time, the enclave APIs cannot be used normally and require special handling.

Distributed TEE API. Considering the interaction process between the client and server, the framework extracts the following user programming interfaces:

1. **init_distributed_tee_context:** Initializes the distributed TEE context based on the configuration and returns a context object.
2. **destroy_distributed_tee_context:** Deregisters from the distributed TE- E and destroys the context.
3. **call_remote_secure_function:** Invokes a function that will be executed in the remote TEE.
4. **publish_secure_function:** Publishes a function as a service.

This framework offers two execution modes: normal mode and transparent mode. The desired execution mode can be specified when creating a new distributed TEE context. In normal mode, the client must explicitly call call_remote_secure_function to remotely invoke a function executed in the TEE, or directly call a function to execute in the local TEE. In transparent mode, developers can call functions in the secure directory as if they were local functions, without needing to manage the underlying distributed and TEE details.

The framework automatically detects the presence of a local TEE environment: if it exists, the secure function is executed locally; if not, it seeks a remote TEE node for execution. If neither local nor remote TEEs are available, the framework can either execute the call in an untrusted manner or treat it as an error.

By using transparent mode, this framework significantly reduces development and debugging complexity, making all secure function calls appear as local function calls to developers. This greatly simplifies the debugging process.

Enclave API. This framework currently provides the following enclave APIs:

1. **eapp_print:** During the development phase, this function is similar to printf. In the deployment phase, this function eventually calls the write system call, which requires the host's assistance to complete.
2. **seal_data(buf, buf_len, data_len):** buf is a pointer to a buffer for storing raw data and storing sealed data after sealing is completed. **buf_len** is the length of the buffer, and **data_len** is the number of bytes of data to be sealed. During the development phase, to some extent simulate the seal function, a predefined key is used for encryption here. In the deployment phase, this function ultimately calls the TEE SDK API to encrypt the data using a key derived from the enclave.
3. **unseal_data(buf, buf_len):** This API corresponds to **seal_data**, used to decrypt previously encrypted data.

3.5 Codegen for Programming Model

The code written by developers using the aforementioned programming model does not inherently possess distributed and TEE (Trusted Execution Environment) capabilities, making it convenient for local debugging. To automatically transform developers' code to gain distributed TEE capabilities, we developed a codegen.

First, codegen collects function definitions from the "insecure" and "secure" directories, obtaining a set of trusted functions $S_{\text{sec}} = \{f \mid f \in \text{secure}\}$ and a set of untrusted functions $S_{\text{insec}} = \{f \mid f \in \text{insecure}\}$. Then, by analyzing the mutual calls between these two sets, two entry sets are derived:

$$\text{ecall\_func} = \{f \mid f \in S_{\text{sec}}, \exists g \in S_{\text{insec}} \text{ such that } f \text{ is called by } g\} \quad (1)$$

and

$$\text{ocall\_func} = \{f \mid f \in S_{\text{insec}}, \exists g \in S_{\text{sec}} \text{ such that } f \text{ is called by } g\} \quad (2)$$

These two sets are used to generate the Enclave Definition Language (EDL) file. The EDL file uses a specific syntax to describe the functions within the enclave and the interfaces that should be exposed to external applications.

3.6 Dteegen

To facilitate development, this work further implements a development suite, including a command-line tool called dteegen and a Docker-based build environment. Developers can easily get started with distributed TEE development using dteegen.

Dteegen supports project initialization, automated code building, and deployment, significantly streamlining the entire lifecycle of distributed TEE project development.

All dependencies required for project building are included in the Docker image, including template files, codegen (for transforming the project), compilers (for compiling the project), TEE-Capability SDK (providing libraries required for distributed TEE), TEE SDK (providing libraries required for TEE like Penglai), and secGear SDK (providing dependencies for secGear). Dteegen will pull the image and use the provided scripts to complete project initialization, building, and deployment, eliminating the hassle of configuring the environment.

Initialization. Project initialization is completed using templates. By providing the project name to replace variables in the template, a project can be quickly generated to support subsequent development. The template project provides basic usage of the framework, lowering the entry barrier.

Building. As previously mentioned, applications developed based on this framework support local building and running. This step can be done manually or using dteegen. The project is built using CMake, so its build steps are fixed.

Deployment. The deployment process transitions the project from the development stage to the deployment stage. Code analysis and generation are completed using the codegen module, and the related dependency libraries encapsulated in the container inject distributed TEE capabilities into the project. Additionally, a signing tool included in the container is used to complete the enclave signing. The entire process is automated, without requiring developers to focus on the underlying tasks. The final output is a signed enclave file (enclave.signed.so) and an executable file for the host part developed in the insecure directory.

4 Implementation

The working process of the `codegen` module of dteegen can be divided into two phases: parsing and generation.

Parsing Phase. First, the source code in the `insecure` and `secure` directories is parsed to obtain:

1. Function definition set in the `insecure` directory (InsecDef).
2. Function definition set in the `secure` directory (SecDef).
3. Function call set in the `insecure` directory (InsecCall).
4. Function call set in the `secure` directory (SecCall).

Then, the intersections are calculated:

$ecall = SecDef \cap InsecCall$, representing functions defined in the `secure` directory called by functions in the `insecure` directory.

$ocall = InsecDef \cap SecCall$, representing functions defined in the `insecure` directory called by functions in the `secure` directory.

To parse function calls, the symbol table in the compiled object files (ELF format on Linux) is used. This table records all symbols (functions and variables) used in the object file. The function definitions are parsed using `libclang`, which provides APIs for analyzing and manipulating C, C++, and Objective-C source code.

Generation Phase. Based on the information obtained in the parsing phase, `codegen` can complete the code generation. This is achieved using templates that follow three simple syntax rules:

1. The first line specifies the path of the generated file using the syntax `path: A/B/C`.
2. The remaining text is placed in the generated file.
3. Strings in the template like `${Variable}` can be replaced with metadata obtained during parsing. Additionally, batch generation using a `foreach`-like syntax is supported with template blocks starting with `**begin**` and ending with `**end**`.

Currently, `codegen` mainly targets the `secGear` project, which includes an `enclave` directory for defining the enclave, and a `host` directory for defining the host. For a secure function (e.g., `add`), dteegen places its definition in the `enclave` directory with a renamed identifier to prevent conflicts and constructs a corresponding definition in the `host` directory for local or distributed invocation based on the presence of a TEE.

5 Evaluation

We conducted three distinct application scenarios to demonstrate the effectiveness and versatility of DTee in real-world contexts. Our evaluation was done on OpenHarmony 3.2 Release, with Penglai [6] RISC-V TEE.

Face recognition is a security-critical task [8]. However, some devices lack the necessary TEE support for confidential computing, thereby facing significant threats [14]. By using DTee, we implement face recognition logic in a secure folder that is executed with a distributed TEE. The face recognition algorithm operates within secure enclaves, significantly reducing the risks of data leakage and unauthorized access. Specifically, we use RetinaNet [9] to extract faces from input images and MobileNet [3] to extract facial features. We run NCNN [17] within the enclave for model deployment.

Fig. 2. The results of the face recognition application utilizing distributed TEE.

The results are shown in Fig. 2: on the left, the client calls the distributed TEE to assist with executing the face recognition computation, while on the right, the server receives requests from clients and performs the computation within the enclave.

5.1 Distributed Speech Recognition System

We also implement distributed speech recognition [10] using the DTee framework. Similarly, we use Zipformer [18] to extract features and deploy the model within the enclave using NCNN. The results are similar to those of the distributed face recognition system, as shown in Fig. 3. When a local TEE is available, the DTee framework can detect it and execute the tasks locally. For convenience, we hardcoded the data in the binary.

Fig. 3. DTee detected local tee to execute auto speech recognition

5.2 Distributed Lua Script Execution

The framework's support for the secure execution of Lua [7] scripts in a distributed environment further illustrates its versatility. Lua scripts, which do not require compilation, offer ease of development and debugging and can be quickly transferred between nodes, enhancing the system's efficiency. Since Lua scripts are data, they can be securely transmitted between the client and the enclave using encryption. By running these scripts within TEEs, the DTee framework mitigates risks associated with code injection and unauthorized modifications, ensuring the integrity and trustworthiness of dynamically executed code.

6 Related Work

Distributed TEE Frameworks. Distributed TEE frameworks aim to extend the security guarantees of TEEs across multiple devices. Town Crier [20] leverages SGX to provide secure oracles for smart contracts. It ensures the integrity and authenticity of data fed to blockchain applications, showcasing the potential of TEEs in distributed systems. VC3 [16] is a system for secure data analytics in the cloud. It uses SGX to protect data and computation, enabling secure processing of sensitive information in untrusted environments.

Programming Frameworks for TEEs. Open Enclave SDK [12] provides a unified API for developing enclave apps across different TEE technologies. SecGear [4] is another framework designed to support the development of secure applications using TEEs. It abstracts the underlying TEE technology and provides a high-level programming interface for developers. DTee utilizes SecGear to support different TEEs, leveraging its capabilities to create a more versatile and robust development environment.

Compared to these existing frameworks, DTee is unique as the first application framework specifically targeting distributed TEE environments. This innovation allows for secure execution and communication across multiple devices, significantly enhancing the scope and applicability of TEE technologies in distributed systems.

7 Conclusion

This paper presents DTee, the first general-purpose programming framework for OpenHarmony that simplifies the development and deployment of applications on a distributed TEE. DTee introduces a novel programming model with easy-to-use APIs and tools to assist developers in designing and debugging their secure apps.

References

1. Agency, X.N.: Data security law of the People's Republic of China. In: Gazette of the Standing Committee of the National People's Congress of the People's Republic of China, no. 005, pp. 951–956 (2021)
2. Agency, X.N.: Personal information protection law of the people's republic of China. In: The 30th Meeting of the Standing Committee of the 13th National People's Congress (2021)
3. Chen, S., Liu, Y., Gao, X., Han, Z.: MobileFaceNets: efficient CNNs for accurate real-time face verification on mobile devices. In: Zhou, J., et al. (eds.) CCBR 2018. LNCS, vol. 10996, pp. 428–438. Springer, Cham (2018). https://doi.org/10.1007/978-3-319-97909-0_46
4. Chenmaodong, [n.d.], H.C.L.: secGear: SDK to develop confidential computing apps based on hardware enclave features (2024). https://gitee.com/src-openeuler/secGear
5. Costan, V., Devadas, S.: Intel SGX explained. Cryptology ePrint Archive (2016)
6. Feng, E., et al.: Scalable memory protection in the {PENGLAI} enclave. In: 15th {USENIX} Symposium on Operating Systems Design and Implementation ({OSDI} 2021), pp. 275–294 (2021)
7. Ierusalimschy, R., De Figueiredo, L.H., Filho, W.C.: Lua-an extensible extension language. Softwa. Pract. Exp. **26**(6), 635–652 (1996)
8. Kortli, Y., Jridi, M., Al Falou, A., Atri, M.: Face recognition systems: a survey. Sensors **20**(2), 342 (2020)
9. Li, Y., Ren, F.: Light-weight retinanet for object detection. arXiv preprint arXiv:1905.10011 (2019)
10. Malik, M., Malik, M.K., Mehmood, K., Makhdoom, I.: Automatic speech recognition: a survey. Multimed. Tools Appl. **80**, 9411–9457 (2021)
11. Meneghello, F., Calore, M., Zucchetto, D., Polese, M., Zanella, A.: IoT: internet of threats? A survey of practical security vulnerabilities in real iot devices. IEEE Internet Things J. **6**(5), 8182–8201 (2019)
12. openEnclave: OpenEnclave: SDK for developing enclaves (2024). https://github.com/openenclave/openenclave
13. Pinto, S., Santos, N.: Demystifying arm trustzone: a comprehensive survey. ACM Comput. Surv. (CSUR) **51**(6), 1–36 (2019)
14. Ramachandra, R., Busch, C.: Presentation attack detection methods for face recognition systems: a comprehensive survey. ACM Comput. Surv. (CSUR) **50**(1), 1–37 (2017)
15. Sabt, M., Achemlal, M., Bouabdallah, A.: Trusted execution environment: what it is, and what it is not. In: 2015 IEEE Trustcom/BigDfataSE/Ispa, vol. 1, pp. 57–64. IEEE (2015)

16. Schuster, F., et al.: VC3: trustworthy data analytics in the cloud using SGX. In: 2015 IEEE Symposium on Security and Privacy, pp. 38–54. IEEE (2015)
17. Tencent: NCNN: A high-performance neural network inference framework optimized for mobile platforms (2024). https://github.com/Tencent/ncnn
18. Yao, Z., et al.: ZipFormer: a faster and better encoder for automatic speech recognition. arXiv preprint arXiv:2310.11230 (2023)
19. Zanella, A., Bui, N., Castellani, A., Vangelista, L., Zorzi, M.: Internet of things for smart cities. IEEE Internet Things J. **1**(1), 22–32 (2014)
20. Zhang, F., Cecchetti, E., Croman, K., Juels, A., Shi, E.: Town crier: an authenticated data feed for smart contracts. In: Proceedings of the 2016 aCM sIGSAC Conference on Computer and Communications Security, pp. 270–282 (2016)

Design and Implementation of a Multi-metric Performance Analysis Tool for Android System and Its Applications

Zhenyu Yang(✉)

Shanghai Jiao Tong University, 800# Dong Chuan Road, Shanghai, China
jolynefr@sjtu.edu.cn

Abstract. Android is a dominant smartphone OS known for its Linux kernel foundation and open-source nature, offering flexibility and customization via its multi-layer design. Understanding system performance under different application loads is essential to improve user experience and optimize devices. However, existing profiling tools fail to combine information from multiple system layers.

The paper introduces PolyMetrica, a framework designed for Android system developers to measure and analyze performance across various layers. PolyMetrica stands out for its extensive data collection, adaptability, and customizability. It collects data through a monitoring module that interfaces with different system layers and allows for the addition of new metrics.

PolyMetrica's testing has proven effective in reflecting Android's performance metrics and providing depth through cross-level data analysis. Its customizable nature enables easy addition and customization of metrics and load scenarios, making it a valuable tool for Android system development and optimization.

Keywords: Android System · Language Runtime · Performance Evaluation · Performance Analysis · Profiling Tools

1 Introduction

1.1 Android System

Android is an open-source mobile operating system based on the Linux kernel [7,8], developed in 2003 under the leadership of Andy Rubin and currently maintained by Google. It aims to create an open, free, and easily extensible mobile platform that supports a wide range of devices and makes it easier for developers to create innovative applications. Android has been updated and iterated many times and is now one of the most popular mobile operating systems in the world [14].

Android Operating System can be categorized into four main layers, from highest to lowest: the application layer, the application framework layer, the system runtime layer, and the Linux kernel layer [3].

1. **Application layer:** Applications can use functions provided by various system applications at this layer [18].
2. **Application Framework Layer:** This layer provides developers with a set of APIs to access system services and resources; these APIs simplify the reuse of core modular system components and services. The Application Framework Layer helps the application build the user interface, manage non-code resources, and the application lifecycle [4].
3. **System Runtime Layer:** This layer provides a virtual machine environment for Android applications: the Dalvik virtual machine or AndroidRuntime. the runtime compiles the Java source code "Ahead of Time" or "Just in Time" and performs garbage collection on the application's memory.
4. **Linux Kernel Layer:** This layer provides the underlying hardware support and drivers, and the Linux kernel layer manages various device drivers, such as audio, video, and network drivers, as well as various system services, such as power management, memory management, process management, and security.

Each of these layers has its own specific functions and responsibilities, and they work in complex interactions to provide developers with a flexible and efficient development environment and to enable Android devices to interact with hardware. It enables developers to build effective applications and provide users with a great experience.

Optimizing the user experience on Android is a wide-ranging effort that includes improving performance, stability, and responsiveness, as well as improving hardware compatibility, conserving energy and battery life, and enhancing security.

1.2 Existing Profiling Tools

Responsiveness Analysis Tool. In mobile devices, the response latency of an application is very important to the user experience, especially when the response latency keeps increasing, which may lead to problems such as lagging in the use of the application and freezing of the interface.

In a responsiveness analysis tool developed for Android applications, researchers at Keio University proposed a static analysis technique to discover responsiveness flaws in Android applications [15]; a context-sensitive call graph generator was used to generate call graphs for Android applications, and a list of potentially blocking APIs was used as a guide to traverse the generated call graphs to discover potential responsiveness defects. In this way, the tool can help developers identify potential long-running operations.

Frame Rate Profiling Tool. Typically, Android phones have a frame rate of 60 frames per second (60 fps) or higher, so the time left for the system to perform rendering calculations is only 16.7 ms or less; however, if the main thread responsible for graphics is blocked for more than this time, the frame (or frames) that could not be rendered in time will be skipped.

Android GPU Inspector (AGI) is a tool developed and maintained by Google to analyze the graphics rendering performance of Android applications [2]. The tool saves the image statistics of each frame drawn into memory by staking in OpenGL ES, the graphics rendering library used by Android, and then exports them to the PC for data analysis.

Integrated Tools. Android comes with a comprehensive performance analysis tool, Simpleperf [1], which is based on the hardware performance counters and performance event sampling mechanism in the Linux kernel [20]. The Simpleperf collects the number of CPU cycles, the number of instructions executed, the number of cache failures, and so on, and also monitors the application's memory and I/O operations on a sampling basis.

Android Profiler is a graphical performance analysis tool built into Android Studio, the official integrated environment for Android development provided by Google [9]. It measures higher-level performance metrics than Simpleperf, helping developers to monitor the application's CPU utilization, memory usage, network connection, disk read/write data. It provides real-time graphs of performance data, and is therefore commonly used in the Android application development community.

1.3 Drawbacks of Existing Tools

Android currently has a good ecosystem of performance profiling tools: not only are there specialized tools for each different metric, but there are also comprehensive application performance analysis platforms like Simpleperf and Android Profiler that integrate a wide range of metrics. However, for Android developers, there are still some problems with existing performance profiling tools.

1. **Performance data processing is not customizable:** System developers need to customize the performance data format to the specific operating scenarios of the device and present results that reflect the state of the system and the user experience.
2. **Optimized for the application rather than the system layer:** The dominant testing tools in the ecosystem are designed for application tuning, and the primary purpose of monitoring data is for application developers to optimize their Java applications.
3. **Integrated tool lacks capability for joint analysis across hierarchical levels:** As the hierarchies (4 layers of system) in Android are complex and collaborative, system developers often need to analyze multiple performance metrics across hierarchical levels in order to find bottlenecks in the operating system that affect the user experience; existing integrated profiling tools

(e.g., Simpleperf) only present the performance data of the different levels independently.

1.4 Contributions

To meet the above requirements, this paper designs and implements PolyMetrica, an automated remote Android system multi-metric and cross-layer performance profiling tool, and builds a working set of real mobile applications for testing.

PolyMetrica collects data on performance metrics in Android from a variety of sources. These include graphics rendering information provided by Android system services, memory page swapping in the Linux kernel and statistics exposed in procfs, and garbage collection logs provided by the Android Runtime.

PolyMetrica also builds workload test sets based on real-world applications, in order to ensure that the test results are consistent with real-world application scenarios. In this paper, we use the Appium framework [19] to control the GUI applications on Android systems.

2 Background

In order to better describe the workings of the framework in this paper, additional explanations of the mechanisms in the Android system and in the Linux kernel are needed.

2.1 Garbage Collection in Android Runtime

For each application process, Android runs a separate Android Runtime for it, and all of the process's Java code is executed in the Android Runtime [17]. The Android Runtime provides automatic memory management for the application, which automatically scans for garbage objects in the Garbage Collection (GC) phase and reclaims their memory [16].

Concurrent copying GC is the default garbage collector used by Android. It concurrently copies tagged objects to another area while briefly stopping the application thread in order to discard garbage objects and defragment the heap.

2.2 Swapping Mechanism in Linux

In a Linux system, the physical memory area is divided into pages of 4KB in size, which are used to store the code and data of a process. When the amount of physical memory used by a process exceeds the limit, the operating system displaces the temporarily unused physical pages into a swap partition (e.g., disk) to free up memory space [13].

The physical page swap-in/swap-out mechanism allows processes to remain available when the size of the memory they need exceeds the size of the physical memory, improving system stability and availability. Testing, analysis and tuning of the swap-in/swap-out mechanism also play an important role in Android user experience optimization [5, 10–12].

3 Design

In order to test and analyze multiple metrics on the complex structure of Android system and Android applications running on it, this paper designs a multi-metric performance profiling framework for Android—PolyMetrica (**Poly**-dimensional **Metric A**nalyzer for **A**ndroid) collects the data of the application during the load running process by various means, such as capturing logs, calling the interface of Android framework, loading kernel modules, reading process file system, etc., and generates a proper performance analysis report to show it.

3.1 Overview

An overview of PolyMetrica's general architecture is shown in Fig. 1:

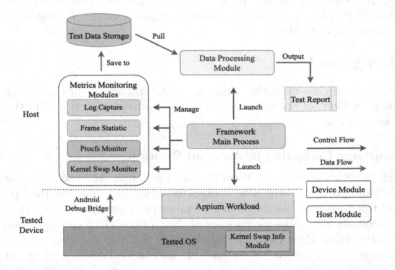

Fig. 1. PolyMetrica Overview

The main body of the test program runs on a host computer connected to the tested device (Android phone), and the operation of the test framework is controlled by a main process, which decides when to run the application workload and ensures that the various metrics monitoring modules continue to acquire and save data during the load run, and finally calls the data processing and report generation module.

The main feature of PolyMetrica is that it simultaneously acquires test data from multiple sources at multiple layers of the Android system, so managing the start and stop of these metrics monitoring modules is an important part of the PolyMetrica testing framework. The metrics monitoring modules work in different ways: they collect logs from Android's framework or kernel layer for the corresponding feature, call APIs provided by the Android framework, or access the Linux kernel directly.

Application Workload based on Appium provides a stable workload for the PolyMetrica framework, ensuring consistent behavior across multiple rounds.

The last running module is used to read the files from the test data storage area, which holds the raw records of all the above mentioned monitoring metrics, after the application load execution is completed, to perform complex data processing and to generate the corresponding test reports.

3.2 Runtime Log Capture Module

In Android, logcat is a command line tool for capturing and viewing log messages from application and system output.

Android's logging system is a global, buffer-based logging system that captures a variety of events and messages throughout the system, including application logs, kernel logs, system service logs, and so on [6]. Log messages are stored in a ring buffer, and when the buffer is full, the oldest log messages are discarded to make room for new ones.

For Garbage Collection events, the Android runtime writes detailed information about the garbage collection just now to the log buffer through the $Heap::LogGC$ function at the end of the object's collection. Therefore, the PolyMetrica log capture module filters out the garbage collection (or other required events) from logcat and writes them to a specified log file.

3.3 Customizable Kernel Swap Information Module

In order to obtain statistics related to physical memory page swapping in the Linux kernel, PolyMetrica has designed a swapping information monitoring mechanism based on function injection and kernel module.

In general, to capture statistics in the Linux kernel that are not exposed to the user mode in a specific way, the kernel code needs to be modified to expose the target statistics to the user in some way (e.g., kernel logs or procfs interfaces), and recompilation is needed.

We observes that the kind of statistical information needed by different developers of the memory page swapping system is not the same, and the framework cannot choose to expose all the information to the user mode: this will drastically slow down the execution speed of each swap in/swap out, causing a user-perceivable performance impact on the application.

Therefore, the kernel module needs to be customizable. Therefore, the developers can conditionally select the statistics for swapping information based on performance tuning needs. As shown in Fig. 2, PolyMetrica uses Linux's kernel module mechanism to achieve this goal: by declaring empty injector function pointers in the kernel, the injector functions are conditionally called on specified paths on the swap-in/swap-out; developers using the test framework write their own customized kernel modules, and when the kernel module is loaded, it points the corresponding function pointer to its customized implementation.

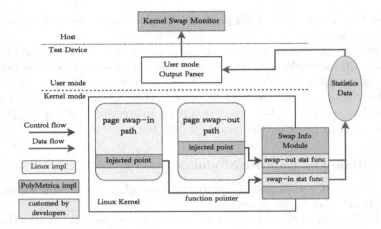

Fig. 2. Kernel Swap Monitor of PolyMetrica

As constraints, the injected functions must not modify sensitive data structures in kernel (e.g., *mm_struct* of a process) and not perform traversal on any list of uncertain length. Otherwise, excessively long function execution will block the swap's critical path.

3.4 Frame Rendering Time Capture Module

Detailed frame rendering information is output by the command line tool gfxinfo, which includes the timestamps of the start of the respective computation phases for each of the last 120 frames of the specified activity.

PolyMetrica runs a background thread in a recurring loop on the host side, querying the current Android foreground activity at 1 s intervals, and accordingly capturing frame rendering stage elapsed time statistics for the last 120 frames of the foreground application.

3.5 Customizable Process Information Capture Module

PolyMetrica obtains statistical information about the target process directly from the Linux layer of the Android system through the process filesystem (*procfs*), which includes the average CPU utilization during the application load operation, the number of minor page-faults and major page-faults occurring in the process etc.

Due to the abstraction of the process file system, these data are represented as virtual file system files in the Linux directory tree, and the framework can read such data using a unified interface after specifying the corresponding file path for each type of metric. For example, in order to collect the CPU utilization of the whole device during the application load operation, the framework reads the user state time, kernel state time and idle time of each CPU since startup from the */proc/stat* file of the process file system.

By providing a python decorator interface, the developer can customize which data metrics in the *procfs* are recorded for the load function.

3.6 Automatic Test Workload

Performance testing frameworks generally need to provide interfaces for test loads in order to ensure consistent test loads for the application under different test configurations, thus controlling variables to draw more informative test conclusions. However, most of the application scenarios of mobile Android phones are users interacting with the graphical interface on the screen, and it is impossible to simulate the real-world application load with simple API calls.

To solve this problem, the PolyMetrica testing framework provides an Appium scripting interface for test workloads. Appium is an open-source automated testing framework fo mobile applications; Appium provides remote access to a mobile device via the WebDriver protocol to control applications on an emulator or a real device.

PolyMetrica has designed three default test workloads via its Appium interface:

1. **Invoke "PCMark for Android":** This workload automates the call to Futuremark's benchmarking application "PCMark for Android" through Appium and captures the results at the end of the call, enabling a single application to cover multiple uses of a mobile device, such as browsing the web, editing images/video, working with documents, and so on.
2. **Swipe-and-Switch:** This workload switches the front-end application between multiple swipe-browsing applications (e.g., online shopping and news applications) and intersperses the switches with multiple screen swipes to simulate the user browsing through the application's content, allowing the application to quickly create a large number of new objects.
3. **Launch-and-Switch:** This workload launches multiple apps and switches to the next one after one is fully launched, terminates all apps when they are all launched, and repeats the process from the first step. Launching large apps causes the Android runtime to allocate a lot of memory to them in a short period of time, which is a more aggressive memory pressure behavior than the **Swipe-and-Switch**.

3.7 Data Processing Module

The data files obtained by the monitoring modules described in the previous section all hold raw data, which is either unprocessed log line text or statistical data strings dumped directly from the process file system. The PolyMetrica framework is designed with a module for data processing in respective output format for each performance metric.

4 Implementation

The implementation of the PolyMetrica framework is divided into three main parts, namely, *Performance Monitors*, *Workload Scripts*, and *Data Processor*. The Monitors and the Data Processor communicate through an intermediate *Data Storage* directory. The main process of the framework is independent of the three main modules and is used to call the module code at the right time to control the overall flow of the test.

PolyMetrica's main process and test loads on the host side are written in Python and PowerShell scripts, and the kernel swap information module is written in C.

For log and process information capture module, we develop the customizable interfaces based on python's decorator mechanism. For the log capture module, developers can pass keywords into the decorator to captured matched log entries; for the process information module, several *procfs* paths can be assigned to a workload function and corresponding process data will be recorded before & after the workload execution.

```
@logcat_collect(keywords=['GC freed', 'ProcessReferences'])
def do_benchmark(benchmark_name: str, output_dir: str):
    # run workload benchmark
```

The kernel swap information module is implemented based on function injection. The framework declares a function prototype for each function pointer used for injection in the kernel, so that developers can point these function pointers to their own function implementations of the same prototype during the initialization of kernel modules written by themselves.

We have placed injection points in the kernel in two key functions, *shrink_page_list* and *do_swap_page*, to capture swapping out and swapping in, respectively. Developers can perform custom operations at the injection point by implementing these two function prototypes:

```
// injection function declaration for swapout
bool polymetrica_swapout_instrumentor(
      struct page *page,     struct vm_area_struct *vma,
      unsigned long address, void *arg);

// injection function declaration for swapin
bool polymetrica_swapin_instrumentor(struct vm_fault *vmf);
```

PolyMetrica has also made efforts to merge data collected by different modules. For example, combining multiple layers of data requires harmonizing the

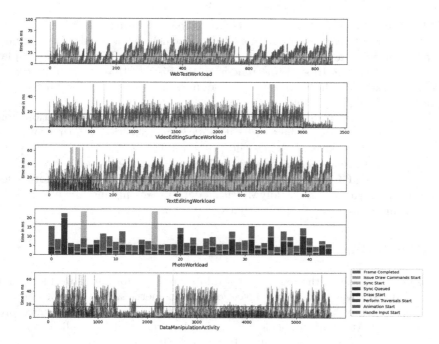

Fig. 3. Frame Rendering Details Generated by PolyMetrica

timestamps of individual events (e.g., GC and swap), but the Runtime and kernel layer logs use different timers. The main thread of PolyMetrica performs a "Sync" operation to record time interval between each system timer, thus the data processing module can merge events from different system layers into a single timeline.

5 Case Study on Workload

This section uses the PolyMetrica framework to test one of the three real-world application cases constructed in this paper, and generates corresponding test analysis reports for each scenario. Based on the charts provided in the report, this paper attempts to analyze the performance metrics during the operation of the workloads as a way of demonstrating the help of the PolyMetrica framework to system developers in perceiving the behavior of devices.

Figure 3 shows the visualization output from the PolyMetrica framework on the PCMark workload. The framework generates a bar chart that records the frame rendering time in chronological order for each of the work sets, from top to bottom: web browsing, video editing, text editing, image processing, and data visualization.

The different colored parts of the individual bars represent the different stages of a frame's rendering, the exact meanings of which are labeled in the legend. For a screen with a fixed refresh rate of 60 Hz, the condition for the display not to lag is that the total frame rendering time is less than 16.7 milliseconds, which is marked by a horizontal line in the image as a reference for whether the application is lagging or not.

The PolyMetrica framework combines the image rendering information from the application framework layer with the garbage collection information from the system runtime layer, labeling the frame rendering that occurs during the application garbage collection process with a **gray background**.

By combining GC information from the Runtime layer and Swap information from the Linux Kernel layer, PolyMetrica's data processing module also has the ability to filter the Swap behavior that occurs during GC.

6 Conclusion

In this paper, because the existing Android system performance profiling tools are difficult to cover the various levels of system metrics, and cannot combine these performance data together, we propose a multi-metrics performance profiling tool, PolyMetrica, targeting the Android system and its applications.

PolyMetrica serves as a profiling framework that collects performance metrics data from Android from a variety of sources, including graphics rendering information, memory page swap-in/out data, process data, and runtime garbage collection data.

PolyMetrica is also flexible and easy to expand. Developers can easily extend the PolyMetrica profiling framework to address different scenarios by decorating custom workload functions, modifying and reloading kernel modules, or adding data processing modules.

In addition, three test workloads were written to simulate real-world scenarios to evaluate the PolyMetrica framework's ability to perform performance analysis across system levels, with loads that can be customized by the developers.

The PolyMetrica framework provides Android system developers with performance statistics across the application framework layer, the system runtime layer, and the Linux kernel layer to find bottlenecks affecting the user experience of Android systems, which can then be used to inspire further system-level optimization.

References

1. AOSP: Introduction of simpleperf (2022). https://android.googlesource.com/platform/prebuilts/simpleperf/+/refs/heads/ndk-r13-release/README.md
2. AOSP: Android GPU inspector (AGI) (2023). https://developer.android.com/agi
3. AOSP: Architecture overview (2023). https://source.android.com/docs/core/architecture

4. Backes, M., Bugiel, S., Derr, E., McDaniel, P., Octeau, D., Weisgerber, S.: On demystifying the android application framework: re-visiting android permission specification analysis. In: Proceedings of the 25th USENIX Conference on Security Symposium, SEC 2016, pp. 1101–1118. USENIX Association, USA (2016)
5. Chae, D., et al.: CloudSwap: a cloud-assisted swap mechanism for mobile devices. In: Proceedings of the 16th IEEE/ACM International Symposium on Cluster, Cloud, and Grid Computing, CCGRID 2016, pp. 462–472. IEEE Press (2016). https://doi.org/10.1109/CCGrid.2016.22
6. contributors, E.L.W.: Android logging system (2022). https://elinux.org/Android_Logging_System
7. Developers, A.: What is android. Dosegljivo **1**, 1–24 (2011)
8. Gargenta, M.: Learning Android. O'Reilly Media Inc., California (2011)
9. Hagos, T., Hagos, T.: Android studio profiler. Android Studio IDE Quick Reference: A Pocket Guide to Android Studio Development, pp. 73–82 (2019)
10. Kim, J., Kim, C., Seo, E.: ezswap: enhanced compressed swap scheme for mobile devices. IEEE Access 1 (2019). https://doi.org/10.1109/ACCESS.2019.2942362
11. Lebeck, N., Krishnamurthy, A., Levy, H.M., Zhang, I.: End the senseless killing: improving memory management for mobile operating systems. In: Proceedings of the 2020 USENIX Conference on Usenix Annual Technical Conference, USENIX ATC 2020. USENIX Association, USA (2020)
12. Li, C., Shi, L., Liang, Y., Xue, C.: SEAL: user experience aware two-level swap for mobile devices. IEEE Trans. Comput.-Aided Design Integr. Circ. Syst. **39**, 1 (2020). https://doi.org/10.1109/TCAD.2020.3012316
13. Mauerer, W.: Professional Linux Kernel Architecture. Wiley, San Francisco (2010)
14. Omoruyi, O.: From 0-70 percent market share, how android gained and maintained the OS market leadership (2022). https://technext24.com/2022/09/09/how-android-gained-os-market-leadership/
15. Ongkosit, T., Takada, S.: Responsiveness analysis tool for android application. In: Proceedings of the 2nd International Workshop on Software Development Lifecycle for Mobile, DeMobile 2014, pp. 1–4. Association for Computing Machinery, New York (2014). https://doi.org/10.1145/2661694.2661695
16. Pufek, P., Grgic, H., Mihaljevic, B.: Analysis of garbage collection algorithms and memory management in java. In: 2019 42nd International Convention on Information and Communication Technology, Electronics and Microelectronics (MIPRO), pp. 1677–1682. IEEE, New York (2019). https://doi.org/10.23919/MIPRO.2019.8756844
17. Sadowska, P.: Android runtime - how dalvik and art work? (2021). https://proandroiddev.com/android-runtime-how-dalvik-and-art-work-6e57cf1c50e5
18. Sarkar, A., Goyal, A., Hicks, D., Sarkar, D., Hazra, S.: Android application development: a brief overview of android platforms and evolution of security systems. In: 2019 Third International conference on I-SMAC (IoT in Social, Mobile, Analytics and Cloud) (I-SMAC), pp. 73–79. IEEE, New York (2019). https://doi.org/10.1109/I-SMAC47947.2019.9032440
19. Singh, S., Gadgil, R., Chudgor, A.: Automated testing of mobile applications using scripting technique: a study on Appium. Int. J. Curr. Eng. Technol. (IJCET) **4**(5), 3627–3630 (2014). https://doi.org/10.1109/ICVRIS.2019.00068
20. Uhsadel, L., Georges, A., Verbauwhede, I.: Exploiting hardware performance counters. In: 2008 5th Workshop on Fault Diagnosis and Tolerance in Cryptography, pp. 59–67. IEEE, New York (2008). https://doi.org/10.1109/FDTC.2008.19

Towards Optimal Leakage Assessment of TVLA

Yuanqiao Bi[1], Weijian Li[2(✉)], and Guiyuan Xie[2]

[1] Engineering Training Center, Jianghan University, Wuhan, China
[2] School of Computer Science, Guangdong Polytechnic Normal University, Guangzhou, China
weijianlee@126.com

Abstract. Since side-channel attacks pose a very serious threat to cryptographic products, leakage assessment plays a key role in the evaluation of the physical security of cryptographic products before putting into the market. However, the widely used Common Criteria (CC) method is both costly and difficult to apply, while the Test Vector Leakage Assessment (TVLA) method fails to accurately detect key leakage points. In this paper, a feature selection technique in machine learning is introduced, with which the leakage points detected by TVLA are treated as features for supervised learning. According to their contribution to the classification, irrelevant features are removed, while the optimal features remained, which identify more important leakage points. We further extend the assessment from two groups to multiple groups, which achieves much better performance. Experimental results show that, compared with existing methods such as TVLA, χ^2-test, TVLA-Bonferroni and ANOVA test, our methods detect leakage points more accurately and stably, and greatly improve the evaluation performance in term of accuracy, leakage hit rate and false positive rate. Besides, our methods require much fewer power traces and shorter time for evaluation than existing methods.

Keywords: Side-channel Attacks · Leakage Evaluation · Feature Selection

1 Introduction

Side-channel attacks exploit physical information such as power consumption and electromagnetic radiation generated during the encryption process of a device to reveal the cryptographic key, posing a serious threat to cryptographic products. Therefore, cryptographic products must undergo side-channel leakage assessments before being released to the market. The CC standard (ISO/IEC 15408 [2]) and TVLA (ISO/IEC 17825 [1]) are the most widely used side-channel leakage assessment methods at present.

The CC standard is attack-oriented, which utilizes various existing side-channel attack methods to evaluate cryptographic products, e.g. Correlation

Power Analysis (CPA) [9], Template Attack (TA) [5], Mutual Information Analysis (MIA) [7] and various ML-Based SCA [3,11,18,19].

However, the CC evaluation method requires detailed knowledge of the cryptographic algorithm's implementation to identify sensitive intermediates, which is both costly and difficult to apply. Evaluation labs and academic researchers began to seek a fast, reliable and robust side-channel black-box evaluation approach. In 2015, Tobias Schneider et al. [13] proposed a clear roadmap for side-channel leakage assessment, including t-test, higher order t-test and their efficient implementations. In 2016, ISO/IEC 17825 [1] specified the TVLA framework as the standard method to assess whether cryptographic products are vulnerable to side-channel attacks. In 2016, Adam Ding et al. [8] proposed a paired t-test to reduce environmental noise effect. In 2018, Amir Moradi et al. [10] described Pearson's χ^2-test as a supplement to Welch's t-test, and combined two tests to mitigated the limitations of the evaluation of higher-order masked implementations. In 2019, Whitnall and Oswald et al. [14] proposed a method to calculate the number of evaluated power traces based on the effect size of the power traces. Afterward, they proposed TVLA-Bonferroni method to reduce the false positive of the evaluation result. In 2021, Yang and Jia et al. [17] proposed a new black-box leakage detection approach to enable the test among multiple groups by using the one-way analysis of variance (ANOVA).

TVLA methodology is based on the fact that side-channel attacks exploit the dependency between the instantaneous power consumption and the intermediate value being processed, any statistic difference between two groups of power traces may point out the potential side-channel vulnerability. The non-specific t-test facilitates black-box evaluation without requiring detailed knowledge of the cryptographic algorithm, making it easier to apply and more robust. However, it is insufficient to advise the designers and manufacturers on best practice in the following aspects.

1) Failure to provide key leakage points: The non-specific TVLA is able to check if a cryptographic implementation is leaking or not, but it is unable to provide key leakage points.
2) Reliability issues of evaluation results: There is a great risk of false positives and negatives in the evaluation results of non-specific TVLA. In ASIACRYPT 2019, Whitnall and Oswald [14] pointed out that the false positives will increase as the number of sample points grows.
3) Efficiency issues of evaluation: A practical evaluation with too few power traces is not able to detect the leakage, while evaluation with too many power traces will reduce the efficiency of evaluation.

This paper introduces feature selection technique to improve the TVLA leakage assessment. The main contributions of this paper are as follows:

1) Feature selection technique is introduced to optimize TVLA methodology. Leakage points detected by TVLA are treated as features for supervised learning and classification. After leakage points are ranked according to their contributions to the classification, irrelevant features are moved, and the optimal

feature subset is obtained as the most important leakage points. Our method greatly improves the accuracy and efficiency of evaluation.
2) TVLA is evaluated between two groups of power traces, which usually results in some undetected leakage points of the same intermediate values during encryption. The ANOVA-based feature selection method is proposed to test among multiple groups of power traces, which provides more precisely and comprehensive leakage information.
3) We propose three performance criteria for leakage assessment, i.e., accuracy, leakage hit rate and false positive rate. Subsequently, we give a detailed comparison with existing evaluation methods including TVLA, χ^2-test, TVLA-Bonferroni and ANOVA test.

The remaining paper is organized as follows: Sect. 2 introduces the existing evaluation methods and feature selection technique. In Sect. 3, we propose both TVLA and ANOVA test optimization framework and implementations based on feature selection technique. Section 4 discusses the experimental results. Section 5 concludes this paper.

2 Preliminaries

2.1 Test Vector Leakage Assessment

Based on the Welch's t-test, TVLA analyses the mean difference between two groups of power traces collected on the device under test (DUT). A sample point will be regarded as leakage point if there is a significant difference between two groups. Let μ_A and μ_B denote the mean values of sample point population A and population B. The null hypothesis and alternative hypothesis of Welch's t-test are written as

$$\begin{cases} H_0 : \mu_A = \mu_B \\ H_1 : \mu_A \neq \mu_B \end{cases}. \tag{1}$$

Let x_A and x_B denote two sample sets from A and B respectively. Let also $\overline{x_A}$ (resp. $\overline{x_B}$) and s_A^2 (resp. s_B^2) stand for the sample mean and sample variance for the set x_A (resp. x_B). Let n_A and n_B indicate the cardinality of each set. The t-test statistic and the degree of freedom v are computed as

$$t = \frac{\overline{x_A} - \overline{x_B}}{\sqrt{\frac{s_A^2}{n_A} + \frac{s_B^2}{n_B}}}, v = \frac{\left(\frac{s_A^2}{n_A} + \frac{s_B^2}{n_B}\right)^2}{\frac{\left(\frac{s_A^2}{n_A}\right)^2}{n_A - 1} + \frac{\left(\frac{s_B^2}{n_B}\right)^2}{n_B - 1}}. \tag{2}$$

Based on the degree of freedom v, the student's t distribution function is drawn as

$$f(t, v) = \frac{\Gamma\left(\frac{v+1}{2}\right)}{\sqrt{\pi v} \Gamma\left(\frac{v}{2}\right)} \left(1 + \frac{t^2}{v}\right)^{-\frac{v+1}{2}}, \tag{3}$$

where $\Gamma(\cdot)$ denotes the gamma function. Based on the two-tailed Welch's t-test the desired probability p is computed as

$$p = 2 \int_{|t|}^{\infty} f(t,v)\, dt. \tag{4}$$

For simplicity in practice, usually a threshold $|t| > 4.5$ is defined to determine whether there is a leakage [13]. This intuition is based on the fact that $p(|t_{v>1000}| > 4.5) < 0.00001$, which leads to a confidence level greater than 0.99999 to reject the null hypothesis. At this time, the confidence level that the means of the two sample sets are significantly different is 99.999%. Therefore, it is considered that there is a leakage at the sample point.

Whitnall and Oswald [14] pointed out that the results of TVLA are affected by the devices, and there are very high false positives in the evaluation results. They proposed a TVLA-Bonferroni correction method to reduce the significance level of Welch's t-test to mitigate false positives by

$$\alpha_{per-test} = \frac{\alpha_{overall}}{N}, \tag{5}$$

where $\alpha_{overall}$ denotes the overall significance level, $\alpha_{per-test}$ indicates the significance level of a single test, and N represents the number of sample points.

2.2 Leakage Detection with ANOVA

Analysis of variance (ANOVA) tests whether the mean values of multiple populations are equal or not by analysing the data error of each power trace. The null hypothesis and alternative hypothesis of n classifications are written as

$$\begin{cases} H_0 : \mu_1 = \mu_2 = \cdots = \mu_n \\ H_1 : \mu_1, \mu_2, \cdots, \mu_n \text{ not all equal} \end{cases}. \tag{6}$$

Let $x_1, x_2, \cdots, x_n$ denote n sample sets from n populations respectively, h_i represents the sample size of the i-th group of samples x_i, where $i = 1, 2, \cdots, n$ and $h = h_1 + h_2 + \cdots + h_n$. Let $x_{i,j}$ stand for the j-th individual of the i-th group of samples. The calculation formulas for the within-group variance SS_w and the between-group variance SS_b are as follows:

$$SS_w = \sum_{i=1}^{n} \sum_{j=1}^{h_i} (x_{ij} - \overline{x_i})^2, \quad SS_b = \sum_{i=1}^{n} \left(\overline{x_i} - \sum_{i=1}^{n} \frac{h_i \overline{x_i}}{h} \right)^2, \tag{7}$$

where $\overline{x_i}$ indicates the sample mean of the i-th group of samples.

The sample point is considered a leakage point if its F-statistic exceeds 1.4 [17], where F is calculated as

$$F = \frac{(h-n)SS_b}{(n-1)SS_w} \sim F(n-1, h-n). \tag{8}$$

2.3 Feature Selection

Feature selection is an important pre-processing technique in the field of machine learning, which greatly reduces the data scale and maintains the information of the original feature set. In order to balance accuracy and efficiency, feature selection practically consists of three stages, i.e., filter feature selection stage, wrapper feature selection stage, and result verification stage.

In stage 1, a filter feature selection method which does not depend on the classifier is used to quickly reduce features of the training dataset. In stage 2, candidate feature subsets are generated based on a specific search strategy. Each candidate feature subset will be evaluated by a classifier, and the former optimal feature subset will be replaced if the new one performs better. The process of subset generation and evaluation iterates until the termination condition is satisfied. In stage 3, the test set is classified based on the optimal feature subset, and the performance of the optimal feature subset is evaluated according to the classification accuracy.

3 The Proposed Methods

One of the reasons that TVLA detects a large number of leakage points is due to the diffusion operations of encryption, which distribute intermediate values after a couple of rounds and provoke a strong diversification in the power consumption. Our methods regard leakage points detected by TVLA as features, and introduce feature selection technique to find the most important leakage points.

3.1 SVM-RFE Leakage Assessment

As shown in Fig. 1, framework of TVLA-SVMRFE leakage assessment includes four stages. In stage 1, two groups of power traces corresponding to a fixed plaintext and random plaintexts respectively are collected. Subsequently, power traces are divided into labelled training samples T and test samples T'. In stage 2, TVLA method is performed as filter feature selection method on the training samples T, which reserves the sample points with statistics $\mid t \mid > 4.5$, and removes other irrelevant sample points. In stage 3, recursive feature elimination (RFE) algorithm is used to generate candidate feature subsets, and support vector machine (SVM) is used as a classifier to evaluate candidate feature subsets. Candidate feature subset with the highest classification accuracy is regarded as the optimal feature subset. In stage 4, the classification accuracy of test samples T' is obtained based on the optimal feature subset. We determine whether there is leakage according to the accuracy and regard the optimal feature subset as the most important leakage points if leaking.

SVM-RFE algorithm is a sequence backward feature selection algorithm based on SVM, which gradually eliminates the features that contribute the least to the classification to find the optimal feature subset. It has been widely used in the pre-processing of various complex classification problems [4,12,15]. RFE

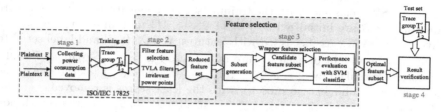

Fig. 1. Framework of TVLA-SVMRFE leakage assessment

method builds the model based on the original feature set and calculates the weight of each feature. Then the feature with the least absolute weight is eliminated, and the process is recursively executed until the optimal feature subset is found.

Numerous highly correlated features exist in the feature set detected by TVLA, which leads to underestimating the importance of these features in SVM-RFE. We use a correlation bias reduction (CBR) strategy [16] to solve this problem.

The main idea of our method is as follows:

1) Collecting power traces. The FRRF crossover order is used in this paper, which is able to reduce the sampling error [8]. The training power traces T and the test power traces T' are collected as

$$\begin{cases} T_n = [t_1^n; t_2^n; \cdots ; t_K^n], n = 1, 2 \\ t_j^n = \{t_{j1}^n, t_{j2}^n, \cdots, t_{jN}^n\}, 1 \leq j \leq K \end{cases}, \quad (9)$$

$$\begin{cases} T'_n = [t_1^{'n}; t_2^{'n}; \cdots ; t_{K'}^{'n}], n = 1, 2 \\ t_j^{'n} = \{t_{j1}^{'n}, t_{j2}^{'n}, \cdots, t_{jN}^{'n}\}, 1 \leq j \leq K' \end{cases}. \quad (10)$$

T contains two groups T_1 and T_2 corresponding to a fixed plaintext and random plaintexts respectively, each group has K power traces, and each power trace t_j^n contains N sample points. T' contains two group T'_1 and T'_2 as well, each group has K' power traces.

2) Evaluation of TVLA. Evaluation of TVLA is performed according to the Eq. (2). The sample points whose t-statistics are greater than the threshold value of 4.5 are reserved to form the feature set X_0. Let M denote the number of the reserved sample points. Y denotes the label set of training power traces, where power traces from fixed group T_1 are labelled 1, while power traces from random group T' are labelled 0. Training set $D = \{X_0, Y\}$ is recorded as

$$\begin{cases} X_0 = [x_1^1; x_2^1; \cdots ; x_K^1; x_1^2; x_2^2; \cdots ; x_K^2], x_j^n = \{x_{j1}^n, x_{j2}^n, \cdots, x_{jM}^n\} \\ Y = [y_1^1; y_2^1; \cdots ; y_K^1; y_1^2; y_2^2; \cdots ; y_K^2], where \ y_j^1 = 1, y_j^2 = 0 \\ n = 1, 2; j = 1, \cdots, K \end{cases}, \quad (11)$$

where X_0 is composed of T_1 and T_2 with M features standardized.

3) Training SVM classifier with the training set D. The feature set f and feature ranking set R are initialized as $f = \{1, 2, \cdots, M\}$ and an empty set respectively. The SVM classifier is fed with features $X = X_0(:, f)$ and label set Y to compute the weight of each feature by equation $w = \sum_{k=1}^{2K} \alpha_k d_k y_k$. Finally, features are ranked by $c_i = w_i^2, i = 1, 2, \cdots, |f|$, where $|f|$ is the size of the feature set f. The feature subset q with the least scores is removed.
4) Reducing the correlation bias. A group size threshold T_g and the correlation threshold T_c are set. Let feature m denote the highest score in q, q' is set to $q - m$ if the cardinality of the set $\{i \in q \mid\mid corr(i, m) \mid > T_c\}$ is greater than T_g and the cardinality of the set $\{j \in f - q \mid\mid corr(j, m) \mid > T_c\}$ is equal to 0, otherwise q' is set to q.
5) Updating iteratively. The feature ranking set is updated by $R = [R, q']$, and the feature set is updated by $f = f/q'$. If f is not empty, go to step 3.
6) Selection of optimal feature subset. Nested feature subsets $F_1 \subset F_2 \subset \cdots \subset F_M$ are generated based on R. Training set is classified by SVM based on these nested feature subsets. The feature subset with the highest classification accuracy is selected as the optimal feature subset S_{best}.
7) Verification of optimal feature subset. The test set D' is constructed from the power traces T' based on S_{best}, and classified by SVM. The high classification accuracy means that the implementation is leaking, and S_{best} points out the most important leakage points.

3.2 ANOVA-SVMRFE Leakage Assessment

The combination of TVLA and feature selection method greatly improves the accuracy of leakage assessment compared with hypothesis testing methods, but its accuracy and leakage hit rate still need to be improved. One of the reasons is that TVLA is evaluated with two groups of plaintexts, which leads to some undetected leakage points of the same intermediate values during encryption. ANOVA is applied to test whether the mean values of multiple populations are equal or not, which is an alternative of TVLA to extend the assessment from two groups to multiple groups. The framework of ANOVA-SVMRFE leakage assessment is shown in Fig. 2.

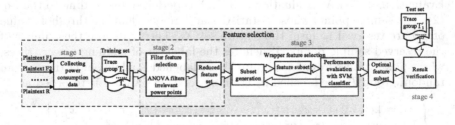

Fig. 2. Framework of ANOVA-SVMRFE leakage assessment

In stage 1, the $F_1, F_2, \cdots, F_{n-1}, R$ crossover order is selected, which effectively improves the convergence speed of the evaluation. n groups of power traces corresponding to $(n-1)$ fixed plaintexts and random plaintexts are alternately collected and divided into training and test power traces as well.

In stage 2, F-statistics are calculated by Eqs. (8) for each sample point, which will be reserved as leakage point if its statistic value $F > 1.4$.

Stages 3 and 4 are performed in the same way as the stages described in Sect. 3.1.

As demonstrated in Sect. 4, the ANOVA-SVMRFE leakage assessment requires fewer power traces for analysis and significantly improves the accuracy, reduces the false positive rate, and enhances the stability of the evaluation.

4 Experimental Results and Discussion

In our experiments, TVLA, χ^2-test, TVLA-Bonferroni adopt the $FRRF$ cross-collection method, with which the fixed plaintext F is set to

- 0x0123456789abcdeffedcba9876543210 specified in ISO/IEC 17825 standard [1], and the random plaintexts R are randomly generated.

The ANOVA method adopts the $F_1 F_2 F_3 R$ cross-collection method, with which F_1, F_2 and F_3 are set to respectively

- 0x0123456789abcdeffedcba9876543210,
- 0xda39a3ee5e6b4b0d3255bfef95601890 recommended by Becker et al. [6],
- 0xc5904800f7b296431b8a8900586ce70f.

All the keys are set to

- 0x0123456789abcdef123456789abcdef0.

The amount of power traces evaluated in our experiments is 250000, which is computed by the method proposed by Whitnall and Oswald [14].

Our experimental setup includes a side-channel standard evaluation board SAKURA-G, an oscilloscope and a computer. The SAKURA-G is designed for evaluation of hardware security and equipped with two independent Spartan-6 FPGA chips.

In order to compare the performance of different evaluation methods, it is necessary to identify the practical leakage points of the cryptographic implementation. In this paper, the leakage points obtained by the CC standard are regarded as the practical leakage points, and the classical CPA method is applied to attack the unprotected AES parallel implementation on SAKURA-G to acquire these practical leakage points, as illustrated in Fig. 3(a).

Leakage points detected by leakage assessment of TVLA, χ^2-test, TVLA-Bonferroni, ANOVA and our methods are compared in Fig. 3(b)–Fig. 3(g), where red points indicate leakage points detected. The numbers of leakage points

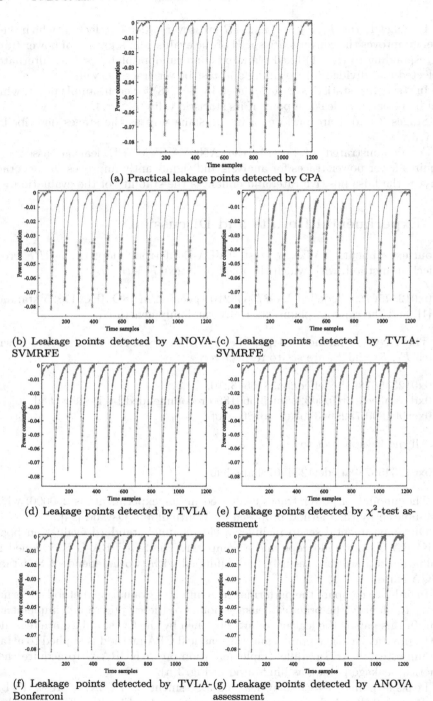

Fig. 3. Practical leakage points detected by CPA and leakage points detected by leakage assessment method with 250,000 power traces of AES implementation

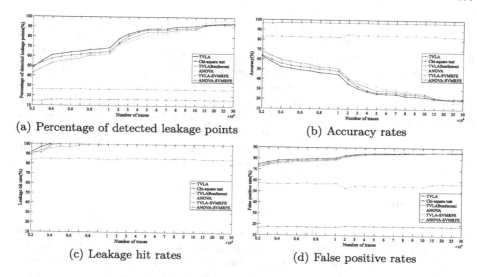

Fig. 4. Performance of different evaluation methods as number of power traces grows

detected by leakage assessment of TVLA, χ^2-test, TVLA-Bonferroni, ANOVA, TVLA-SVMRFE and ANOVA-SVMRFE are 1127, 1116, 1112, 1164, 317 and 196 respectively. Both the leakage points detected by TVLA-SVMRFE and ANOVA-SVMRFE are very close to the practical leakage points, whose performance is far better than other methods.

As illustrated in Fig. 4(a), the percentages of leakage points detected by leakage assessment methods based on hypothesis testing, i.e., TVLA, χ^2-test, TVLA-Bonferroni and ANOVA, gradually increase to almost 100% as the number of power traces grows. Therefore, the methods are unable to provide key leakage information. While the percentage of leakage points detected by TVLA-SVMRFE and ANOVA-SVMRFE always stabilize at about 26% and 16%.

Subsequently, the accuracy, leakage hit rate and false positive rate are used to compare the performance of leakage assessment methods. As shown in Fig. 4(b), the accuracy rates of leakage assessment methods based on hypothesis testing gradually decrease to 21% as the number of power traces grows. However, the accuracy rates of TVLA-SVMRFE and ANOVA-SVMRFE are always stable around 82% and 97%. Our proposed methods achieve respectively 61% and 76% improvement in accuracy compared to other methods.

As illustrated in Fig. 4(c), the leakage hit rates of leakage assessment methods based on hypothesis testing reaches 100% after more than 4000 traces are evaluated, because large number of leakage points, approximately 50%-100% of sample points, are detected to include all the practical leakage points, which means extra high false positive rate. However, TVLA-SVMRFE achieves 85% leakage hit rate under only 26% leakage points detected. While ANOVA-SVMRFE detects 16% sample points as leakage points and still reaches 100% leakage hit rate.

As shown in Fig. 4(d), false positive rates of leakage assessment methods based on hypothesis testing are higher than 70% after more than 2000 traces are evaluated, while false positive rates of TVLA-SVMRFE and ANOVA-SVMRFE are lower than 60% and 20%. Our proposed methods achieve respectively 28% and 67% reduction in false positive rate compared to other methods.

Based on the analysis of the above experimental results, the leakage points detected by our methods are similar to the practical leakage points. The performance is much better than existing leakage assessment methods in term of accuracy, leakage hit rate and false positive rate. Besides, our methods require few power traces, and keep stable as the number of traces grows.

5 Conclusions

In this paper, to provide best practice advice to designers and manufacturers of cryptographic products, we introduce the feature selection technique in machine learning into the field of side-channel leakage assessment to address the issues with leakage assessment methods based on hypothesis testing. A TVLA-SVMRFE leakage assessment is proposed to detect more precise leakage points, which also offers the advantages of being easy to apply, highly accurate, and efficient.

This paper further extends the assessment from two groups to multiple groups by ANOVA-SVMRFE test to greatly improve the performance. In addition, performance criteria including accuracy, leakage hit rate and false positive rate are proposed to compare the performance of different evaluation methods. Experimental results show that our proposed TVLA-SVMRFE and ANOVA-SVMRFE methods achieve respectively 61% and 76% improvement in accuracy, 28% and 67% reduction in false positive rate compared to existing leakage assessment methods. Meanwhile, our methods achieve 85% and 100% leakage hit rates under only 26% and 16% leakage points detected, while conventional methods reach 100% leakage hit rate under more than 50% detected leakage points.

Acknowledgments. This work was supported by Guangdong Provincial Department of Education Key Field Project for Ordinary Universities under Grant No.2023ZDZX1008 and No.2020ZDZX3059, Guangzhou key field research and development plan project under Grant No. 202206070003, General Project of the National Natural Science Foundation of China under Grant No. 62072123, Key Research and Development Program of JiangXi Province under Grant No. 20212BBE53002, Key Research and Development Program of YiChun City under Grant No. 20211YFG4270.

References

1. Information technology - security techniques - testing methods for the mitigation of non-invasive attack classes against cryptographic modules. International Organization for Standardization (2016)
2. Common criteria for information technology security evaluation part 1: Introduction and general model version 3.1 release 5. NIST (2017)

3. Alam, M., Bhattacharya, S., Mukhopadhyay, D.: Victims can be saviors: a machine learning-based detection for micro-architectural side-channel attacks. ACM J. Emerg. Technol. Comput. Syst. (JETC) **17**(2), 1–31 (2021)
4. Albashish, D., Hammouri, A.I., Braik, M., Atwan, J., Sahran, S.: Binary biogeography-based optimization based SVM-RFE for feature selection. Appl. Soft Comput. **101**, 107026 (2021)
5. Choudary, M.O., Kuhn, M.G.: Efficient, portable template attacks. IEEE Trans. Inf. Forensics Secur. **13**(2), 490–501 (2017)
6. Cooper, J., et al.: Test vector leakage assessment (TVLA) methodology in practice. In: International Cryptographic Module Conference, vol. 20 (2013)
7. de Chérisey, E., Guilley, S., Rioul, O., Piantanida, P.: Best information is most successful. Cryptology ePrint Archive (2019)
8. Ding, A.A., Chen, C., Eisenbarth, T.: Simpler, faster, and more robust t-test based leakage detection. In: Standaert, F.-X., Oswald, E. (eds.) COSADE 2016. LNCS, vol. 9689, pp. 163–183. Springer, Cham (2016). https://doi.org/10.1007/978-3-319-43283-0_10
9. Fahd, S., Afzal, M., Abbas, H., Iqbal, W., Waheed, S.: Correlation power analysis of modes of encryption in AES and its countermeasures. Future Gener. Comput. Syst. **83**, 496–509 (2018)
10. Moradi, A., Richter, B., Schneider, T., Standaert, F.-X.: Leakage detection with the x2-test. IACR Trans. Cryptographic Hardware Embed. Syst. 209–237 (2018)
11. Perin, G., Chmielewski, Ł., Picek, S.: Strength in numbers: Improving generalization with ensembles in machine learning-based profiled side-channel analysis. IACR Trans. Cryptographic Hardware Embed. Syst. 337–364 (2020)
12. Sanz, H., Valim, C., Vegas, E., Oller, J.M., Reverter, F.: SVM-RFE: selection and visualization of the most relevant features through non-linear kernels. BMC Bioinform. **19**(1), 1–18 (2018)
13. Schneider, T., Moradi, A.: Leakage assessment methodology. In: International Workshop on Cryptographic Hardware and Embedded Systems, pp. 495–513. Springer (2015)
14. Whitnall, C., Oswald, E.: A critical analysis of ISO 17825 ('testing methods for the mitigation of non-invasive attack classes against cryptographic modules'). In: Galbraith, S.D., Moriai, S. (eds.) ASIACRYPT 2019. LNCS, vol. 11923, pp. 256–284. Springer, Cham (2019). https://doi.org/10.1007/978-3-030-34618-8_9
15. Xue, Y., Zhang, L., Wang, B., Zhang, Z., Li, F.: Nonlinear feature selection using gaussian kernel SVM-RFE for fault diagnosis. Appl. Intell. **48**(10), 3306–3331 (2018)
16. Yan, K., Zhang, D.: Feature selection and analysis on correlated gas sensor data with recursive feature elimination. Sens. Actuators B Chem. **212**, 353–363 (2015)
17. Yang, W., Jia, A.: Side-channel leakage detection with one-way analysis of variance. Secur. Commun. Netw. **2021** (2021)
18. Zaid, G., Bossuet, L., Dassance, F., Habrard, A., Venelli, A.: Ranking loss: maximizing the success rate in deep learning side-channel analysis. IACR Trans. Cryptographic Hardware Embed. Syst. 25–55 (2021)
19. Zhang, L., Dejun, M., Wei, H., Tai, Yu.: Machine-learning-based side-channel leakage detection in electronic system-level synthesis. IEEE Netw. **34**(3), 44–49 (2020)

Joint UAV Trajectory Optimization and Task Offloading in Integrated Air-Ground Networks

Yuanwei Zhang, Jiaqi Shuai, Weichang Wen, and Haixia Cui[✉]

School of Electronics and Information Engineering, South China Normal University, Foshan 528200, China
2022024890@m.scnu.edu.cn

Abstract. As an essential component of the integrated air-ground-space networks, the unmanned aerial vehicle (UAV) is indispensable for the mobile edge computing (MEC) processing. In this paper, considering the highly dynamic nature of information exchange, we apply the multi-agent reinforcement learning (MARL) for the UAV-based communication networks and introduce a low-complexity K-means algorithm for the pre-cluster/partition users to improve the system performance. We devise a general reward and punishment strategy within the multi-agent reinforcement learning process to generate the optimal trajectories and promote the coverage maximization. Furthermore, we utilize NOMA technology to provide stable data rate gains and allocate more power to weaker users under optimizing energy consumption so as to maximize the overall system throughput. Simulation results demonstrate that the proposed joint optimization algorithm achieves excellent convergence performance in UAV based integrated air-ground-space networks.

Keywords: UAV · K-means algorithm · multi-agent reinforcement learning

1 Introduction

With the application of large-scale edge computing tasks and the deployment of AI large language models, challenges arise. Fixed edge servers have limited computing resources and coverage. As a network device in mobile edge computing [1] deployments, UAV can effectively complement the growing computational demands as highly flexible mobile small base stations and offer lots of advantages, such as more economical, easier to maintain, and environment friendly. Particularly, they provide highly on-demand services for maritime and aerial users, which play a crucial role in densely populated ground user scenarios and facilitate applications such as emergency communication and enhanced coverage. As an integral part of the integrated space-air-ground network [2], the UAV

Y. Zhang and J. Shuai—They contribute equally to this work.

© The Author(s), under exclusive license to Springer Nature Singapore Pte Ltd. 2025
W. Zhang et al. (Eds.): APWeb-WAIM 2024, CCIS 2246, pp. 212–226, 2025.
https://doi.org/10.1007/978-981-96-0055-7_18

communication network (UCN) [3] involves numerous dynamic factors, including trajectories, user mobility, and channel instability [4] which make the traditional optimization methods challenging to be applied. However, the reinforcement learning [5] can adapt to the dynamic environments through continuous interaction. Due to the complex interaction between UAVs and ground users, the traditional optimization methods often struggle to address such highly interactive systems. Additionally, we combine the UAV network with the V2X scenario in vehicular networks to verify the effectiveness of UAV mobility coverage and to improve the timeliness of computational tasks [6].

Reinforcement learning allows agents to learn optimal strategies through interactions with the environment in complex problems, such as resource allocation, route selection, and power control which can be learned directly from data without the need for complex mathematical models. Furthermore, multi-agent reinforcement learning (MARL) [7] allocates learning tasks to individual UAVs, allowing each UAV to train its strategy based on its local observations and lightweight information exchange with other UAVs [8]. Cui et al. [9] studied a joint user selection, power, and subchannel allocation problem using Multi-Agent Q-learning (MA-QL), where UAVs indirectly interact with each other through user selection. Wang et al. [10] proposed a multi-UAV coverage problem under a multi-agent Deep Q-learning (DQL) framework, where rewards are shared among UAVs, and compared convergence under static and dynamic user scenarios. Authors in [11] compared orthogonal and non-orthogonal spectrum utilization with multiple UAVs involved, maximizing the minimum throughput of cellular edge users by jointly optimizing spectrum allocation, coverage radius, and the number of UAVs.

Differing from the existing work, this paper considers a general policy design about rewards and punishments for UAVs as individual agents in multi-agent reinforcement learning, on top of addressing the challenges of ground user mobility and utilizing NOMA technology to enhance the network spectrum efficiency [12]. This implies that most research strategies are determined based on specific network environment problems and goals, and some works even lack specific punishment strategies. Effective reward and punishment policies can significantly improve the overall performance of the network and the utilization of network resources. We comprehensively consider factors such as communication performance, system stability, and energy efficiency to achieve the optimization of system performance. Main contributions are summarized as follows.

- We investigate the deployment of UAVs in three-dimensional space of integrated air-ground network communication models. In environments characterized by dense and dynamic ground users, UAVs provide computational services for task offloading to clusters of associated users. Additionally, when the volume of task data requested by user clusters exceeds the computational limits of UAVs, we consider UAVs acting as relay nodes to transmit data to low Earth orbit satellites and cloud processing centers to ensure long-distance communication. The goal is to maximize throughput of the system network under stable operation of the UAV fleet.

- We employ a combination of the heuristic algorithm k-means and the reinforcement learning algorithm MA-DQN to jointly optimize the dynamic trajectories and energy consumption of multiple UAVs. The k-means clustering algorithm periodically associates UAVs with user clusters and partitions to reduce interference, decrease computational complexity, and save unnecessary exploration time and kinetic energy consumption of UAVs. Concurrently, NOMA technology is introduced to allocate user transmit power, ensuring energy efficiency of UAV communication. The MA-DQN algorithm guides UAVs to make optimal action decisions, maximizing effective coverage of the UAV fleet.
- By balancing the rewards and penalties in MA-DQN algorithm, the rewards are given to the agents when the data needs to be transmitted to low Earth orbit satellites or cloud servers, while the penalties are imposed when the distance between UAVs is too small, avoiding service loss for users with smaller task demands due to the large task demands of some users. Simulation results indicate that the proposed approach effectively reduces energy consumption during task execution, accelerates learning training convergence, achieves optimal trajectory design, and enhances the robustness of the system network compared to the traditional DQN algorithm.

2 System Model

The illustration in Fig. 1 presents an example of computation offloading for IoT user devices in the integrated air-ground network. The green dashed lines represent data exchanges where there is a high level of data exchange, while the red dashed lines represent areas where there is less data exchange due to high computational resource utilization. The data processing methods for real-time communication can generally be divided into two parts: the first part involves local processing by IoT user devices, while the second part entails offloading the processing to servers with more powerful computing resources, which then return the processed tasks to the IoT user devices requesting them. Drones serve as auxiliary cellular offloading in the latter part, offloading task data for associated users within their coverage area via drone networks, and then executing offloading and resource allocation strategies for users based on the current network resources and task demands. Given the broad distribution of users served by drone services and the varying distances between users and drones, the channel conditions for each user may vary significantly. Therefore, we employ Non-Orthogonal Multiple Access (NOMA) technology here to provide stable data rate gains, ensuring that more power is allocated to weaker users to enhance power efficiency for drones under energy-limited conditions. Additionally, due to inherent hardware performance limitations of drones, if computational tasks exceed their payload limits, they need to be retransmitted to low Earth orbit satellites with more abundant computing resources or to more distant cloud server centers for processing, serving as relay nodes to ensure drone endurance and load balancing. This type of multi-terminal collaborative distributed computing can dynamically allocate tasks to the optimal execution endpoints based on the capabilities of the devices and the requirements of the current tasks, thereby improving overall system efficiency.

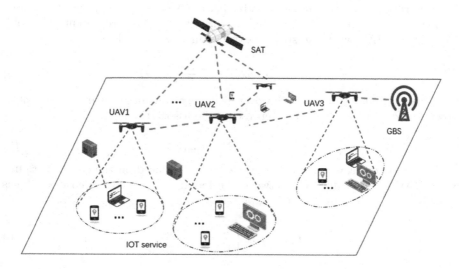

Fig. 1. Network Model

2.1 Communication Model

In our communication model, we adopt a cellular network with Non-Orthogonal Multiple Access (NOMA) technology, where each unmanned aerial vehicle (UAV) is equipped with a single antenna, catering to the downlink transmission in outdoor scenarios with dynamically distributed users. Hence, the channel gain between each user p and UAV u is calculated based on the positions of the UAVs and users that is:

$$g_p^u(t) = \beta \cdot 10^{-L_p^u/10} \tag{1}$$

In the equation, β represents the fading coefficient between UAV u and user p, and L_p^u denotes the path loss in free space, formulated as follows:

$$L_p^u = 32.45 + 20F_c + 20D_{up} \tag{2}$$

In the above equation, F_c represents the carrier frequency, and D_{up} denotes the Euclidean distance between UAV u and user p. The symbol D_{up} expressed as,

$$D_{up} = \sqrt{[x_u(t) - x_p^u(t)]^2 + [y_u(t) - y_k^u(t)]^2 + h_u^2(t)} \tag{3}$$

2.2 Computation Model

The calculation model is divided into two parts: data transmission volume and energy consumption. In particular, $r_{up}(t)$ denotes the data transmission amount, where B represents the channel bandwidth. $SINR_{up} = \frac{P_{user}g_{up}(t)}{\delta^2}$ The transmission power of each UAV is denoted as user P_{user}, and δ represents the noise power of the UAV, with the specific expression given as follows:

$$r_{up}(t) = Blog2(1 + \frac{P_{user}g_{up}(t)}{\delta^2}) \tag{4}$$

For each user's time slot, the overall energy consumption expression is as shown in Eq. (5), where up T_{up} represents the association time.

$$E_{up}(t) = P_{user}T_{up}(t) \tag{5}$$

The successful coverage of the mobile device is based on the coverage radius of the UAV and the distance between the mobile device, so the coverage basis of the device in the time slot t is expressed as

$$C_{p,u}(t) = \begin{cases} 0, & d_p > r_u \\ 1, & d_p < r_u \end{cases} \tag{6}$$

2.3 Mobility Models and User Association

The user mobility model [13] in this paper adopts a random walk model, where users move aimlessly. The movement direction and speed within each time slot t are completely random. Vmin and Vmax represent the minimum and maximum speeds of user terminals, $v(t) \in [Vmin, Vmax]$. respectively. The movement direction φ is randomly chosen from the range $[0, 2\pi]$. Therefore, for each user p, the velocity vector is expressed as $[v(t)cos(\varphi(t)), v(t)sin(\varphi(t))]$. We denote the set of user-UAV associations as $\{O = o_{u,p}(t), t = 1, \ldots, T\}$.

3 Problem Formulation

Our objective is to maximize the overall throughput by implementing optimal trajectory control strategies for UAVs under spatial constraints, minimum energy consumption constraints, and Quality of Service (QoS) constraints. We assume that the UAVs' velocities are fixed. Therefore, the optimization problem can be formulated as follows:

$$\max_{O} R = \sum_{t=0}^{T}\sum_{u=1}^{U}\sum_{p=1}^{p} r_{up}(t) \tag{7}$$

$$\text{s.t. } x_{min} \leq x_{u(t)} \leq x_{max}, \forall u, \forall t \quad \text{(C1)}$$
$$y_{min} \leq y_{u(t)} \leq y_{max}, \forall u, \forall t \quad \text{(C2)}$$
$$h_{min} \leq h_{u(t)} \leq h_{max}, \forall u, \forall t \quad \text{(C3)}$$
$$u_i \neq u_j, \ i,j \in U, \forall t \quad \text{(C4)}$$
$$P_{user}(t) \leq P_u, \forall user, \forall t \quad \text{(C5)}$$

In Eq. (6), constraints C1-C3 represent the three-dimensional position constraints of the UAVs, constraint C4 ensures that there are no collisions between UAVs, and constraint C5 ensures that the power allocation of each UAV does not exceed the upper limit of the transmission power. The UAV and user movements involved in Problem (6) are highly dynamic. Therefore, this problem is a non-linear constraint mixed integer non-linear non-convex problem. Traditional convex optimization algorithms are not suitable for solving such NP-hard problems. For such problems, we typically need to use Deep Reinforcement Learning (DRL) methods, which are capable of making sequential decisions in dynamic environments, to address them. It is worth noting that while multi-agent reinforcement learning methods can guarantee convergence, they do not guarantee optimality. Therefore, we can combine heuristic algorithms to further optimize this multi-agent framework, ensuring that the distributed execution of learning and strategies has better scalability.

4 Algorithm Design

In this section, we outline the implementation process of the K-means algorithm and the MA-DQN algorithm, as well as how to integrate these two algorithms to address our target problem. The approach is divided into two steps: firstly, employing the K-means clustering algorithm to determine the user association clusters for UAV-assisted offloading, and subsequently proposing the MA-DQN algorithm to realize the action trajectories of UAVs. The resulting trajectories represent the process of achieving effective coverage and energy efficiency optimization for UAVs.

4.1 K-Means Based User Clustering

The K-means clustering algorithm [14], as a heuristic approach, is characterized by its low computational complexity, timely discovery of clustering results, and the absence of the need for prior knowledge training. Consequently, it has found wide application in wireless communications. Its objective is to partition samples into K clusters, where the distances between data points within clusters are minimized while maximizing the distances between different clusters. The initial cluster centers are randomly chosen. In this paper, we treat ground user locations as data points, with the set of user locations $L = (l_1, l_2...l_p)$ serving as the observation set for unmanned aerial vehicles (UAVs). Clustering partitions are denoted by $(J_1, J_2...J_u)$, and the minimum sum-of-squares error for partitions is expressed as follows:

$$d_{up} = \sum_{l \in J_i} \left\| l - \frac{\sum_{l \in J_i} l}{J_u} \right\|^2 \tag{8}$$

Algorithm 1. K-means clustering algorithm

Input: users location $L = (l_1, l_2...l_p)$, UAV maximum load ϕ
Output: $(J_1, J_2...J_i)$
    Initialize cluster number $(J_1, J_2...J_i)$, Randomly select U samples in L as initial centroid, iterations N,
1: **for** $n = 1, 2...N$ **do**
2:     Calculate d_{up} according to(7)
3:     Allocate l_p to J_i with minimum d_{up}
4:     update centroid
5:     **if** iterations $N \geq n$ **then**
6:         end loop
7:     **end if**
8: **end for**
9: **while** $J_u > \phi$ **do**
10:     Remove l_p from J_u
11:     Add l_p to another J_i
12: **end while**

4.2 MA-DQN Algorithm for Deployments

We typically formulate decision problems involving large-scale dynamic mobility as a Markov Decision Process (MDP) [15], and then address them using a multi-agent DQN algorithm. Each independent agent interacts with the environment to continuously optimize its decisions. In the MA-DQN framework, each unmanned aerial vehicle (UAV) is treated as an intelligent agent, and then each agent is trained and executes its own strategy based on local states and information from other agents. In this section, we mainly elaborate on the state space, action space, and reward function of this algorithm.

1) State Space: The MA-DQN model calculates the Q-values of actions based on the input state space S, which is defined by the positions of the unmanned aerial vehicles (UAVs) in the model.

$$S_i = (x_u(t), y_u(t), h_u(t), g_p^u(t)), t = (1, \cdots T) \tag{9}$$

In the equation, $(x_u(t), y_u(t), h_u(t))$ represents the spatial coordinates of each unmanned aerial vehicle's three-dimensional position at time slot t. $g_p^u(t)$ represents the Channel gain of associated users.

2) Action Space: one UAV has five possible horizontal movements: forward, backward, right, left, and hover. We use 0, 1, 2, 3, and 4 respectively to represent horizontal movement A. We adopt a greedy strategy for selecting action strategies, and the formula is as follows:

$$A_i = \{0, 1, 2, 3, 4\} \tag{10}$$

$$A = \begin{cases} random \quad action, & \epsilon \\ argmax_A Q(S, A, \theta_i), & 1 - \epsilon \end{cases} \tag{11}$$

3) Reward: Our goal is to maximize the system throughput, where the reward obtained by each agent based on state information and subsequent actions sums up to the system throughput. In practical applications, to achieve maximum reward, specific reward and penalty policies need to be designed. This is because considering the states of all UAVs and user states before each UAV makes a decision about its motion would lead to higher computational complexity. Additionally, the number of users served by each UAV is limited. Therefore, in scenarios with uneven distribution of dynamic users, when considering a significant amount of task data to be offloaded by a user cluster, assistance from multiple UAVs is required. If this exceeds the maximum QoS requirements of UAVs, the UAVs need to send the excess data to satellites and cloud service centers for processing, resulting in a larger actual system throughput. In such cases, we double the reward function. The reward factor is expressed as λ. However, to prevent UAVs from becoming overly concentrated in pursuit of throughput rewards, we introduce a distance penalty into the reward function, The penalty factor is represented as φ. This encourages UAV dispersion to reduce overlapping coverage areas, enhance the number of serviced users, and ensure fairness among users to some extent. The reward function is expressed as,

$$R_u = \sum_{u \in i} R_i(t) \cdot 2^{\lambda - \varphi} \tag{12}$$

As shown in Fig. 2, we integrate K-means with DQN to address the problem. In this framework, agents need to be appropriately connected to neural networks (NNs). Connected unmanned aerial vehicles (UAVs) first input abstract state information into the evaluation network to determine the optimal action. Subsequently, rewards are computed, and actions are executed in the environment. After all UAVs complete their actions, the data rate for that time slot is calculated. The specific algorithmic procedure is outlined in Algorithm 2.

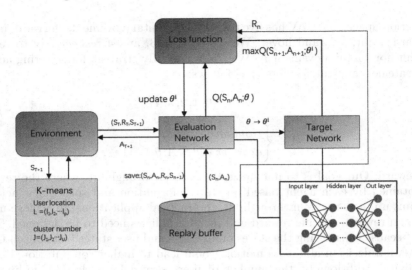

Fig. 2. Optimization process of joint K-means and MA-DQN algorithm

---

Algorithm 2. MA-DQN Algorithm for Deployments

1: **for** each episode **do**
2:   initialize the main DQN $Q(S_i, A_i|\theta_i)$
3:   Initialize the target DQN $Q'(S_i, A_i|\theta_i') with \theta_i$
4:   Initialize experience replay buffer Bi;
5:   **for** time step t := 0 : T **do**
6:     **for** each UAV **do**
7:       Generate state array S and Calculate $g_p^u(t)$
8:       take action A according to ϵ-greedy policy
9:       observe R and S', Calculate Ri(t + 1) accordingto Eq.10
10:      Store e = (S, A, R, S') into Buffer
11:      Randomly sample a mini-batch from Buffer
12:      Compute target $y = R + \alpha max Q(S', A', \theta_i)$
13:      Calculate Loss function
14:      $\sigma(\theta_i) = E[Q_{target} - Q(S', A'|\theta_i)]^2$
15:      Train parameter θ_i with a gradient descent step
16:      $(y - Q(S, A, \theta_i))^2$
17:      **if** update = true **then**
18:        $\theta_t \longleftarrow \theta_i$
19:      **end if**
20:    **end for**
21:    Users move
22: **end for**
23: **end for**

5 Simulation Results

This section validates the effectiveness of the proposed solution in an integrated aerial-ground network scenario through simulation experiments. Each UAV covers an area of 500*500 square units, with a maximum altitude of 150 m. Users are randomly distributed within the coverage area of each UAV. Each UAV can connect up to 6 users, assuming the UAV's connection to users is always maximized. The neural network used in the MA-DQN algorithm consists of 3 layers, with 40 nodes in the hidden layer and rectified linear unit activation function. The greedy action selection strategy for action selection is set to linearly decrease from 0.9 to 0. The loss function is chosen as mean squared error, and the Adam optimizer is utilized to train the neural network. Table 1 lists the other default simulation parameters.

Table 1. Simulation Parameters

Parameters	Values
Carrier frequency f_c	2 GHz
RB bandwidth f_c	15 kHz
Number of UAVs and Users f_c	(3,18)
Maximum and Minimum UAV Altitude H f_c	(150 m, 20 m)
Maximum transmitting power P	29 dBm
User maximun moving speed V_{max}	0.5 m/s
UAV speed	5 m/s
Max QOS require	100 kb/s
Service area boundary (x_{min}, y_{min})	0
Service area boundary (x_{max}, y_{max})	500 m
Discount factor β	0.99
The noise power at the UAV δ^2	-100 dBm/Hz
Learning rate	0.001
Batch size	512
Epsilon ϵ	0–0.9

Figure 3 illustrates the system throughput under different numbers of UAVs. Generally, with an increase in the number of UAVs, communication coverage is enhanced, leading to an improvement in the overall system throughput. In the case of a single UAV, the maximum throughput is limited by its own performance due to the absence of cooperation and interference among clusters. With the communication gain of three or four UAVs, the overall system throughput can be significantly increased, and the training convergence is faster.

Figure 4 depicts the impact of the maximum number of associated users per UAV on the system throughput. As we gradually increase the number of

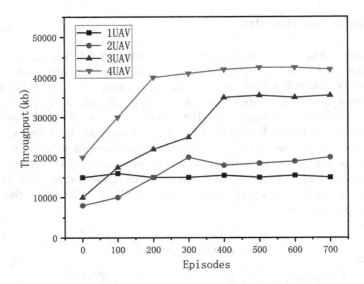

Fig. 3. Convergence of the throughput per episode for different fleet size.

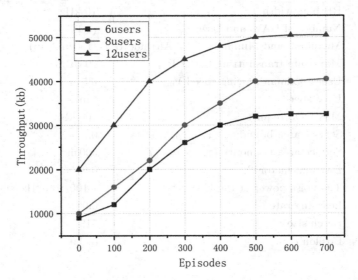

Fig. 4. Convergence of the throughput per episode for different user numbers.

user clusters associated with each UAV, the overall system throughput naturally improves. However, considering the limitations of UAV hardware performance and the complexity of training data, we do not significantly increase the number of associated user clusters for offloading. Nevertheless, it is evident that our approach provides stable and effective performance for serving different numbers of users. Figure 5 presents our example illustrating the optimal trajectories for serving 6 and 8 users per UAV.

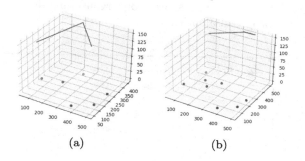

Fig. 5. The optimal trajectory of UAV for different users distribution.

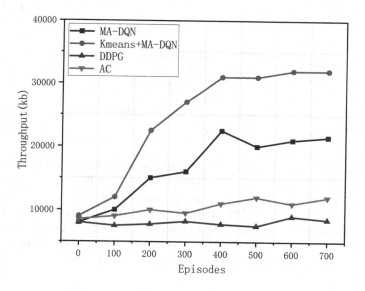

Fig. 6. Comparison of different method.

Figure 6 compares the convergence performance of different reinforcement learning algorithms in terms of system throughput. We assume a scenario with three UAVs. It is evident that, with the integration of the k-means clustering algorithm into the DQN algorithm, the UAVs can quickly connect to the users within their respective partitions and then take optimal actions by following the users. This approach allows the system to maintain an optimal state before the training of the MA-DQN algorithm, resulting in a rapid improvement in system performance and accelerated training convergence. Although the initial performance difference is not significant without reclustering, the gap widens over time, and convergence is slower. The other two algorithms, DDPG and AC, generally perform worse, with slower convergence, highlighting the superior effectiveness of the proposed algorithm.

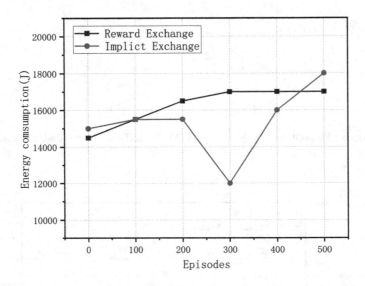

Fig. 7. Convergence of the number of connected users per episode for different information exchange.

Figure 7 illustrates the effect of balancing rewards on the system's overall throughput. In scenarios with dynamic user mobility, designing appropriate reward mechanisms for agents ensures the stability of learning algorithms, thereby effectively enhancing the overall system performance. From the graph, we observe that our designed reward strategy leads to better convergence during training. Conversely, intermittent reward exchange, i.e., training without specific reward and penalty policies, results in slower convergence and greater fluctuations.

6 Conclusions

With the continuous advancement of UAV technology and the advent of the 6G era, the era of ubiquitous connectivity is drawing closer. The efficient deployment of integrated air-ground networks is essential to facilitate data transmission for the vast array of IoT devices. This paper focuses on the dynamic scenarios of densely populated users in UAV-based small-scale aerial base stations. We propose a trajectory control algorithm combining k-means with MA-DQN, leveraging NOMA technology for multi-user connections. Specifically, we treat each

UAV as an individual agent and design corresponding reward and penalty strategies to enhance the robustness of the entire system network through continuous learning. Simulation results demonstrate that our approach not only reduces energy consumption but also accelerates training convergence in highly dynamic environments, thereby ensuring a more stable and efficient system network compared to the traditional DQN algorithms.

Acknowledgments. This work was supported in part by the National Natural Science Foundation of China under grants 61871433, 61828103, and 61201255, in part by the GuangDong Basic and Applied Basic Research Foundation under Grant 2024A1515012052, and in part by the Research Platform of South China Normal University and Foshan.

References

1. Chen, Y., Zhao, J., Zhou, X., Qi, L., Xu, X., Huang, J.: A distributed game theoretical approach for credibility-guaranteed multimedia data offloading in MEC. Inf. Sci. **644**, 119306 (2023)
2. Li, H., Ota, K., Dong, M.: AI in sagin: building deep learning service-oriented space-air-ground integrated networks. IEEE Netw. **37**(2), 154–159 (2022)
3. ur Rahman, S., Kim, G.H., Cho, Y.Z., Khan, A.: Positioning of UAVs for throughput maximization in software-defined disaster area UAV communication networks. J. Commun. Netw. **20**(5), 452–463 (2018)
4. Mozaffari, M., Saad, W., Bennis, M., Nam, Y.H., Debbah, M.: A tutorial on UAVs for wireless networks: applications, challenges, and open problems. IEEE Commun. Surv. Tutor. **21**(3), 2334–2360 (2019)
5. Pham, H., La, H., Feil-Seifer, D., Nefian, A.: Cooperative and distributed reinforcement learning of drones for field coverage. arXiv (2018). arXiv preprint arXiv:1803.07250
6. Zeng, H., Zhu, Z., Wang, Y., Xiang, Z., Gao, H.: Periodic collaboration and real-time dispatch using an actor–critic framework for UAV movement in mobile edge computing. IEEE Internet Things J. (2024)
7. Rapetswa, K., Cheng, L.: Towards a multi-agent reinforcement learning approach for joint sensing and sharing in cognitive radio networks. Intell. Converged Netw. **4**(1), 50–75 (2023)
8. Hu, Y., Chen, M., Saad, W., Poor, H.V., Cui, S.: Distributed multi-agent meta learning for trajectory design in wireless drone networks. IEEE J. Sel. Areas Commun. **39**(10), 3177–3192 (2021)
9. Cui, J., Liu, Y., Nallanathan, A.: Multi-agent reinforcement learning-based resource allocation for UAV networks. IEEE Trans. Wirel. Commun. **19**(2), 729–743 (2019)
10. Tripathi, S., Zhang, R., Wang, M.: Distributed user connectivity maximization in UAV-based communication networks. In: 2023 IEEE Global Communications Conference, GLOBECOM 2023, pp. 3753–3758. IEEE (2023)
11. Song, Q., Zheng, F.C., Jin, S.: Multiple UAVs enabled data offloading for cellular hotspots. In: 2019 IEEE Wireless Communications and Networking Conference (WCNC), pp. 1–6. IEEE (2019)

12. Zhong, R., Liu, X., Liu, Y., Chen, Y.: Multi-agent reinforcement learning in noma-aided UAV networks for cellular offloading. IEEE Trans. Wirel. Commun. **21**(3), 1498–1512 (2021)
13. Barbosa, H., et al.: Human mobility: models and applications. Phys. Rep. **734**, 1–74 (2018)
14. Wilkin, G.A., Huang, X.: K-means clustering algorithms: implementation and comparison. In: Second International Multi-symposiums on Computer and Computational Sciences (IMSCCS 2007), pp. 133–136. IEEE (2007)
15. Deng, S., et al.: Dynamical resource allocation in edge for trustable internet-of-things systems: a reinforcement learning method. IEEE Trans. Ind. Inf. **16**(9), 6103–6113 (2020)

Research on Ethical Issues of Classroom Dialogue in the Era of Large Language Model

Jianxia Ling[1], Jia Zhu[1(✉)], Jianyang Shi[1], and Pasquale De Meo[2]

[1] College of Education, Zhejiang Normal University, Jinhua, China
jiazhu@zjnu.edu.cn
[2] Department of Ancient and Modern Civilizations, University of Messina, Messina, Italy

Abstract. As society moves towards the intelligence era, the large language model ChatGPT has garnered widespread attention, particularly regarding its applications in education and the ethical issues it raises. This study examines the ethical dimensions of classroom dialogue for novice and expert teachers using a method that combines human verification with ChatGPT's judgment. The content of classroom dialogue from expert teachers was used as prompts for ChatGPT to predict and generate subsequent classroom dialogue, which was then compared with those of the expert teachers. Results show that expert teachers exhibit significantly higher ethical levels in classroom dialogue compared to novice teachers. While there are differences in the characteristics of dialogue content predicted by ChatGPT and those of expert teachers, there is no significant difference in ethical levels between the two types of dialogue.

Keywords: ChatGPT · Classroom dialogue · Ethics · Expert teachers · Novice teachers

1 Introduction

In recent years, the field of artificial intelligence has witnessed remarkable progress, especially with the continuous evolution of deep learning and natural language processing technologies, making large language models a prominent achievement in this field. Against this backdrop, OpenAI, an American company, released ChatGPT at the end of 2022, representing a standout example of large language models [12]. ChatGPT finds applications across a wide range of fields, including automatic translation, natural language understanding, intelligent customer service, and more. Particularly noteworthy is its significance in the field of education, where ChatGPT can provide timely assistance to students, answer questions, enhance learning outcomes, and provide educators with additional tools and resources to better meet students' needs. However, despite the vast potential for ChatGPT's application in education, it comes with a series of ethical concerns [4,14]. Some scholars are concerned that the content generated by

ChatGPT may contain inaccurate information, potentially misleading students [16]. It could also include discriminatory language that harms certain groups, or even lead to the inadvertent disclosure of sensitive information from educational institutions, which could have detrimental effects on educational practices [1].

Assessing the overall impact of the content generated by ChatGPT on education and the ethical issues it raises, as well as exploring whether it possesses ethical judgment capabilities, has become a crucial topic. Therefore, this study focuses on the perspective of classroom dialogue in the education sector. It analyzes ChatGPT's ability to judge ethical issues and its ability to generate classroom dialogue under simulated classroom situations. The study also validates that ChatGPT-predicted classroom dialogue exhibits no significant differences in ethical standards when compared to expert teachers' classroom dialogue.

2 Related Work

In the past 20 years, international attention to classroom dialogue has gradually increased [19]. This study adopts Christine Howe's definition, which states that "one individual addresses another individual or individuals, and at least one addressed individual replies" [5]. Frequent dialogue interactions can attract students' attention [2], but improving the quality of classroom dialogue is more crucial [17]. Statistical evidence provided by Muhonen and others indicates that high-quality dialogue predicts better student learning outcomes [11]. However, although classroom dialogue can enhance teaching effectiveness, ethical issues can arise. Questioning by teachers often varies depending on students' academic performance, leading to educational inequality [13]. Moreover, classroom dialogue may neglect cultural diversity, focusing primarily on mainstream culture and local students' experiences, potentially disadvantaging minority groups [6]. Therefore, attention to ethical issues is essential to ensure educational equity when implementing effective classroom dialogue.

ChatGPT, a sophisticated chatbot built upon a large language model, leverages supervised fine-tuning and reinforcement learning from human feedback [15]. Trained on a vast database of human conversations, it can generate human-like responses [18]. ChatGPT has demonstrated utility in education, including language translation, coding assistance, and article generation [10]. It can help educators develop course outlines, create lecture topics, prepare presentations, and simulate classroom dialogue [9]. Additionally, ChatGPT assists students in tackling complex problems, explaining terms, drafting papers, and completing assignments [14]. However, the application of ChatGPT in education, while making significant strides, has also raised a series of ethical issues. Training data may contain biases related to race, gender, age, and other factors, and ChatGPT can facilitate cheating, undermining academic integrity [3,8]. Current researchers have explored the application of ChatGPT and its ethical issues, yielding some valuable insights. However, there is a relative lack of research on ChatGPT's application in classroom dialogue.

Given the background and analysis mentioned above, this study raises several questions: What are the similarities and differences in the ethical issues of classroom dialogue between novice teachers and expert teachers? What are the similarities and differences in the characteristics of classroom dialogue predicted by ChatGPT and those of expert teachers? What are the similarities and differences in the ethical issues of classroom dialogue between ChatGPT's predicted dialogue and that of expert teachers?

3 Methodology

In this section, we will primarily introduce the data sources and the methods of data processing and analysis. Firstly, the classroom dialogue texts were transcribed and underwent manual content checks. Secondly, the text was encoded, categorizing dimensions of ethical judgment, and ethical judgments and dialogue predictions were made using ChatGPT. Finally, the encoded text data underwent analysis.

3.1 Dataset

The data for this study were obtained from the National Education Resources Public Service Platform provided by the Ministry of Education of China (website: https://www.eduyun.cn/). This platform aggregates curriculum records from primary and secondary schools across China, covering various grade levels and subjects. For this study, we selected the ninth-grade mathematics curriculum chapter "Determining a Square" as the subject. From this chapter, we randomly selected one unrated and one highly-rated lesson recording from the platform as sample data. The teachers associated with the unrated recording represented novice teachers, while those from the highly-rated recording represented expert teachers. Each class has 36 students, with a nearly equal gender distribution, and lasts for 45 min.

3.2 Data Processing

The data processing mainly includes video-to-text conversion, dialogue text encoding, and categorization of ethical judgment dimensions. The specific process is as follows.

Speech to Text and Text Encoding. In this study, we used the iFlytek interface to implement the conversion of speech to text in the videos. To enhance data accuracy, the converted text data underwent manual inspection and adjustment. There were 340 dialogue statements for novice teachers and 204 for expert teachers in their respective classrooms. The dialogue content was divided into 17 segments based on the dialogue rounds. For novice teacher classrooms, each segment comprised ten rounds of dialogue, while for expert teacher classrooms, each segment comprised six rounds of dialogue.

The coding framework for this study was referenced and adapted from the classroom dialogue coding tool developed by Song Y [20] and the dialogue coding framework developed by Zhao W [21]. Using this framework, two researchers independently coded the classroom dialogue content, achieving an inter-coder reliability of 89%.

Ethical Judgment Dimensions and Scoring. In this study, we categorized ethical issues in classroom dialogues into five dimensions: information reliability, fairness and justice, privacy, responsibility, and transparency, drawing from ethical principles extracted by Huang C [7] from 146 AI guideline documents. Each dimension is scored on a scale of five, totaling 25 points. "Information reliability" pertains to the accuracy and reliability of information sources in classroom dialogue. "Fairness and justice" ensures equal participation and expression without bias or discrimination. "Privacy" focuses on ensuring that classroom dialogue does not involve personal privacy or sensitive personal information. "Responsibility" emphasizes the responsibility of both teachers and students. "Transparency" concerns the clarity of discourse objectives. Human evaluators and ChatGPT collaboratively assess ethical issues in classroom dialogue content. Each round of dialogue is input into ChatGPT three times, and human evaluators judge which assessment is most appropriate based on ethical concepts, using it as the final score.

3.3 Data Analysis

In response to the research questions raised, the encoded data was analyzed as follows. First, Classroom dialogues from novice and expert teachers separately into the ChatGPT dialogue interface to assess ethical issues, as illustrated in Fig. 1. Independent sample t-tests in SPSS were used to examine differences in ethical issues between the classroom dialogues of novice and expert teachers. Secondly, used the teaching design of expert teacher classrooms as a reference, with each segment consisting of six rounds of classroom dialogues. ChatGPT was then used to sequentially predict dialogues. Descriptive statistics were employed to analyze the frequency of various types of classroom dialogue by expert teachers and ChatGPT predictions, exploring their similarities and differences. Finally, an ethical assessment was conducted on the classroom dialogue of expert teachers and the predicted classroom dialogue of ChatGPT, exploring the specific ethical issues in classroom dialogue.

4 Results

The results of this study are divided into three parts. The first part compares the ethical issues in classroom dialogues between novice teachers and expert teachers. The second part explores the similarities and differences in the characteristics of classroom dialogue between expert teachers and ChatGPT predictions. Finally, it compares the level of ethical issues in classroom dialogues between expert teachers and ChatGPT predictions.

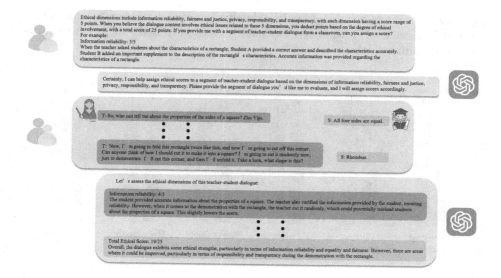

Fig. 1. Example of ethical level ratings for ChatGPT

4.1 The Level of Ethical Issues in Classroom Dialogues Between Novice Teachers and Expert Teachers

Figure 2 displays the t-test results, indicating that the overall ethical level of expert teacher classroom dialogue ($M = 3.66, SD = 1.03$) is significantly higher than that of novice teacher classroom dialogue ($M = 4.75, SD = 0.49, p < 0.001$). In the dimension of "information reliability" ($t(32) = -2.05, p = 0.05$, Cohen's d = 0.58), there is a significant difference between novice and expert teacher classroom dialogues. However, both novice and expert teachers' classroom dialogues scored high in "information reliability", indicating that teachers and students based their statements and discussions on factual and verifiable evidence. For "fairness and justice" ($t(32) = -7.05, p < 0.001$, Cohen's d = 0.83), novice teachers scored significantly lower than expert teachers. This indicates that in novice teachers' classrooms, there may be occasional unfriendly attitudes between teachers and students, and the range of participants in dialogue is limited. In contrast, expert teachers' classrooms prioritize an environment of equality and inclusivity, ensuring that everyone is treated equally and fairly. In the "privacy" dimension ($t(32) = -1.48, p = 0.15$, Cohen's d = 0.81), there is no significant difference between novice and expert teacher classroom dialogue. Both have relatively high scores, indicating that dialogues generally do not involve sensitive personal information, although ChatGPT deducts points for mentioning names. In the dimensions of "responsibility" ($t(32) = -7.14, p < 0.001$, Cohen's d = 0.67) and "transparency" ($t(32) = -4.15, p < 0.001$, Cohen's d = 0.70), there are significant differences between novice and expert teacher classroom dialogue. In novice teachers' classrooms, there is a lack of responsibility in guiding answers to questions and providing timely clear explanations when correcting students' mistakes. Additionally, students occasionally make offensive remarks when responding to the teacher.

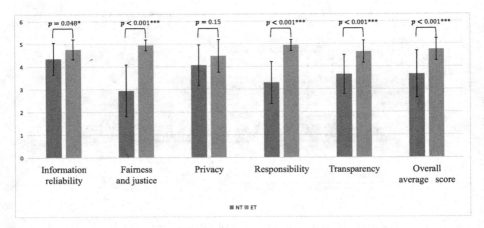

Fig. 2. Ethical issues in classroom dialogue between novice teachers (NT) and expert teachers (ET)

4.2 Similarities and Differences Between Characteristics of Expert Teacher Classroom Dialogue and ChatGPT Predicted Classroom Dialogue

Table 1 displays the types of classroom dialogues in expert teachers and ChatGPT-predicted dialogues, showing some common characteristics. The proportions of "agreement and challenge" (5.5% vs. 5%, $p = 0.823$) and "coordination" (6.5% vs. 6%, $p = 0.365$) are relatively low, while "analysis" (16.5% vs. 20%) is more prevalent, and "construction" (12% vs. 16%) types are moderately represented.

Comparing the proportions of the four coding types, namely "agreement and challenge", "coordination", "analysis", and "construction", between ChatGPT-predicted dialogues and expert teacher classroom dialogues, there were no significant differences in the numerical values. This indicates that, under the same conditions of dialogue text quantity and teaching design, there is consistency in these four coding types between ChatGPT-predicted and expert teacher dialogues. Despite different methods for dialogue generation, both emphasize analysis, with teachers guiding students in problem analysis. The lower proportion of "agreement and challenge" types is due to the longer dialogue utterances, where explanations or inquisitive statements are included. Consequently, in both expert classroom dialogues and ChatGPT-predicted classroom dialogues, the proportion of "agreement and challenge" is relatively low. Additionally, since this lesson's teaching design focuses on the new lesson content of square determination,

Table 1. The distribution of dialogue categories in expert teachers' classroom dialogues and ChatGPT-predicted classroom dialogues

Coding	ChatGPT(%)	ET(%)	Z statistic	p value
Basic knowledge**	15	26	2.725	0.006
Personal information***	23	8.5	−3.982	<.001
Analysis	16.5	20	0.906	0.365
Coordination	6.5	6	−0.207	0.836
Construction	12	16	1.153	0.249
Agreement and challenge	5.5	5	−0.224	0.823
Instruction and guidance*	18.5	10.5	−2.272	0.023
Other*	3	8	2.193	0.028

***p < 0.001, **p < 0.01, *p < 0.05

rather than a practice exercise class, there is also relatively less inductive discourse.

The encoding proportions of ChatGPT-predicted and expert teacher classroom dialogues differ significantly in the category of "personal information" (23% vs. 8.5%, $p < 0.001$). ChatGPT-generated dialogues contain a significantly higher proportion of "personal information" compared to expert teacher dialogues, due to its data sources primarily from internet text. There is also a significant difference in the encoding proportions of "basic knowledge" (15% vs. 26%, $p = 0.006$), with expert teacher dialogue having a higher proportion. This is because ChatGPT has limited knowledge of textbook content without contextual cues, resulting in less dialogue content encoded under "basic knowledge". Differences in the encoding proportions of "instruction and guide" (18.5% vs. 10.5%, $p = 0.023$) show fewer instances in expert teacher classroom dialogue. Real classrooms tend to share similar cognitive frameworks and background knowledge between teachers and students, resulting in relatively fewer instances of instruction and guide discourse. The "Other" category (3% vs. 8%, $p = 0.028$) represents uncoded content, which is not dialogue but rather refers to student behavior, such as practicing, with a significant difference between the two methods. Expert teacher classrooms focus more on students' thinking and discussions, while ChatGPT primarily involves teacher-student dialogue, with limited involvement in behavior and language alternation processes.

4.3 The Level of Ethical Issues in Classroom Dialogues Between Expert Teachers and ChatGPT-Predicted Dialogues

As shown in Table 2, an independent samples t-test revealed that the overall ethical level of ChatGPT-predicted classroom dialogue ($M = 4.86$, $SD = 0.38$) was slightly higher than that of expert teacher classroom dialogue ($M = 4.75$,

$SD = 0.49, p = 0.116$), but the difference was not statistically significant. Except for the "transparency" dimension, there were no significant differences in ethical issues between ChatGPT-predicted and expert teacher classroom dialogues across the other four dimensions. Both ChatGPT-predicted and expert teacher classroom dialogues provided reasonably accurate information, logical reasoning, and explanations in the "information reliability" and "transparency" dimensions. In the "fairness and justice" dimension ($t(32) = 0.00$, $p = 1.000$, Cohen's d = 0.24), both ChatGPT-predicted and expert teacher classroom dialogues scored near the maximum, indicating exceptionally low probability of ethical issues. The communication between teachers and students is equal, and there is mutual respect for each other's opinions. In the "privacy" dimension ($t(32) = 0.00$, $p = 1.000$, Cohen's d = 0.67), the scores of both are consistent. Neither received full marks in this dimension because ChatGPT considers any mention of names to be within the realm of privacy. In ChatGPT-predicted dialogues, the "responsibility" dimension received a full score, demonstrating active engagement from both teachers and students, without any irresponsible or aggressive words.

Table 2. The ethical issues in expert teachers'(ET) classroom dialogues and ChatGPT-predicted classroom dialogues

Implicit variable	Experimental condition	
	ChatGPT(M ± SD)	ET(M ± SD)
Information reliability*	4.94 ± 0.24	4.76 ± 0.44
Fairness and justice***	4.94 ± 0.24	4.94 ± 0.24
Privacy	4.47 ± 0.62	4.47 ± 0.72
Responsibility***	5 ± 0	4.94 ± 0.24
Transparency***	4.94 ± 0.24	4.65 ± 0.49
Overall average score***	4.86 ± 0.38	4.75 ± 0.49

***p < 0.001, **p < 0.01, *p < 0.05

5 Conclusion

Since the widespread popularity of ChatGPT, the education sector has started to investigate its application in teaching and associated ethical issues. From the perspective of classroom dialogue, this study reveals through analysis that the ethical level of expert teachers' classroom dialogue is significantly higher than that of novice teachers. There were significant differences between ChatGPT-predicted and expert teacher classroom dialogues in categories such as "basic knowledge", "personal information", "instruction and guide", and "other", but not in categories like "analysis", "coordination", "construction", and "agreement and challenge". Furthermore, there was no significant difference in the ethical level between ChatGPT-predicted and expert teacher classroom dialogues.

This study has two main limitations that could serve as starting points for future research. Firstly, this study primarily focuses on the ethical issues within classroom dialogue and ChatGPT's ability to evaluate ethical standards, without delving deeply into the fundamental causes of ethical issues that arise in classroom dialogue. Secondly, the research results of this study are based on the comparison between ChatGPT and Chinese mathematics subjects. In other subjects and national contexts, different ethical issues may exist, and our conclusions remain exploratory until further confirmation through additional research.

Acknowledgments. This work was supported by the Zhejiang Provincial Philosophy and Social Sciences Planning Project under Grant (No. 24NDJC191YB), the National Key R&D Program of China under Grant (No. 2022YFC3303600), and the Zhejiang Provincial Natural Science Foundation of China under Grant (No. LY23F020010).

Disclosure of Interests. The authors have no competing interests to declare that are relevant to the content of this article.

References

1. Baidoo-Anu, D., Ansah, L.O.: Education in the era of generative artificial intelligence (AI): understanding the potential benefits of chatgpt in promoting teaching and learning. J. AI **7**(1), 52–62 (2023)
2. Chin, C.: Classroom interaction in science: teacher questioning and feedback to students' responses. Int. J. Sci. Educ. **28**(11), 1315–1346 (2006)
3. Costa-juss, C.B.M.R., Casas, N.: Evaluating the underlying gender bias in contextualized word embeddings. In: GeBNLP 2019, p. 33 (2019)
4. Eysenbach, G., et al.: The role of chatgpt, generative language models, and artificial intelligence in medical education: a conversation with chatgpt and a call for papers. JMIR Med. Educ. **9**(1), e46885 (2023)
5. Howe, C., Abedin, M.: Classroom dialogue: a systematic review across four decades of research. Camb. J. Educ. **43**(3), 325–356 (2013)
6. Howe, C., Hennessy, S., Mercer, N., Vrikki, M., Wheatley, L.: Teacher-student dialogue during classroom teaching: does it really impact on student outcomes? J. Learn. Sci. **28**(4–5), 462–512 (2019)
7. Huang, C., Zhang, Z., Mao, B., Yao, X.: An overview of artificial intelligence ethics. IEEE Trans. Artif. Intell. (2022)
8. Hutchinson, B., Prabhakaran, V., Denton, E., Webster, K., Zhong, Y., Denuyl, S.: Social biases in NLP models as barriers for persons with disabilities. arXiv preprint arXiv:2005.00813 (2020)
9. Javaid, M., Haleem, A., Singh, R.P., Khan, S., Khan, I.H.: Unlocking the opportunities through chatgpt tool towards ameliorating the education system. BenchCouncil Trans. Benchmarks Stand. Eval. **3**(2), 100115 (2023)
10. Kasneci, E., et al.: Chatgpt for good? On opportunities and challenges of large language models for education. Learn. Individ. Differ. **103**, 102274 (2023)
11. Muhonen, H., Pakarinen, E., Poikkeus, A.M., Lerkkanen, M.K., Rasku-Puttonen, H.: Quality of educational dialogue and association with students' academic performance. Learn. Instr. **55**, 67–79 (2018)
12. OpenAI, C.: Optimizing language models for dialogue (2022). https://openai.com/blog/chatgpt

13. Pehmer, A.K., Gröschner, A., Seidel, T.: How teacher professional development regarding classroom dialogue affects students' higher-order learning. Teach. Teach. Educ. **47**, 108–119 (2015)
14. Qadir, J.: Engineering education in the era of chatgpt: promise and pitfalls of generative AI for education. In: 2023 IEEE Global Engineering Education Conference (EDUCON), pp. 1–9. IEEE (2023)
15. Rahman, M.M., Watanobe, Y.: Chatgpt for education and research: opportunities, threats, and strategies. Appl. Sci. **13**(9), 5783 (2023)
16. Sallam, M.: Chatgpt utility in healthcare education, research, and practice: systematic review on the promising perspectives and valid concerns. In: Healthcare, vol. 11, p. 887. MDPI (2023)
17. Schwab, G.: From dialogue to multilogue: a different view on participation in the English foreign-language classroom. Classroom Discourse **2**(1), 3–19 (2011)
18. Shen, Y., et al.: Chatgpt and other large language models are double-edged swords (2023)
19. Song, Y., Chen, X., Hao, T., Liu, Z., Lan, Z.: Exploring two decades of research on classroom dialogue by using bibliometric analysis. Comput. Educ. **137**, 12–31 (2019)
20. Song, Y., Lei, S., Hao, T., Lan, Z., Ding, Y.: Automatic classification of semantic content of classroom dialogue. J. Educ. Comput. Res. **59**(3), 496–521 (2021)
21. Zhao, W., Ma, J., Cao, Y.: What is effective classroom dialog? A comparative study of classroom dialog in Chinese expert and novice mathematics teachers' classrooms. Front. Psychol. **13**, 964967 (2022)

Adaptive Exploration: Elevating Educational Impact of Unsupervised Knowledge Graph Question Answering

Xi Yang[1], Zhangze Chen[2](✉)[iD], Hanghui Guo[3](✉)[iD], and Tetiana Shestakevych[1]

[1] Lviv Polytechnic National University, Lviv, Ukraine
[2] College of Education, Zhejiang Normal University, Jinhua, China
zjnuczz@zjnu.edu.cn
[3] School of Computer Science and Technology, Zhejiang Normal University, Jinhua, China
ghh1125@zjnu.edu.cn

Abstract. Question answering (QA) systems have been widely used in educational domain, significantly contributing to immediate information access and enhancing learning experiences. This paper introduces an efficient educational Knowledge Graph Question Answering (KGQA) framework that operates without relying on annotated training data. Our method enables the swift deployment of new knowledge graphs by simulating human-like information acquisition through a symbolic exploration module. Leveraging diverse program generation and large language models (LLMs), we formulate natural language questions to optimize the query for each program. To address semantic accuracy challenges in LLMs due to the absence of contextual training data, we propose adaptive strategies, including dynamic contextual re-ranking. These techniques significantly enhance question generation precision, showcasing robust performance in unsupervised settings. The framework demonstrates exceptional adaptability across varied queries and outperforms state-of-the-art models in zero-shot queries across various knowledge graph scales and datasets, underscoring its efficacy and scalability.

Keywords: Knowledge Graph · Question Answering · Large Language Models

1 Introduction

Improving natural language interaction with structured data repositories, like KGQA, is crucial for better information accessibility [3]. Despite advancements, the challenge of handling complex queries remains due to the reliance on annotated training data. Collecting such data is often impractical in fields like biomedicine and education [20], where it can be prohibitively expensive.

Models fine-tuned on one dataset often struggle to generalize across different datasets within the same Knowledge Graph (KG), particularly in education where data varies greatly [11]. The dynamic nature of educational datasets makes it hard to create universally adaptable QA systems [14]. This study introduces a framework for reasoning over new KGs without prior query knowledge, focusing on efficiently understanding KG semantics. This is especially beneficial for educational environments where the adaptability and relevance of information can significantly impact learning outcomes.

When encountering a new Knowledge Graph (KG), people familiarize themselves by inspecting random nodes and their properties, gaining a holistic understanding. Our framework mimics this process using an exploration module that employs diversified sampling, viewing KGs as symbolic graphs. This module enhances program diversity and improves QA systems' ability to navigate educational knowledge graphs, providing more accurate and relevant information for learning.

After acquiring diverse programs, we use large language models (LLMs) to generate natural language questions. Ensuring semantic accuracy is only possible with in-context learning data. To improve zero-shot generation, we introduce three strategies: adaptive prompting based on query complexity, dynamic contextual re-ranking for semantic consistency, and runtime-efficient pruning with candidate re-ranking to reduce processing time.

Integrating these strategies enhances question generation precision, which is crucial for unsupervised contexts. Leveraging the Pangu framework [9], our approach iteratively synthesizes projected programs guided by contextual exemplars, improving accuracy, adaptability, and robustness. Our contributions include adaptive prompting, dynamic contextual re-ranking, and runtime-efficient pruning with candidate re-ranking. These strategies boost zero-shot generation using LLMs, enhancing performance across various datasets and scales. Two public datasets show significant improvements, surpassing state-of-the-art models in zero-shot queries. By enhancing the accuracy and relevance of question answering systems in educational contexts, our framework aims to support more effective and personalized learning experiences.

2 Related Works

Generalization in Knowledge Graph Question Answering (KGQA) has become a key focus in research, with initiatives introducing benchmarks and frameworks. Recent work extends generalization to unseen KGs [7], but often relies on curated training data, a limitation our unsupervised approach addresses. Galkin et al. [5] presented a model for transferable KGQA representations, lacking natural language query resolution.

Another research path focuses on developing semantic parsers for new KGs using automatic training data generation [10], often involving human annotation for final data curation. Our work emphasizes using string similarity for token-KG schema matching, ranking, and query graph construction. Reasoning in KGQA involves iteratively planning to navigate the KG based on a test query. Several approaches use reinforcement learning for path-finding in KGQA [4], though they are designed for something other than natural language queries. Recent studies leverage language models as navigational planners for various environments beyond KGs [15].

Integrating KGQA systems has shown the potential to enhance personalized learning and provide detailed explanations for complex topics in educational contexts [16]. For instance, educational platforms can utilize KGQA to create intelligent tutoring systems that adapt to individual student needs by providing context-aware answers and explanations [8]. Additionally, KGQA can support curriculum development by identifying knowledge gaps and suggesting relevant content to address those gaps [18,19]. These applications are particularly beneficial in settings where annotated training data is limited, allowing for creating adaptive learning environments where students receive immediate, accurate responses to their queries, promoting a more engaging and effective educational experience [13,17].

We adopt a Case-Based Reasoning (CBR) approach, applicable to tasks like link prediction, semantic parsing, and reading comprehension. We utilize CBR by drawing from observed similar cases during exploration as in-context demonstrations. Recent endeavors have aimed at integrating language models and knowledge graphs [12]. However, the majority of previous approaches rely on a training corpus for retrieving in-context demonstrations, a necessity absent in our context. An illustrative work operating in a completely zero-shot setting is KAPING [1], employing a retrieve-and-generate methodology where triples are retrieved from the KG to generate the final answer. Nevertheless, this method falls short in independently providing the answer text.

3 The Adaptive Exploration and Reasoning Framework for Educational KGQA

3.1 The Structure of the Educational KGQA Framework

The adaptive exploration and reasoning framework for educational KGQA seamlessly integrates Educational KGs into a QA system, eliminating the need for training data and enabling swift deployment of a natural language interface. It begins with the Graph Exploration Stage, where the framework employs symbolic, graph-based random walks with diversified sampling strategies to enumerate executable programs, ensuring efficient navigation and comprehensive coverage of the Educational KG. This stage lays the foundation for subsequent processes by systematically exploring and understanding the educational knowledge graph.

Next, the Question Generation Stage transforms these executable programs into natural language questions using an LLM. This stage involves an adaptive prompting strategy that dynamically adjusts the model's prompts based on query complexity, enhancing the model's adaptability to various query types.

Additionally, dynamic contextual re-ranking is employed to fine-tune the ranking of generated queries, ensuring semantic consistency and task alignment. This approach enables fast and accurate answers based on extensive knowledge and provides students with personalized, contextual support, significantly enhancing the learning experience.

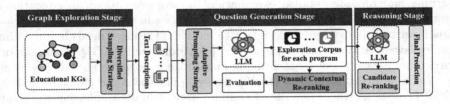

Fig. 1. The adaptive exploration and reasoning framework for Educational KGQA

Finally, the Reasoning Stage uses knowledge from the exploration phase to drive a bottom-up inference process. This stage implements a candidate re-ranking strategy to handle large sets of potential answers efficiently. By pruning and re-ranking candidates, the framework reduces interference from irrelevant information and enhances reasoning efficiency. This iterative process of ranking candidates generates the final program for accurate query responses, as shown in Fig. 1.

The following sections detail the key functions of this framework, emphasizing the seamless integration of graph exploration, question generation, and reasoning stages to achieve accurate and efficient educational KGQA.

3.2 The Graph Exploration Stage of Educational KGQA

This stage begins by initializing a sub-program with a randomly sampled class (e.g., courses, educational resources). Through guided random walks, the agent extends the sub-program using reachable schema items to ensure diversity. This iterative process continues until the desired program complexity is achieved. Optionally, operators from a predefined set, encompassing functions like COUNT or comparatives, may be applied to the sub-programs. To ensure diversification, sub-programs undergo scrutiny and are discarded if they already exist in the set of explored programs. The grounding of classes involves random sampling from entities(e.g., students, teachers, learning resources), guaranteeing non-empty answer sets upon program execution. The grounded program is then included in the set of explored programs, ensuring approximate coverage of query patterns. This symbolic, diversified sampling approach enhances control during exploration, empowering the framework to efficiently navigate the KG while maintaining a balance between complexity and diverse information representation.

3.3 The Question Generation Stage of Educational KGQA

The question generation stage is important for converting executable programs obtained from the knowledge graph into natural language questions. This stage ensures that the generated questions are contextually relevant and semantically accurate, which is particularly important in educational QA systems. It consists of two main parts: adaptive prompting strategy and dynamic context re-ranking.

Adaptive Prompting Strategy. In our approach, we construct an exploration corpus by generating questions using the extensive knowledge acquired during LLMs pretraining, complemented by textual descriptions of relevant schema items(e.g., course content, syllabi, learning objectives). Operating in an unsupervised setting poses challenges due to the lack of in-context exemplars. To address this, we introduce an adaptive prompting strategy for accurate single-shot predictions in multi-step problems, especially multi-hop queries. It dynamically adjusts the model's prompting strategy based on query complexity.

Each program begins with a dynamic complexity assessment. Direct prompting for low-complexity queries and staged prompting for high-complexity ones. The strategy evolves with the model's output, allowing real-time adjustments based on accuracy and completeness. We generate a question for each program and append it to the prompt, enabling adept It incorporates a feedback loop for continuous improvement, with an evaluation mechanism assessing performance across query complexities, refining strategies iteratively. This ensures the framework evolves and optimizes over time, enhancing the model's reasoning proficiency for intricate queries. Its adaptability is crucial for addressing student queries, from simple to complex.

Dynamic Contextual Re-ranking. While LLMs produce satisfactory surface-form outputs, ensuring task alignment remains challenging, especially for smaller models. To address this, we introduce dynamic contextual re-ranking (DCR), which fine-tunes inverse consistency ranking based on query context. This approach optimizes semantic consistency in question generation, adapting to diverse query contexts and ensuring more accurate, task-aligned predictions.

The generative task involves predicting a target sequence given query tokens, textual instruction, and in-context demonstrations. Initial predictions are derived from length-normalized log probability scores from candidate sequences. DCR refines the ranking by dynamically adjusting the inverse consistency ranking based on the query context. This adjustment uses a similarity measurement between the contextual information of the generated query and the task context, incorporating an adjustment factor.

Scoring for a single candidate involves a forward pass to compute the log probability score, with efficiency enhanced by batched forward passes for the entire candidate set. In summary, DCR refines predictions by considering the contextual alignment of generated queries, leading to more accurate and task-aligned outcomes in question generation tasks. This ensures that the questions generated are aligned with the teaching content, supporting more effective learning and assessment.

3.4 The Reasoning Stage of Educational KGQA

Our reasoning methodology mirrors previous work, but handling a large candidate set can lead to inefficiencies. We introduce a candidate pruning step, limiting the set to 10 per iteration, based on the similarity between anonymized candidate programs and the natural language test question using a sentence embedding model. In few-shot scenarios, unseen schema items may cause the model to over-score candidates with irrelevant relations and functions from demonstrations.

To overcome this challenge, we implement a candidate re-ranking strategy. Specifically, we use a weighted combination of the original and inverse scores to make predictions. This process balances these scores to refine the ranking and optimize predictions.

These strategies have effectively improved reasoning efficiency and reduced interference from irrelevant information, particularly when handling large candidate sets and unseen schema items. The application of candidate pruning and re-ranking strategies enhances the practicality of our approach and provides a viable solution for tackling complex educational reasoning problems.

4 Experiments

4.1 Datasets and Evaluation Metric

The evaluation of educational KGQA performance involves the utilization of off-the-shelf open-source LLMs juxtaposed against competitive baseline models. Our comprehensive experimentation spans a diverse spectrum of graphs, encompassing datasets ranging from small-scale to extensive collections such as MoviesKG and Freebase.

In the domain of graph exploration, we meticulously manage temporal and computational resources dedicated to question generation. To streamline the development of a QA system within a singular day, we allocate an exploration budget of 10,000 programs. Furthermore, we cap the maximum number of programs per distinct query pattern at 5, maintaining consistency for both Freebase and MoviesKG. These extensively explored programs form the bedrock for query generation, collectively constituting the exploration corpus X for both datasets.

Our evaluation criterion of choice is the F1-score, a metric that quantifies prediction precision by gauging entity and literal overlap within the reference set. This metric ensures a robust and comprehensive assessment of Educational KGQA performance.

4.2 Experimental Results

We employ two notable LLMs, MPT-Instruct and GPT-3.5, as the foundational models for our approach. MPT-Instruct is a decoder-style transformer pre-trained on a substantial corpus of 1 trillion tokens encompassing both English text and code. It undergoes instruction fine-tuning on the Databricks-Dolly-15k [2] and Anthropic Helpful and Harmless datasets [6]. Our primary experiments utilize the 7B model, supplemented by a small-scale experiment using the 30B variant to assess scalability. GPT-3.5 stands as a state-of-the-art, closed-source model developed by OpenAI. In a small-scale experiment, due to budget constraints, we opt for the text-davinci-003 variant to demonstrate the scalability of our methodology.

Our experimental setup rigorously evaluates models without query supervision. In the zero-shot scenario, we use a bottom-up reasoning approach without in-context demonstrations to score sub-programs at each step.

KGQA Results on MetaQA and GrailQA. Figure 2 shows the effectiveness of our exploration strategy across diverse settings. Significant improvements over the zero-shot baseline in MetaQA with MoviesKG and GrailQA with FreebaseKG highlight our approach's robustness. Notably, F1 scores improved nearly sixfold on GrailQA and fivefold on MetaQA, especially in the complex 3-hop setting, showcasing our framework's proficiency with intricate multi-hop queries. In most cases, students' questions often require connections across multiple knowledge points, and our approach can effectively handle these complex queries and provide precise answers.

The n-hop scenarios (1-hop, 2-hop, and 3-hop) reveal our approach's scalability and adaptability. Consistent trends across these scenarios indicate our framework's efficacy with varying question complexities. This is critical for educational question answering systems, which need to process educational data from various sources and forms. These capabilities enable our framework to provide more efficient and personalized support for education, promoting better learning outcomes and educational experiences.

Prompting for Question Generation. We analyze the impact of the adaptive prompting strategy on question generation compared to the standard approach. By annotating 100 questions, we examine the semantic accuracy of generated questions relative to logical programs. Figure 3(a) shows the performance of MPT-7B and MPT-30B, highlighting significant improvements with adaptive prompting. MPT-7B improves from 61.55 to 78.20, and MPT-30B increases from 66.75 to 86.45. These results underscore the efficacy of adaptive prompting in enhancing question generation quality over standard prompting.

Dynamic Contextual Re-ranking. We explored the impact of dynamic contextual re-ranking on question generation quality. Figure 3(b) compares the top-1 results from a standard beam-search procedure with our re-ranked output,

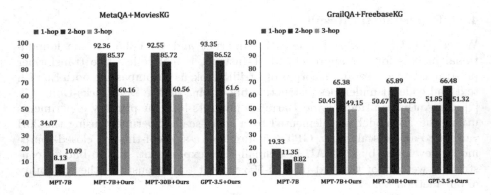

Fig. 2. Educational KGQA Performance Comparison

based on 2,000 randomly chosen questions from the GrailQA development set. Using ROUGE-1 and BLEU metrics for evaluation, with MPT-7B as the underlying model, the results show that dynamic contextual re-ranking significantly improves generation quality. Both metrics exhibited substantial enhancements, achieving approximately 25% higher accuracy compared to the standard beam-search method.

Candidate Re-ranking. Figure 3(c) offers an extensive comparative analysis of F1 accuracy between standard scoring and re-ranked outputs. In this evaluation, we randomly sub-sample 2,000 questions from the GrailQA development set and MetaQA test set. Re-ranking results show significant performance improvements: a 4.72 F1 gain on GrailQA and a 1.71 increase on MetaQA, highlighting the positive impact of candidate re-ranking on model accuracy.

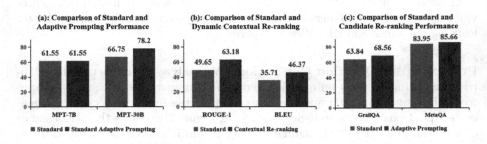

Fig. 3. Comparison of Various Strategies

In summary, the experimental results show that our framework significantly improves the semantic accuracy and relevance of generated questions, ensuring answers' accuracy and contextual relevance. This is of great help in the development of educational question-answering systems. By providing more accurate and relevant answers, students can understand the essence of the problem more quickly and find solutions quickly. At the same time, high-quality answers can also further assist teachers in their teaching work, saving teachers time in preparing and finding materials, thereby improving teaching effectiveness.

5 Conclusion

Our study presents a versatile educational KGQA framework for swift deployment across diverse knowledge graphs without human-annotated training data. Inspired by curiosity-driven learning, our framework explores unknown KGs, leveraging insights for question-answering. We combine LLMs with adaptive exploration to facilitate unsupervised KG traversal and iterative reasoning. Novel methods like candidate re-ranking enhance zero-shot LLM performance. Extensive evaluations on prominent datasets and KGs affirm the framework's robustness and effectiveness, highlighting the impact of various design choices.

In the future, we will use the educational KGQA framework to empower education. We can not only use this framework to build intelligent tutoring systems that support personalization. It can also be used for educational evaluation and improving teaching methods and can even promote educational equity. In short, our framework demonstrates the potential for educational change, improves the accuracy, efficiency, and relevance of question-answering systems, and helps achieve better learning outcomes and a more personalized educational experience.

Acknowledgments. This work was supported by the National Natural Science Foundation of China under Grant (62077015), the Zhejiang Provincial Philosophy and Social Sciences Planning Project under Grant (24NDJC191YB), the National Key R&D Program of China under Grant (2022YFC3303600), and the Natural Science Foundation of Zhejiang Province under Grant (LY23F020010).

Disclosure of Interests. The authors have no competing interests to declare that are relevant to the content of this article.

References

1. Baek, J., Aji, A.F., Saffari, A.: Knowledge-augmented language model prompting for zero-shot knowledge graph question answering. arXiv preprint arXiv:2306.04136 (2023)
2. Conover, M., et al.: Free dolly: introducing the world's first truly open instruction-tuned LLM (2023)
3. Das, R.: Nonparametric contextual reasoning for question answering over large knowledge bases. Ph.D. thesis, Doctoral dissertation (2022)
4. Das, R., et al.: Go for a walk and arrive at the answer: reasoning over paths in knowledge bases using reinforcement learning. In: International Conference on Learning Representations (2018)

5. Galkin, M., Yuan, X., Mostafa, H., Tang, J., Zhu, Z.: Towards foundation models for knowledge graph reasoning. arXiv preprint arXiv:2310.04562 (2023)
6. Ganguli, D., et al.: Red teaming language models to reduce harms: methods, scaling behaviors, and lessons learned. arXiv preprint arXiv:2209.07858 (2022)
7. Gao, J., Zhou, Y., Ribeiro, B.: Double permutation equivariance for knowledge graph completion. arXiv preprint arXiv:2302.01313 (2023)
8. Gao, M., Li, J.Y., Chen, C.H., Li, Y., Zhang, J., Zhan, Z.H.: Enhanced multi-task learning and knowledge graph-based recommender system. IEEE Trans. Knowl. Data Eng. **35**(10), 10281–10294 (2023)
9. Gu, Y., Deng, X., Su, Y.: Don't generate, discriminate: A proposal for grounding language models to real-world environments. In: Proceedings of the 61st Annual Meeting of the Association for Computational Linguistics (Volume 1: Long Papers), Toronto, Canada, pp. 4928–4949. Association for Computational Linguistics (2023)
10. Gu, Y., et al.: Beyond IID: three levels of generalization for question answering on knowledge bases. In: Web Conference, pp. 3477–3488 (2021)
11. Khosla, S., Dutt, R., Kumar, V.B., Gangadharaiah, R.: Exploring the reasons for non-generalizability of KBQA systems. In: The Fourth Workshop on Insights from Negative Results in NLP, Dubrovnik, Croatia, pp. 88–93. Association for Computational Linguistics (2023)
12. Li, T., Ma, X., Zhuang, A., Gu, Y., Su, Y., Chen, W.: Few-shot in-context learning on knowledge base question answering. In: Proceedings of the 61st Annual Meeting of the Association for Computational Linguistics (Volume 1: Long Papers), pp. 6966–6980. Association for Computational Linguistics (2023)
13. Raj, N.S., Renumol, V.: A systematic literature review on adaptive content recommenders in personalized learning environments from 2015 to 2020. J. Comput. Educ. **9**(1), 113–148 (2022)
14. Sajja, R., Sermet, Y., Cwiertny, D., Demir, I.: Platform-independent and curriculum-oriented intelligent assistant for higher education. Int. J. Educ. Technol. High. Educ. **20**(1), 42 (2023)
15. Shinn, N., Labash, B., Gopinath, A.: Reflexion: an autonomous agent with dynamic memory and self-reflection. arXiv preprint arXiv:2303.11366 (2023)
16. Tang, Y., Yuan, E., Chen, W., Zhang, S., Liu, L., Wu, Y.: Analysis of learning effectiveness based on knowledge graph. In: 2023 4th International Conference on Education, Knowledge and Information Management (ICEKIM 2023), pp. 1742–1749. Atlantis Press (2023)
17. Wei, Q., Yao, X.: Personalized recommendation of learning resources based on knowledge graph. In: 2022 11th International Conference on Educational and Information Technology (ICEIT), pp. 46–50. IEEE (2022)
18. Xia, X., Qi, W.: learning behavior interest propagation strategy of MOOCs based on multi entity knowledge graph. Educ. Inf. Technol. **28**(10), 13349–13377 (2023)
19. Zhang, H., Shen, X., Yi, B., Wang, W., Feng, Y.: KGAN: knowledge grouping aggregation network for course recommendation in MOOCs. Expert Syst. Appl. **211**, 118344 (2023)
20. Zheng, L., Niu, J., Long, M., Fan, Y.: An automatic knowledge graph construction approach to promoting collaborative knowledge building, group performance, social interaction and socially shared regulation in CSCL. Br. J. Edu. Technol. **54**(3), 686–711 (2023)

xLSTM-FER: Enhancing Student Expression Recognition with Extended Vision Long Short-Term Memory Network

Qionghao Huang[✉] and Jili Chen

Zhejiang Key Laboratory of Intelligent Education Technology and Application,
Zhejiang Normal University, Jinhua, Zhejiang, China
{qhhuang,irelia}@zjnu.edu.cn

Abstract. Student expression recognition has become an essential tool for assessing learning experiences and emotional states. This paper introduces xLSTM-FER, a novel architecture derived from the Extended Long Short-Term Memory (xLSTM), designed to enhance the accuracy and efficiency of expression recognition through advanced sequence processing capabilities for student facial expression recognition. xLSTM-FER processes input images by segmenting them into a series of patches and leveraging a stack of xLSTM blocks to handle these patches. xLSTM-FER can capture subtle changes in real-world students' facial expressions and improve recognition accuracy by learning spatial-temporal relationships within the sequence. Experiments on CK+, RAF-DF, and FER-plus demonstrate the potential of xLSTM-FER in expression recognition tasks, showing better performance compared to state-of-the-art methods on standard datasets. The linear computational and memory complexity of xLSTM-FER make it particularly suitable for handling high-resolution images. Moreover, the design of xLSTM-FER allows for efficient processing of non-sequential inputs such as images without additional computation.

Keywords: Facial Expression Recognition · Student Academic Performance · Memory Network · Vision xLSTM

1 Introduction

Student facial expression recognition is a burgeoning field with significant implications for educational technology. By analyzing students' facial cues, educators can gain insights into their emotional states, engagement levels [25], cognitive load [12], and academic performance [8,11] during learning activities [9]. The current student face recognition systems primarily include those based on traditional CNN-based and Vision Transformer [5] (ViT)-based approaches. The lightweight and efficient characteristics of CNNs have attracted the attention of early education researchers, leading to the development of a series of face recognition systems and teaching environments based on CNNs [22]. The ViT has replaced CNN as a more robust backbone network for student facial expression recognition. The Vision Transformer, leveraging self-attention for global image modeling, has surpassed the performance of CNNs in both teaching feedback systems [28] and the assessment of learning outcomes [10,29].

However, for CNNs, the main limitation is their lack of global receptive fields and dynamic weighting capabilities, which can restrict their ability to capture long-range dependencies and integrate information from the entire image [13]. Besides, this advantage of ViTs comes at the cost of quadratic complexity in terms of image sizes, which leads to a significant computational overhead when dealing with dense prediction tasks such as object detection and semantic segmentation [33].

To address the aforementioned issues, we propose the xLSTM-FER. xLSTM-FER begins by segmenting the input image into a series of non-overlapping patches, converting the 2D image into a 1D token sequence with added learnable 2D positional encodings to retain spatial information. These sequences are then fed into an xLSTM encoder composed of stacked xLSTM blocks. The xLSTM blocks maintain a linear complexity while capturing long-range dependencies and spatial-temporal dynamics within the image sequence. Each xLSTM block employs a modified LSTM layer (mLSTM) that uses matrix values for memory retrieval, enhancing the model's capacity to discern subtle facial movements. To overcome the inherent difficulty of parallel processing in LSTM, the mLSTM utilizes a memory matrix to enhance parallel capabilities. By integrating different path traversals, the model achieves a comprehensive image representation. The summary of our contributions is as follows:

- We propose xLSTM-FER, which segments input images into a series of patches and processes them through a stack of xLSTM blocks, allowing the model to capture subtle facial expression changes and improve recognition accuracy by learning the spatial-temporal dynamics within the sequence.
- The xLSTM-FER has the capabilities of parallelization and scalability through the memory matrix calculation. With its linear computational and memory complexity, which is essential for capturing clear and detailed student expressions and making xLSTM-FER a more practical solution for real-world applications.
- The extensive empirical evaluations of the xLSTM-FER model on multiple standard datasets demonstrate its superior performance in facial expression recognition tasks, including a perfect score on the CK+ [18] dataset, and shows substantial improvements over previous state-of-the-art methods on both RAF-DB [16] and FERplus [2] datasets.

2 Related Work

2.1 Student Facial Expression Recognition in Learning Environment

Early work utilizes Convolutional Neural Networks (CNNs) as the backbone for facial expression recognition tasks. Mohamad *et al.* [21] use a VGG-B network to calculate the level of student engagement in MOOCs based on their facial expressions. Lasri et al. [15] demonstrate a CNN-based automatic facial recognition system in educational settings can assist teachers in adjusting their teaching strategies and materials according to the emotional responses of students. Wang et al. [26] introduce a framework integrating an enhanced MobileViT [19] model with an online platform for real-time student emotion analysis. To analyze student expressions and inform teaching strategies, Ling et al. [17] present a classroom-based FER system using YOLO and ViT. The computational demands of ViTs grow quadratically with the self-attention mechanism, which can be prohibitive for applications requiring high-resolution processing. To make facial recognition more efficient in educational scenarios, xLSTM-FER demonstrates linear computational and memory complexity, making it more suitable for training and practical deployment.

2.2 Long Short-Term Memory Network

LSTM [7] is a type of recurrent neural network (RNN) architecture that is particularly good at learning order dependence in sequence prediction problems. Recently, Beck [3] propose improvements to LSTM, including exponential gating and novel memory structures, to address the limitations of LSTM and enable it to scale to larger model sizes. Alkin et al. [1] verify that xLSTM is also applicable as a visual backbone network. Compared to CNNs, xLSTM has the characteristic of being scalable, and compared to Vision Transformers, it has a more linear complexity which makes it easier to deploy in practice. However, its application in student facial expression recognition remains unexplored. Therefore, we propose xLSTM-FER to explore the potential application of LSTM-based models and overcome the challenge in student expression recognition.

3 Methodology

3.1 Patch Embedding

The overall architecture of our network is shown in Fig. 1a. We first perform patchification on the image. The input image $x \in R^{H \times W \times C}$ is divided into a grid of non-overlapping patches. Each patch is a small square or rectangle of pixels with a width of P. Then, each patch is flattened into a sequence of pixel values $X_p \in R^{N \times (P^2 \times C)}$, where $N = HW/(P^2)$. The flattened patch sequences are then linearly projected to a higher-dimensional space. To provide the model with information about the relative positions of the patches, we add learnable 2D positional embeddings to the patch sequences.

3.2 xLSTM Encoder

xLSTM Block. The xLSTM encoder is a structure composed of L-layer stacked xLSTM Blocks, as shown in Fig. 1b. The xLSTM Block begins by layer-normalizing and then inverting the input. One branch doubles the channels

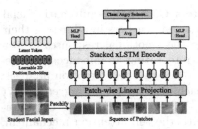
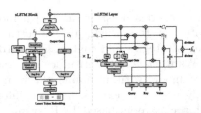

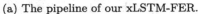

(a) The pipeline of our xLSTM-FER.

(b) Left: The model diagram of the xLSTM Block. Right: The model diagram of the xLSTM Layer.

Fig. 1. Framework of our xLSTM-FER.

(F = 2) to construct the output gating, while the other branch uses a causal convolution layer (Kernel = 3) to build the input for the mLSTM layer, which includes the query and key branch vectors for the linear attention mechanism [14], with the value vector bypassing the causal convolution. The output of the mLSTM layer is passed through a group normalization layer [30] and is then summed with the output of the causal convolution via a weighted residual connection to obtain $\tilde{h}_t$, and $\tilde{h}_t$ is gated with the result of the output gate o_t to obtain the output of the hidden state h_t. Finally, the channels are halved (F = 1/2) and sum with the input embeddings of the block through a residual connection to obtain the entire xLSTM block's output. This high-capacity storage capability enables the model to distinguish between subtle differences in facial expressions, crucial for identifying even the most nuanced emotions. This scalability is essential for creating robust systems capable of operating in diverse real-world environments.

mLSTM Layer. The mLSTM employs a FlashAttention mechanism, which is simulated using query, key, and value to guide the updates of both the cell state and the normalizer state, and subsequently outputs the results of the hidden layer as illustrated in Fig. 1b. Specifically, the mLSTM layer first performs linear projections on the query, key, and value vectors:

$$
\begin{aligned}
\text{QueryMapping} \quad & q_t = W_q\, x_q + b_q, \\
\text{ScaledKeyMapping} \quad & k_t = \frac{1}{\sqrt{d}} W_k\, x_k + b_k, \\
\text{ValueMapping} \quad & v_t = W_v\, x_v + b_v,
\end{aligned}
\quad (1)
$$

where x_q, x_k, and x_v, represent the input query, key, and value vectors respectively, while W_q, W_k, and W_v are the corresponding mapping matrices. b_q, b_k, and b_v are the corresponding bias terms. By concatenating the mapped query, key, and value vectors, the input x_t is obtained for the memory network to perform memory updates. The xLSTM uses an input gate and a forget gate to control the situation of memory updates and employs exponential gating and OR gating to facilitate the matrix memory calculation:

$$\text{Input Gate:} \quad i_t = \exp(\tilde{i}_t), \quad \tilde{i}_t = w_i^\top x_t + b_i,$$
$$\text{Forget Gate:} \quad f_t = \exp(\tilde{f}_t) \text{ OR } \sigma(f_t), \quad \tilde{f}_t = w_f^\top x_t + b_f, \quad (2)$$

where $w_i^\top$, $w_f^\top$, b_i, b_f denote the weight vectors and bias terms corresponding to the input gate and forget gate, respectively. The σ denotes the activation function, and $\exp(\cdot)$ signifies the exponential operation. The mLSTM expands the memory cell into a matrix. By integrating the update mechanism of LSTM with the information retrieval scheme from Transformers, mLSTM introduces an attention-integrated cell state and hidden state update scheme, enabling the extraction of memories from different time steps:

$$\text{Cell State:} \quad C_t = f_t C_{t-1} + i_t v_t k_t^\top,$$
$$\text{Normalizer State:} \quad n_t = f_t n_{t-1} + i_t k_t,$$
$$\text{Output Gate:} \quad o_t = \sigma(\tilde{o}_t), \quad \tilde{o}_t = W_o x_t + b_o, \quad (3)$$
$$\text{Hidden State:} \quad h_t = o_t \odot \tilde{h}_t, \quad \tilde{h}_t = C_t q_t / \max\{|n_t^\top q_t|, 1\},$$

inspired by [23], the cell state uses a weighted sum according to proportions, where the forget gate corresponds to the weighted proportion of memory, and the input gate corresponds to the weighted proportion of the key-value pair to satisfy the covariance-based update rule. The mLSTM employs a normalizer that weights key vectors. Ultimately, through normalization and weighted control by the output gate, the hidden state h_t of the network is obtained. The mLSTMs employ matrix values to process memory retrieval, which allows the retrieval process in mLSTMs to be conducted directly through matrix multiplication. The hidden state from timestep $t-1$ is not included in the processing flow, which greatly enhances the parallelization capability of the mLSTM. The mLSTM introduction of matrix memory and parallelization brings a new level of sophistication to facial expression recognition systems. By employing a matrix memory cell, the mLSTM can store a richer feature representation, capturing the intricate details and variations that define different emotional expressions. Moreover, the parallelization feature of the mLSTM block enables the model to process this complex facial data more efficiently, significantly reduce the computational load.

3.3 Path Transfer

By integrating the outcomes from these various views [24], a more accurate modeling of the sequence can be achieved. Traditional sequence modeling typically has two path traversal schemes: forward traversal and backward traversal. We have integrated four path scanning schemes: forward and backward bidirectional in the column direction and forward and backward bidirectional in the row direction. The xLSTM incorporates a flip module to achieve a more comprehensive image representation by weighting four paths of the image data.

3.4 Classification Head

The output of the xLSTM module will be mapped to the classification dimensions. The current main methods of token aggregation are as follows: 1. Using a learnable [CLS] token placed at the beginning [5] or middle [34] of the sequence. 2. Applying average pooling to the entire sequence. 3. Using the average of the first token and the last token as the input for the classification head. In the vast majority of datasets, objects are typically centered around the middle token by default. To avoid this bias and enhance the generality of our model, our experiments adopt the last scheme mentioned. Our loss function is the cross-entropy loss function:

$$\mathcal{L} = -\sum_{n=1}^{N} y \log(\hat{y}) \qquad (4)$$

4 Experiments

4.1 Datasets and Metrics

We conduct experiments on three datasets in FER research: CK+ [18], RAF-DB [16], and FERplus [2]. We report the Top-1 accuracy on the seven-category task as the evaluation metric. Here is a brief introduction to the datasets. **CK+**. The CK+ dataset includes annotations for the following emotions: Anger, Contempt, Disgust, Fear, Happy, Sadness, and Surprise. The CK+ dataset comprises 784 training samples and 197 test samples. **RAF-DB.** The RAF-DB encompasses seven basic emotional categories: surprise, fear, disgust, happiness, sadness, anger, and neutrality. The training subset encompasses 12,271 images, while the test subset consists of 3,068 images. **FERplus.** The FERplus dataset is an enhanced version of the original FER dataset. The FERplus dataset categorizes expressions into eight distinct emotions: anger, disgust, fear, happy, sad, surprise, neutral, and contempt. The dataset comprises a total of 28,709 images for training, along with 3,589 images allocated for validation and 3,589 designated for testing purposes.

4.2 Experiment Settings

We conduct experiments with a patch size set to 16 × 16, the number of stacked xLSTM layers being 26, and the base dimension of the model being 384, which means the dimensions of the query, key, and value vectors are 768. The number of our attention heads is 192.

4.3 Results

Table 1. Results on CK+, RAF-DB, and FERplus. The previous state-of-the-art (SOTA) values are marked with underlines, while the current SOTA values are marked in bold. All reported values are based on the **"from scratch"** setting.

Method	CK+	RAF-DB	FERplus
FER-GCN [6]	99.54%	-	-
FMPN [4]	98.06%	-	-
FAN [20]	<u>99.70%</u>	-	-
SelfCureNet [27]	-	<u>78.31%</u>	<u>83.42%</u>
ViT [5]	96.88%	63.75%	73.36%
MA-Net [32]	-	67.48%	-
EAC [31]	-	73.73%	75.77%
xLSTM-FER(ours)	**100%**	**87.06%**	**88.94%**
Rank	1	1	1

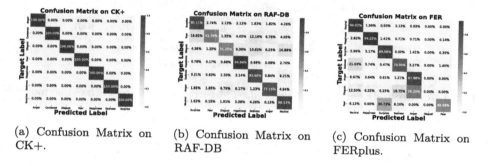

(a) Confusion Matrix on CK+.

(b) Confusion Matrix on RAF-DB.

(c) Confusion Matrix on FERplus.

Fig. 2. Confusion Metrics on Datasets.

We compare xLSTM-FER with the recent CNN-based models such as FER-GCN [6], EAC [31] and ViT-based face recognition models including ViT [5], MA-Net [32], and others to verify the effectiveness of xLSTM-FER. All experiments are conducted from scratch. The experimental results are shown in Table 1.

Results on CK+. In the CK+ dataset, the outcomes presented in Table 1 reveal that our technique, xLSTM-FER, has pioneered a perfect accuracy rate of 100% for classifying facial expressions. The confusion matrix in Fig. 2a indicates that xLSTM-FER has achieved 100% accuracy across all categories. These outcomes excel over the sota methods in the realms of both video and image-based facial expression analysis. Because xLSTM-FER successfully captures the interdependencies among different patch blocks.

Results on RAF-DB. xLSTM-FER achieves an impressive overall accuracy of 87.06% and shows a 14% improvement over the previous sota values, demonstrating superior performance compared to other models. In contrast, ViT only achieves a lower accuracy of 63.75%, and another visual transformer-based FER model, MA-Net, does not perform well on this in-the-wild dataset. xLSTM-FER achieves a competitive accuracy of 87.06%, indicating its robust performance compared to current state-of-the-art methods.

Results on FERplus. On the FERplus dataset, our model has outperformed all contemporary methods, attaining an accuracy rate of 88.94%. xLSTM-FER shows a 4.5% improvement compared with previous sota values. This confirms that the synergistic effect of the memory gating and attention mechanisms within xLSTM-FER can achieve an accurate representation of facial images.

4.4 Case Analysis

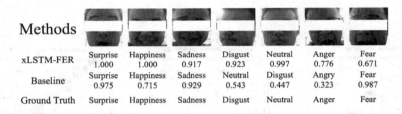

Methods							
xLSTM-FER	Surprise 1.000	Happiness 1.000	Sadness 0.917	Disgust 0.923	Neutral 0.997	Anger 0.776	Fear 0.671
Baseline	Surprise 0.975	Happiness 0.715	Sadness 0.929	Neutral 0.543	Disgust 0.447	Angry 0.323	Fear 0.987
Ground Truth	Surprise	Happiness	Sadness	Disgust	Neutral	Anger	Fear

Fig. 3. A case study on the accuracy of xLSTM-FER compared to the baseline model (EAC [31]) in real-world examples.

To further verify the advantages of xLSTM-FER over the baseline, we test several photos in the learning environment. The test results are shown in Fig. 3. We find that, except for the "Fear" and "Sadness" categories, xLSTM-FER can provide more accurate predictions with higher confidence compared to EAC in other categories. This indicates that the memory network of xLSTM in its image extraction approach can adapt to the real-world needs of students' FER tasks even with the linear complexity.

5 Conclusion

To overcome the quadratic complexity in traditional student facial expression recognition, this paper presents xLSTM-FER, which has profound implications for the assessment of learning experiences and emotional states. The innovative approach of xLSTM-FER in segmenting input images into patches and processing them through a stack of xLSTM blocks. Our experimental results on CK+, RAF-DB, and FERplus not only validate the potential of xLSTM-FER in student facial expression recognition tasks but also highlight its competitive performance when compared to state-of-the-art methods on standard datasets. The linear computational and memory complexity of xLSTM-FER is a significant advantage, making it exceptionally well-suited for processing high-resolution images, which is essential for the clear and detailed capture of student expressions. We are confident that with further optimization and fine-tuning, xLSTM-FER will evolve as a significant tool in student expression recognition.

Acknowledgement. The research project is supported by the National Natural Science Foundation of China (No. 62207028), partially by Zhejiang Provincial Natural Science Foundation (No. LY23F020009), and the Key R&D Program of Zhejiang Province (No. 2022C03106), and Scientific Research Fund of Zhejiang Provincial Education Department (No. 2023SCG367).

References

1. Alkin, B., Beck, M., Pöppel, K., Hochreiter, S., Brandstetter, J.: Vision-LSTM: xLSTM as generic vision backbone. arXiv preprint arXiv:2406.04303 (2024)
2. Barsoum, E., Zhang, C., Ferrer, C.C., Zhang, Z.: Training deep networks for facial expression recognition with crowd-sourced label distribution. In: Proceedings of the 18th ACM International Conference on Multimodal Interaction, pp. 279–283 (2016)
3. Beck, M., et al.: xLSTM: extended long short-term memory. arXiv preprint arXiv:2405.04517 (2024)
4. Chen, Y., Wang, J., Chen, S., Shi, Z., Cai, J.: Facial motion prior networks for facial expression recognition. In: 2019 IEEE Visual Communications and Image Processing (VCIP), pp. 1–4. IEEE (2019)
5. Dosovitskiy, A., et al.: An image is worth 16×16 words: transformers for image recognition at scale. arXiv preprint arXiv:2010.11929 (2020)
6. Fan, Y., Lam, J.C., Li, V.O.: Video-based emotion recognition using deeply-supervised neural networks. In: Proceedings of the 20th ACM International Conference on Multimodal Interaction, pp. 584–588 (2018)
7. Hochreiter, S., Schmidhuber, J.: Long short-term memory. Neural Comput. **9**(8), 1735–1780 (1997)
8. Huang, Q., Chen, J.: Enhancing academic performance prediction with temporal graph networks for massive open online courses. J. Big Data **11**(1), 52 (2024)
9. Huang, Q., Huang, C., Huang, J., Fujita, H.: Adaptive resource prefetching with spatial-temporal and topic information for educational cloud storage systems. Knowl.-Based Syst. **181**, 104791 (2019)
10. Huang, Q., Huang, C., Wang, X., Jiang, F.: Facial expression recognition with grid-wise attention and visual transformer. Inf. Sci. **580**, 35–54 (2021)
11. Huang, Q., Zeng, Y.: Improving academic performance predictions with dual graph neural networks. Complex Intell. Syst. 1–19 (2024)
12. Jagadeesh, M., Baranidharan, B.: Facial expression recognition of online learners from real-time videos using a novel deep learning model. Multimedia Syst. **28**(6), 2285–2305 (2022)
13. Jiang, F., et al.: Face2nodes: learning facial expression representations with relation-aware dynamic graph convolution networks. Inf. Sci. **649**, 119640 (2023)

14. Katharopoulos, A., Vyas, A., Pappas, N., Fleuret, F.: Transformers are RNNs: fast autoregressive transformers with linear attention. In: International Conference on Machine Learning, pp. 5156–5165. PMLR (2020)
15. Lasri, I., Solh, A.R., El Belkacemi, M.: Facial emotion recognition of students using convolutional neural network. In: 2019 Third International Conference on Intelligent Computing in Data Sciences (ICDS), pp. 1–6. IEEE (2019)
16. Li, S., Deng, W., Du, J.: Reliable crowdsourcing and deep locality-preserving learning for expression recognition in the wild. In: Proceedings of the IEEE Conference on Computer Vision and Pattern Recognition, pp. 2852–2861 (2017)
17. Ling, X., Liang, J., Wang, D., Yang, J.: A facial expression recognition system for smart learning based on yolo and vision transformer. In: Proceedings of the 2021 7th International Conference on Computing and Artificial Intelligence, pp. 178–182 (2021)
18. Lucey, P., Cohn, J.F., Kanade, T., Saragih, J., Ambadar, Z., Matthews, I.: The extended Cohn-Kanade dataset (CK+): a complete dataset for action unit and emotion-specified expression. In: 2010 IEEE Computer Society Conference on Computer Vision and Pattern Recognition-Workshops, pp. 94–101. IEEE (2010)
19. Mehta, S., Rastegari, M.: Mobilevit: light-weight, general-purpose, and mobile-friendly vision transformer. arXiv preprint arXiv:2110.02178 (2021)
20. Meng, D., Peng, X., Wang, K., Qiao, Y.: Frame attention networks for facial expression recognition in videos. In: 2019 IEEE International Conference on Image Processing (ICIP), pp. 3866–3870. IEEE (2019)
21. Mohamad Nezami, O., Dras, M., Hamey, L., Richards, D., Wan, S., Paris, C.: Automatic recognition of student engagement using deep learning and facial expression. In: Joint European Conference on Machine Learning and Knowledge Discovery in Databases, pp. 273–289. Springer (2020)
22. Ozdemir, M.A., Elagoz, B., Alaybeyoglu, A., Sadighzadeh, R., Akan, A.: Real time emotion recognition from facial expressions using CNN architecture. In: 2019 Medical Technologies Congress (TIPTEKNO), pp. 1–4. IEEE (2019)
23. Schlag, I., Irie, K., Schmidhuber, J.: Linear transformers are secretly fast weight programmers. In: International Conference on Machine Learning, pp. 9355–9366. PMLR (2021)
24. Schuster, M., Paliwal, K.K.: Bidirectional recurrent neural networks. IEEE Trans. Signal Process. **45**(11), 2673–2681 (1997)
25. Tonguç, G., Ozkara, B.O.: Automatic recognition of student emotions from facial expressions during a lecture. Comput. Educ. **148**, 103797 (2020)
26. Wang, J., Zhang, Z.: Facial expression recognition in online course using lightweight vision transformer via knowledge distillation. In: Pacific Rim International Conference on Artificial Intelligence, pp. 247–253. Springer (2023)
27. Wang, K., Peng, X., Yang, J., Lu, S., Qiao, Y.: Suppressing uncertainties for large-scale facial expression recognition. In: Proceedings of the IEEE/CVF Conference on Computer Vision and Pattern Recognition, pp. 6897–6906 (2020)
28. Wang, K., Cheng, M.: Teaching feedback system based on VIT expression recognition in distance education. In: 2024 13th International Conference on Educational and Information Technology (ICEIT), pp. 93–97. IEEE (2024)
29. Wu, X., et al.: FER-CHC: facial expression recognition with cross-hierarchy contrast. Appl. Soft Comput. **145**, 110530 (2023)
30. Wu, Y., He, K.: Group normalization. In: Proceedings of the European Conference on Computer Vision (ECCV), pp. 3–19 (2018)

31. Zhang, Y., Wang, C., Ling, X., Deng, W.: Learn from all: erasing attention consistency for noisy label facial expression recognition. In: European Conference on Computer Vision, pp. 418–434. Springer (2022)
32. Zhao, Z., Liu, Q., Wang, S.: Learning deep global multi-scale and local attention features for facial expression recognition in the wild. IEEE Trans. Image Process. **30**, 6544–6556 (2021)
33. Zhou, S., Wu, X., Jiang, F., Huang, Q., Huang, C.: Emotion recognition from large-scale video clips with cross-attention and hybrid feature weighting neural networks. Int. J. Environ. Res. Public Health **20**(2), 1400 (2023)
34. Zhu, L., Liao, B., Zhang, Q., Wang, X., Liu, W., Wang, X.: Vision mamba: efficient visual representation learning with bidirectional state space model. arXiv preprint arXiv:2401.09417 (2024)

RDNeRF: Radiance Distribution Guided NeRF for Floaters Removing

Yizhou Chen[1], Zixuan Huang[1], Weijun Wu[1], Weihao Yu[2], and Jin Huang[1](✉)

[1] South China Normal University, Guangzhou 08544, China
{chengyz,huangzx,wwj,huangjin}@m.scnu.edu.cn
[2] Research Institute of China Telecom Corporate Ltd., Guangzhou 510630, China
yuwh3@chinatelecom.cn

Abstract. In the field of education, virtual reality and augmented reality are significant for realizing online education and creating a multi-sensorial learning environment. Virtual online education requires effective support for reconstructing 3D representations of real-world objects. In recent years, the Neural Radiance Field (NeRF) is renowned for its high-quality 3D reconstruction capabilities across various fields. However, NeRF still suffers from artifacts named floaters when extended to large-scale scenes such as campuses or museums, which reduces its general applicability in the online education sector. In this paper, we analyze the possible causes of floater generation and reuse the NeRF pipeline to compute scene priors for guiding the geometric reconstruction of large-scale scenes. Experimental results show that our method can effectively remove floating artifacts and provide better background details. Our method achieves competitive performance on metrics compared to baseline models without additional computational overhead.

Keywords: Neural radiance fields · Virtual reality · Augmented reality

1 Introduction

The application of virtual reality (VR) and augmented reality (AR) technologies in the learning process will enhance deeper perception through 3D visualization. In VR and AR, certain 3D models, such as large architectural structures and human anatomy, can help students understand and acquire knowledge more efficiently. However, efficiently reconstructing and rendering from real objects still a major challenge in creating virtual learning environments. The method of rendering novel views based on existing object photos is called Image-Based Rendering (IBR), which was proposed by Chen and Williams [3]. However, IBR methods [2,7,8,14] usually cannot extract 3D information from objects. In recent years, the Neural Radiance Field (NeRF) [9] has gained significant attention because of the promising ability for 3D modeling from pure captured RGB images.

Although NeRF can synthesize photorealistic novel views in small-scale scenes, it may exhibit artifacts called "floaters" when the target is extended

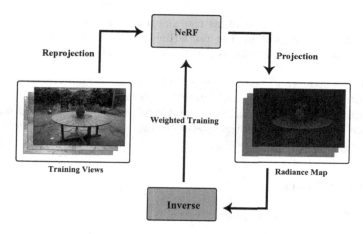

Fig. 1. The pipeline of our method: pixels are reprojected into the neural radiance field to calculate the radiance distribution and then projected back to the pixel space to produce the radiance map. We then multiplicative inverse the radiance map for weighted training.

to large-scale unbounded scenes. Recent methods employ additional regularization to avoid floaters, include discourage transparent geometry [1], appearance regularization [11] and gradient scaling [12]. However, we found that regularization is sensitive to the scenes, and larger regularization weights may create over-smooth surface. Recent depth-based methods [4,5,16,18,19] leverage depth from pre-trained models or sensors as additional supervision to address the issue of sparse inputs. DS-NeRF [5] utilizes sparse 3D point clouds from structure-from-motion (SFM) as surface key points for additional geometric supervision. MonoSDF [19] and SparseNeRF [16] utilize coarse depth maps from pre-trained depth models or consumer-level depth sensors. However, the upstream depth priors often contain potential noise and lack multi-view consistency, thus lacking sufficient robustness.

We believe that one significant reason for the occurrence of floaters is the sparse sampling of local spaces in large scenes, meaning certain background areas are captured less frequently. Based on the above reasoning, we calculated an intermediate geometric feature to guide NeRF in reconstructing accurate geometry. As shown in our pipeline (Fig. 1), we reuse the NeRF pipeline to render pixel depths and reproject them back into 3D space. Subsequently, we use a hash grid to effectively cluster the pixels in 3D space to compute the captured radiance distribution. Finally, the radiance distribution is projected back onto the pixel screen as additional supervision. Unlike previous depth-based methods, our approach achieves good multi-view consistency and low noise, and is free from other priors due to reusing the NeRF pipeline. Finally, to further enhance view synthesis quality, we propose a novel coordinate reparameterization method for better representation of distant surfaces.

The experimental results demonstrate that our approach can effectively eliminate floaters while improving the rendering quality and geometric fidelity of reconstructions. Our contributions can be summarized as follows:

- We propose a method that reutilize NeRF pipeline gained spatial radiance distribution with low-noise and multi-view consistency.
- We employ weighted regularization and gradient balancing based on spatial radiance distribution, effectively eliminating artifacts such as floaters and improving rendering quality.

2 Related Work

2.1 NeRF for Sparse Inputs

The original NeRF's performance degrades significantly when the number of input views is sparse (fewer than 10) due to the lack of multi-view information. Recent work attempts to use depth priors as additional supervision. DS-NeRF [5] utilizes sparse point clouds estimated from Structure-from-Motion (SfM) as key points to supervise surface depths across different views. MonoSDF [19] utilizes depth and normal cues from geometric monocular priors improve performance to improve geometric performance. SparseNeRF [16] utilizes coarse depth from pre-trained models and aligns the estimated depth from NeRF with locally sorted depth to avoid potential depth noise. However, depth estimates from upstream suffer from noise and multiple view inconsistency, which is the first challenge that depth-based methods need to address.

2.2 NeRF for Unbounded Scenes

Mip-NeRF 360 [1] is the first to mention the phenomenon of floater artifacts in unbounded scenes and it introduces a regularization term that encourages compact and solid surfaces along rays to avoid the appearance of semi-transparent artifacts. Nerfbusters [17] found that hand-crafted regularization terms are unable to completely remove floaters and employ a method based on 3D diffusion to discourage artifacts. NeRF++ and similar works [1,13,20] adopt coordinate reparameterization to represent the peripheral regions of scenes.

3 Method

3.1 Preliminaries of NeRF

NeRF represents a scene as a continuous radiance field in Cartesian coordinates, allowing for the querying of color and volume density at any given coordinate. The continuous radiance field is encoded as an implicit function within an MLP network. It takes 3D spatial coordinates $\mathbf{x} = (x, y, z)$ and a 2D viewing direction $\mathbf{d} = (\theta, \phi)$ as input, producing corresponding view-dependent color $\mathbf{c}$ and volume density σ as outputs. During the training stage, NeRF renders views from the same perspectives as the training images and minimizes the RGB differences between rendered and ground truth images. Specifically, rays are projected $\mathbf{r}(t) = \mathbf{o} + t\mathbf{d}$ from the camera origin to the center of each pixel (ray casting) and strategically samples N points along each ray. Afterward, the color of the pixel is obtained by integrating the color c_i and volume density σ_i of sampled points through the volume rendering technique:

$$\hat{C}(\mathbf{r}) = \sum_i^N \alpha_i T_i \mathbf{c}_i, \text{ where } T_i = \exp(-\sum_{i-1}^{j=1} \sigma_j \delta_j), \quad (1)$$

where T_i denotes the transmittance and $\alpha_i = 1 - \exp(-\sigma_i \delta_i)$, and δ_i is the sampling step size.

In each iteration, NeRF samples a set of pixels r from the training images and minimizes the difference between the predicted RGB values and the actual RGB values with MSE loss:

$$\mathcal{L}_{\text{color}} = \sum_r \|\mathbf{C}(\mathbf{r}) - \hat{\mathbf{C}}(\mathbf{r})\|^2 \quad (2)$$

When NeRF is extended to larger scenes, regularization terms are often added to mitigate the generation of floaters. A commonly used distortion regularization term was proposed by Mip-NeRF 360 [1]:

$$\mathcal{L}_{\text{dist}}(\mathbf{s}, \boldsymbol{\alpha}) = \int\int_{-\infty}^{\infty} \boldsymbol{\alpha_s}(u) \boldsymbol{\alpha_s}(v) |u - v| \, d_u d_v, \quad (3)$$

where $\mathbf{s}$ represent a set of regularized ray distances, $\boldsymbol{\alpha_s}(u)$ is the interpolation of the step function defined by $(\mathbf{s}, \boldsymbol{\alpha})$ at u: $\boldsymbol{\alpha_s}(u) = \sum_i \boldsymbol{\alpha}_i \mathbb{I}_{[s_i, s_{i+1})}(u)$.

3.2 Spatial Radiance Distribution

In this section, we reuse the NeRF pipeline to extract scene priors and use them for the geometric guidance in Sect. 3.3. We regard the pixels from the training views as samples of surface radiance to model the distribution of captured radiation. Initially, we reproject each pixel back into the pre-trained radiance field to find the corresponding surface point. For each pixel, there is a ray $\mathbf{r}(t) = \mathbf{o} + t\mathbf{d}$ casted from camera origin $\mathbf{o}$ with direction $\mathbf{d}$ and passing through pixel's center. We perform depth volume rendering over the ray $\mathbf{r}$ to obtain the relative depth $\hat{D} \in \mathbb{R}$. The depth can be obtained as follows:

$$\hat{D}(\mathbf{r}) = \sum_{i=1}^N T_i \alpha_i |\mathbf{o} - \mathbf{x}_i|, \quad (4)$$

where T_i denotes the transmittance in (1), $\mathbf{x}_i \in \mathbb{R}^3$ is the sample point along the ray $\mathbf{r}$, $\mathbf{o} \in \mathbb{R}^3$ is the camera origin, and $|\mathbf{o} - \mathbf{x}_i|$ denotes the depth of the sample point from viewpoints $\mathbf{o}$.

Based on the depth $\hat{D}$ and camera origin $\mathbf{o}$, we reproject the pixel back into 3D space to find the corresponding surface point. Let $\mathbf{y}_i \in \mathbb{R}^3$ represents the spatial coordinates after reprojection for the i-th pixel, the reprojection formula is as follows:

$$\mathbf{y}_i = \mathbf{o} + \hat{D}_i \mathbf{d} \tag{5}$$

To obtain the number of spatial pixel points $\mathbf{y}$ corresponding to a small region in the scene, we discretize the continuous radiance field into a voxel grid. Specifically, we discretize the space into a voxel grid with a resolution of h^3 where $h = b/s$, b denotes the scene boundary length, and s denotes the voxel edge length. We count the number of pixel points in each voxel grid to acquire the spatial radiance distribution. Considering memory and computational efficiency, we use a hash grid to compute the number of pixels in each voxel grid. After that, we project the spatial radiation distribution onto the image plane. We refer to the projection of the radiance distribution as the **radiance map**, as shown in Fig. 1 (right). In the radiance map, each pixel's value is $n \in \mathbb{R}$, where n represents the number of pixel points in the corresponding surface voxel. In next section, we use the reciprocal of the radiance map as the weight w:

$$w_i = \frac{1}{n_i}, \tag{6}$$

where w_i represents the weight of the i-th pixel in the training view.

3.3 Weighted Loss Training

In this section, we train NeRF using weighted reconstruction loss and regularization terms based on the weight w that obtained from previous section. We found that when the regularization strength is too low, floating artifacts cannot be prevented, while excessively high strength lead to the model failing to converge. Since floaters only appear in sparsely sampled regions, it is necessary to apply strategic weighting. Therefore, we assign higher regularization weights to correlated pixels during optimization. We perform per-pixel weighting for the regularization:

$$\mathcal{L}'_{\text{dist}} = \sum_{i=1}^{N} w_i \mathcal{L}_{\text{dist}}, \tag{7}$$

where N represents the total number of pixels, and w_i represents the weight of the i-th pixel.

Recall that NeRF uniformly samples pixels from all images, leading to gradients being dominated by the main object of the scene, which is more likely to be sampled. Such imbalanced gradient assignment may lead to the loss or blurring of details on the surfaces of distant objects, prompting us to propose a weighted reconstruction loss to scale the gradients of sparsely sampled regions and compensate for the sampling imbalance. We incorporate the weight $\mathbf{w}$ into the reconstruction loss, amplifying the gradients on surfaces at the distance for improved surface quality. The weighted reconstruction loss can be expressed as

$$\mathcal{L}'_{color} = \sum_{i=1}^{N} w_i \|\mathbf{C}(\mathbf{r}) - \hat{\mathbf{C}}(\mathbf{r})\|^2, \qquad (8)$$

Our overall loss comprises the weighted reconstruction loss and regularization loss:

$$\mathcal{L} = \mathcal{L}'_{color} + \lambda \mathcal{L}'_{dist} \qquad (9)$$

4 Experiment

4.1 Implementation Details

Our method is implemented based on Instant-NGP [10], a hash-based NeRF variant that balances efficiency and performance. Since Instant-NGP does not address floaters, we chose it as the base model to demonstrate the effectiveness of our method. Our training process consists of two stages. In the first stage, we train original Instant-NGP for 20,000 iterations until the radiance field converges. In the second stage, we fine-tune the radiance field based on weighted loss training and coordinate truncation for 30,000 iterations. We found that an additional 30,000 iterations may lead to a decrease in quality for Instant-NGP, so we use the best result at convergence (20,000 iterations) for comparison.

4.2 Datasets and Baseline

Our experiments were conducted on the large-scale unbounded dataset introduced in Mip-NeRF 360 [1], which consists of three outdoor scenes and four indoor scenes (the two undisclosed outdoor scenes are excluded). Each scene consists of a collection of RGB images captured from different angles. Our baseline includes recent NeRF variants that balance efficiency and performance: instant-NGP [10], DVGO [15], and Plenoxels [6], as well as the traditional NeRF with a pure MLP structure: NeRF [9], NeRF++ [20] and Mip-NeRF [1].

4.3 Quality Comparison

The quantitative evaluation results of the comparative experiments are displayed in Table 1. Considering the strong correlation between reconstruction quality and scene scale, we divided the results into two sections for indoor and outdoor comparisons. NeRF++ uses two sets of MLPs to represent the foreground and background, resulting in relatively higher quality. However, this approach requires significant training time and may lead to a corresponding decrease in rendering speed. As the state-of-the-art (SOTA) pure MLP nerf method, Mip-NeRF 360 achieve higher metrics than other methods (Fig. 2).

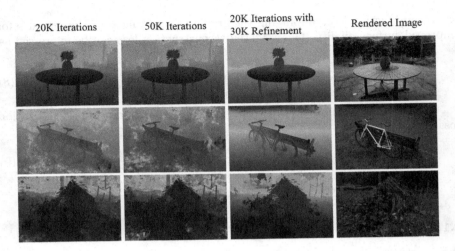

Fig. 2. From left to right for each row: Rendered depth map of 20K iterations of NGP, 20K iterations of NGP, 50K iterations of NGP with 30K ours refinement method (total 50K Iteration) and ours rendered image. Continuing training after NGP convergence (first column) could introduce more artifacts (second column). Employing our proposed method (third column) for further training proves effective in removing artifacts, leading to cleaner and higher-quality geometry.

Our method achieves the best results in real-time rendering methods for two main reasons. First, the weighted regularization enables the model to apply stronger regularization terms to reduce floaters without sacrificing overall rendering quality. Second, we amplify the gradient weights of distant surfaces, enabling the network to better optimize the global region rather than the central region. However, due to the model's use of an explicit hash structure, it cannot optimize the scene as well as the pure MLP NeRF, resulting in a gap in rendering quality compared to Mip-NerF 360.

4.4 Ablation Study

In this section, we perform an ablation study on the weighted loss, keeping the experimental setup consistent with previous sections. As shown in Table 2, the overall experimental results demonstrate that each proposed module contributes to the final rendering quality.

Background Refinement. In this section, we conducted ablation by removing weighted reconstruction loss fine-tuning, which means training the model for more iterations. Figure 3 shows that after refinement, the texture on the distant wall is restored. We also conducted experiments on outdoor scenes, and the results are presented in Table 3. Although refinement can slightly improve the PSNR and SSIM, more iterations can lead to the appearance of more artifacts, which results in a decrease in the LPIPS metric. While our weighted fine-tuning method can restore background details, it still cannot mitigate artifacts, so there is no significant improvement in the metrics.

Table 1. Quantitative results on public scenes from the Mip-NeRF 360 dataset. †indicates that the results are quoted from the original paper. The highlighted method enables real-time rendering.

	Outdoor			Indoor		
	PSNR↑	SSIM↑	LPIPS↓	PSNR↑	SSIM↑	LPIPS↓
NeRF [9]	22.79	0.492	0.518	28.31	0.800	0.330
NeRF++ [20]	23.67	0.554	0.460	29.87	0.858	0.256
Mip-NeRF 360† [1]	**25.91**	**0.747**	**0.244**	**31.72**	**0.917**	**0.180**
Plenoxels [6]	22.47	0.493	0.518	26.31	0.766	0.344
DVGO [15]	22.16	0.497	0.507	27.435	0.792	0.316
Instant-NGP [10]	23.93	0.565	0.457	29.45	0.870	0.241
Ours	**24.65**	**0.614**	**0.425**	**29.93**	**0.883**	**0.221**

Table 2. The quantitative results and training time for the overall ablation.

Method	PSNR↑	SSIM↑	LPIPS↓	Time (mins)
w/o Weighted Regularization	23.91	0.555	0.431	14
w/o Weighted Reconstruction	23.93	0.565	0.457	15
Full Model	**24.65**	**0.614**	**0.425**	15

Weighted Regularization. In the weighted regularization ablation study, we compared the rendering quality between strategy regularization, global regularization, and the absence of regularization. We set the values of λ in Eq. (9) to 0.001 and 0.004, by conducting experiments separately with global and strategy regularization. In Fig. 3, the qualitative results of the experiment are presented, demonstrating that our method can effectively remove artifacts. Table 4 shows that global regularization lead to a decrease in rendering quality. Additionally, the extreme regularization term produces a sharp decline in rendering quality. In contrast, our approach can eliminate artifacts while ensuring decreased the rendering quality. An extreme regularization term does not significantly decline in rendering quality, demonstrating the robustness of strategy regularization to regularization weights.

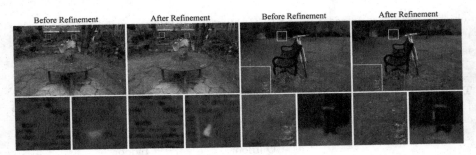

Fig. 3. Qualitative analysis of background refinement: Since the training images are captured around a specific object, the areas farther from the center exhibit lower gradients within a batch, resulting in the production of blurry surfaces and the loss of details.

Table 3. Results for ablation study of background refinement.

Method	PSNR↑	SSIM↑	LPIPS↓	Iteration (K)
w/o refinement	23.68	0.535	0.508	20
w/o weighted refinement	**23.91**	0.560	0.484	50
w/weighted refinement	**23.91**	**0.563**	**0.483**	50

Table 4. Results of the weighted regularization ablation.

Setting	Weight λ	PSNR↑	SSIM↑	LPIPS↓
w/o Reg.	0	**24.21**	0.579	0.447
Global	0.001	23.30	0.548	0.470
Global	0.004	15.08	0.279	0.724
Weighted	0.001	24.10	**0.582**	**0.446**
Weighted	0.004	24.04	0.581	0.452

5 Conclusion

In this work, we have demonstrated that reusing NeRF pipeline can obtain captured spatial radiance distribution with low-noise and multi-view consistency. This free and effective feature could provide additional geometric guidance, such as weighted loss, for NeRF-based methods.

References

1. Barron, J.T., Mildenhall, B., Verbin, D., Srinivasan, P.P., Hedman, P.: Mip-NeRF 360: unbounded anti-aliased neural radiance fields. In: Proceedings of the IEEE/CVF Conference on Computer Vision and Pattern Recognition, pp. 5470–5479 (2022)
2. Buehler, C., Bosse, M., McMillan, L., Gortler, S., Cohen, M.: Unstructured lumigraph rendering. In: Proceedings of the 28th Annual Conference on Computer Graphics and Interactive Techniques, pp. 425–432 (2001)
3. Chen, S.E., Williams, L.: View interpolation for image synthesis. In: Proceedings of the 20th Annual Conference on Computer Graphics and Interactive Techniques, pp. 279–288 (1993)
4. Deng, C., et al.: NeRDi: single-view nerf synthesis with language-guided diffusion as general image priors. In: Proceedings of the IEEE/CVF Conference on Computer Vision and Pattern Recognition, pp. 20637–20647 (2023)
5. Deng, K., Liu, A., Zhu, J.Y., Ramanan, D.: Depth-supervised NeRF: fewer views and faster training for free. In: Proceedings of the IEEE/CVF Conference on Computer Vision and Pattern Recognition, pp. 12882–12891 (2022)
6. Fridovich-Keil, S., Yu, A., Tancik, M., Chen, Q., Recht, B., Kanazawa, A.: Plenoxels: radiance fields without neural networks. In: Proceedings of the IEEE/CVF Conference on Computer Vision and Pattern Recognition, pp. 5501–5510 (2022)
7. Hedman, P., Philip, J., Price, T., Frahm, J.M., Drettakis, G., Brostow, G.: Deep blending for free-viewpoint image-based rendering. ACM Trans. Graph. (ToG) **37**(6), 1–15 (2018)
8. Levoy, M., Hanrahan, P.: Light field rendering. In: Seminal Graphics Papers: Pushing the Boundaries, vol. 2, pp. 441–452 (2023)
9. Mildenhall, B., Srinivasan, P.P., Tancik, M., Barron, J.T., Ramamoorthi, R., Ng, R.: NeRF: representing scenes as neural radiance fields for view synthesis. Commun. ACM **65**(1), 99–106 (2021)
10. Müller, T., Evans, A., Schied, C., Keller, A.: Instant neural graphics primitives with a multiresolution hash encoding. ACM Trans. Graph. (ToG) **41**(4), 1–15 (2022)
11. Niemeyer, M., Barron, J.T., Mildenhall, B., Sajjadi, M.S., Geiger, A., Radwan, N.: RegNeRF: regularizing neural radiance fields for view synthesis from sparse inputs. In: Proceedings of the IEEE/CVF Conference on Computer Vision and Pattern Recognition, pp. 5480–5490 (2022)
12. Philip, J., Deschaintre, V.: Floaters no more: radiance field gradient scaling for improved near-camera training (2023)
13. Reiser, C., et al.: MERF: memory-efficient radiance fields for real-time view synthesis in unbounded scenes. ACM Trans. Graph. (TOG) **42**(4), 1–12 (2023)
14. Riegler, G., Koltun, V.: Stable view synthesis. In: Proceedings of the IEEE/CVF Conference on Computer Vision and Pattern Recognition, pp. 12216–12225 (2021)
15. Sun, C., Sun, M., Chen, H.T.: Direct voxel grid optimization: super-fast convergence for radiance fields reconstruction. In: Proceedings of the IEEE/CVF Conference on Computer Vision and Pattern Recognition, pp. 5459–5469 (2022)
16. Wang, G., Chen, Z., Loy, C.C., Liu, Z.: SparseNeRF: distilling depth ranking for few-shot novel view synthesis. In: Proceedings of the IEEE/CVF International Conference on Computer Vision (ICCV), pp. 9065–9076 (2023)
17. Warburg, F., Weber, E., Tancik, M., Holynski, A., Kanazawa, A.: Nerfbusters: removing ghostly artifacts from casually captured NeRFs. arXiv preprint arXiv:2304.10532 (2023)

18. Xu, D., Jiang, Y., Wang, P., Fan, Z., Wang, Y., Wang, Z.: NeuralLift-360: lifting an in-the-wild 2D photo to a 3D object with 360deg views. In: Proceedings of the IEEE/CVF Conference on Computer Vision and Pattern Recognition, pp. 4479–4489 (2023)
19. Yu, Z., Peng, S., Niemeyer, M., Sattler, T., Geiger, A.: MonoSDF: exploring monocular geometric cues for neural implicit surface reconstruction. In: Advances in Neural Information Processing Systems 35, pp. 25018–25032 (2022)
20. Zhang, K., Riegler, G., Snavely, N., Koltun, V.: NeRF++: analyzing and improving neural radiance fields. arXiv preprint arXiv:2010.07492 (2020)

Multi-user VR Content Wireless Delivery Using Motion Prediction and Adaptive Multicasting

Ke Wang[✉], Yuqi Li, Kaikai Chi, and Liang Huang

School of Computer Science and Technology, Zhejiang University of Technology,
Hangzhou, China
202103150616@zjut.edu.cn

Abstract. Efficient multi-user wireless collaborative Virtual Reality (VR) systems are becoming more and more widely used in the field of education, especially during the epidemic period, teachers and students can teach and watch movies together through VR devices. However, the following challenges need to be addressed: 1. How to achieve efficient VR content delivery and high-quality user experience under limited wireless network resources? 2. How can we efficiently realize the scenario of multiple users watching the same VR video? To address these challenges, this paper introduces a novel wireless collaborative VR system. It proposes a 360-degree video transmission algorithm based on motion prediction, and an adaptive multicast algorithm using a hybrid clustering method to improve transmission efficiency and user experience. Specifically, we apply a Long Short-Term Memory Network (LSTM) model using the real-time position of the primary user's head to achieve efficient motion prediction, and only transmit the visual image data to the other following users, thus improving the utilization efficiency of network resources. Additionally, in the adaptive multicast algorithm, based on real-time network information, a clustering approach using the Self-Organizing Map (SOM) network and the k-means method realizes the effective partition of multicast groups. Then, VR content quality selection is carried out according to the status of the multicast group network, further reducing the transmission content within the system. We experimentally verified the accuracy of the LSTM model in achieving motion prediction, and verified our algorithm by deploying it to commodity mobile devices. Our system improves the quality of experience (QoE) index by nearly 74% and the frame rate index by more than 30% compared to the state-of-the-art (SOTA) technology.

Keywords: Virtual Reality in Education · Motion Prediction · Video Streaming · QoE Optimization · Multicast

1 Introduction

Virtual Reality (VR) is becoming more and more diverse. No longer confined to gaming, it is reshaping the way we collaborate and share experiences in various fields. In astronomy class, Mintz et al. [9] introduced a dynamic 3D model of the solar system that invites learners on a journey through space, significantly sparking students' interest and providing a hands-on learning experience. In healthcare, VR has become a crucial tool for training surgeons [11], enhancing their skills before they perform actual surgeries. The visualized 3D organ models [10] are revolutionizing medical education. These cases all show that VR technology is booming in education.

To deliver an immersive VR experience through wireless transmission, the VR content often needs to support resolution above 4K and fluency up to 60 frames per second (FPS), which is extremely demanding on network bandwidth. However, users can only watch a small part of the content in the VR headset at any given time. This results in a huge waste of bandwidth. In the case of the Oculus Quest 2, the video viewing angle is 100° horizontally and 90° vertically, and the transmission of video content beyond this viewing angle is invisible. Such unnecessary data transmission also increases the power consumption of mobile devices.

In addition, 360-degree video enables multiple viewers to watch content together. This viewing mode is widely used in situations such as multi-person video conferences and virtual reality classrooms. However, the traditional Dynamic Adaptive Streaming technology (DASH) [14] requires servers to stream the same content individually for each viewing device, which is very inefficient. In this case, an efficient multicast algorithm is particularly important, which can significantly reduce bandwidth consumption and show great potential for technical application.

Facing these existing challenges, this paper proposes a novel efficient multicast system. The system first collects the clients' network information, so that the clients with similar network conditions form a group by a hybrid clustering method to receive the same video stream content. In the process of video stream transmission, the system uses a 360-degree video transmission algorithm based on motion prediction to predict the next frame video content range after receiving the real-time head position of the primary user, then adjusts an additional transmission field of view according to the prediction accuracy, making it part of the transmission content. After that, the system will transmit the highest quality video stream that can be played smoothly under the bandwidth condition of each multicast group. In addition, the system will constantly update the network information of clients in real-time, and adjust the multicast group accordingly. Our main contributions are as follows:

1. We propose a LSTM-based motion prediction method to better perform 360-degree motion prediction tasks.
2. We propose a clustering approach using the SOM network and the k-means method, which is more accurate for multicast group division.
3. We conduct experimental verification on commercial devices. Our approach improves the Quality of Experience (QoE) index by nearly 74% and the frame rate index by over 30% on average compared to the state-of-the-art (SOTA) methods.

2 Related Work

2.1 Motion Prediction for 360-Degree Video Streaming

In the study of 360-degree video streaming, Facebook's dynamic streaming protocol [6] preferentially transmits high-resolution data in the center of the user's field of view while adopting low-resolution in the edge area to improve transmission efficiency. However, the lack of motion prediction can cause users to experience low-resolution content when moving quickly. Feng et al. [3] developed an innovative hybrid viewport prediction algorithm, which integrated the PARIMA model [2] to accurately predict the user's viewing focus. Son et al. [12] proposed a 360 video tiled streaming method that transmits 360-degree videos using high-efficiency video coding (HEVC) and the scalability extension of HEVC (SHVC), which generated a bitstream that can transmit tiles independently. Subsequently, Chen et al. [1] further optimized the transmitted content by using a Recurrent Neural Network (RNN) tile-based transfer algorithm, which is also one of the SOTA methods for this task.

The above works discuss the view angle optimization for single-user 360-degree video transmission, while our algorithm is more suitable for wireless multi-user collaborative application scenarios.

2.2 Adaptive Multicast Algorithm For 360-Degree Video Streaming

To cope with the rapidly changing network environment, DASH technology [13] based on HTTP is adopted, mainly targeting single-user environments. Studies [7] have shown that in scenarios where multiple users compete for 360-degree video network resource streams, these schemes may lead to bitrate fluctuations and QoE unfairness. In mobile networks, this is particularly evident. In response to the above challenges, Vinay et al. [5] propose the PAVQ algorithm to achieve a fair distribution of perceptual video quality. Liu et al. [8] propose the FireFly algorithm, which is an offline content preparation technique and demonstrates the feasibility of supporting 10 users to experience VR at 60 FPS on a single server. However, in experiments, we found that although these algorithms can offer great viewing quality, they have high transmission delay and low bandwidth utilization.

Compared with previous 360-degree video transmission schemes, we form an efficient multicast network by collecting client network status and updating it in real-time according to the dynamic changes in the network. Our experimental results have been deployed and verified on commercial devices, and have better practicability and robustness than the above schemes.

3 System Overview

In this section, we design a wireless collaborative VR system and apply it to a common educational scenario. In the classroom, the teacher plays the teaching video and explains the lecture through the VR device, while the students watch the teaching video through their VR devices. In this system, we choose the teacher as the primary user and the students as the following users. The system structure design is shown in Fig. 1.

It should be noted that the video content for the students needs to be consistent with that of the teacher to achieve the effect of listening to the class. Therefore, we can transmit the same video content to all members in the multicast group through the multicast algorithm, without the need for separate content evaluation for each user. If the transmission mode is unicast, nine processing cycles are required, while the system in Fig. 1 only needs three processing cycles. Moreover, since we add motion prediction and multicast optimization, the total processing time of the system for each frame will be reduced. This reflects the optimization effect of our system on the data transmission content.

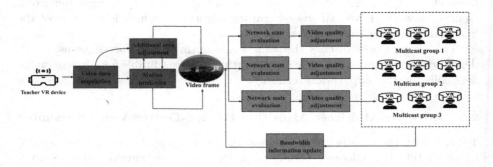

Fig. 1. Collaborative VR system architecture

First, during the system initialization phase, the client devices actively report their network status to the server. The server collects network information and divides multicast groups through the hybrid clustering algorithm. When the system runs, the server first collects the teacher's real-time head position and then uses the LSTM-based motion prediction model to predict the potential viewing range in the next frame. Based on this prediction, video content processing is performed to retain the video content within the teacher's viewing range in the next frame, and additional range adjustments are made by assessing the accuracy of the model prediction. After that, the system evaluates the network status of each multicast group and transmits the appropriate quality video. To ensure the quality of service, the server regularly reevaluates and updates the network status of the clients, and then adjusts the partition of the multicast groups according to the latest network information to ensure the stability and consistency of user viewing.

4 Algorithm Design and Analysis

In this section, we will introduce in detail the 360-degree video transmission algorithm based on motion prediction and the adaptive multicast algorithm using a hybrid clustering method.

4.1 360-Degree Video Transmission Based on Motion Prediction

Given the limitations of existing methods, we propose a 360-degree video transmission algorithm based on motion prediction. Specifically, the algorithm's execution process is as follows: In each frame cycle, the system first collects the primary user's perspective data, which includes the position on each axis and its synchronization direction. Then, the system uses the motion prediction method to independently predict the visual angle range $P_{pred}(t+1)$ of the next frame. The video data of the user's visual angle range $P_{pred}(t+1)$ will be streamed in higher resolution. In addition, the system evaluates the accuracy of the model and transmits an additional range of viewing angles at lower resolution P_θ. The total viewing range of the transmission is

$$P_{total}(t+1) = P_{pred}(t+1) + 2P_\theta \qquad (1)$$

as the left and right perspectives need to be expanded. The size of P_θ is dynamically adjusted based on model accuracy to ensure that the transmitted video content can fully cover the user's visible range.

It is important to note that in our algorithm, the prediction accuracy of the regression model greatly affect the efficiency of the whole algorithm. Therefore, we need to compare the performance of three commonly used time series prediction models, namely the Linear Regression model (LR), the Recurrent Neural Network model (RNN), and the Long Short-Term Memory model (LSTM). Three parameters of Mean value, Standard Deviation (SD), and Root Mean Square Error (RMSE) were selected to evaluate the prediction accuracy of the models.

In terms of the dataset, we adopted the AVTrack360 dataset [4]. The dataset includes data from users who viewed panoramic videos using a head-mounted display device (HMD). The data records the rotation of the user's head around the X, Y, and Z axes, including pitch angle, yaw angle, and roll angle. The dataset included behavioral data from 248 participants (125 women and 123 men) watching 20 different panoramic videos for 30 s each. The experimental results are shown in the Table 1.

Combined with the data from Table 1, it can be seen that the prediction accuracy of models will decrease with the increase of the prediction cycle time, so we need to choose the prediction cycle within 0.3 s. In addition, the LSTM model shows obvious advantages in Mean, RMSE, and SD indicators, and has more stable prediction accuracy than the LR and RNN models. Thus we use the LSTM model as the regression model for motion prediction.

Table 1. Performance of different models (LR, RNN, LSTM) at various time intervals for Mean, RMSE, and SD. Red indicates best performance, yellow indicates second-best.

Metric	Model	0.1 s	0.2 s	0.3 s	0.4 s	0.5 s
Mean	LR	3.42	4.08	7.53	11.92	15.52
	RNN	1.23	3.52	4.47	6.12	8.27
	LSTM	0.88	2.24	3.89	6.33	8.40
RMSE	LR	1.98	3.66	8.92	11.73	16.91
	RNN	1.65	3.78	5.10	8.92	10.77
	LSTM	1.33	2.12	4.67	8.72	10.89
SD	LR	1.72	2.55	5.97	7.88	10.72
	RNN	0.82	1.89	4.16	6.35	8.67
	LSTM	0.87	1.33	3.92	5.22	6.53

4.2 Adaptive Multicast Using the Hybrid Clustering Method

In the initial stage of the system, client network evaluation is carried out first. Through active measurement, the network bandwidth information BW_{c_i} of each client c_i is collected and analyzed. This process involves sending packets of predefined size when the client connects and measuring the round trip time of these packets to estimate bandwidth.

In the running stage of the system, the 360-degree video transmission based on motion prediction is used to obtain the processed video content, adjust the quality of the video content according to the network status of each multicast group, and send the adjusted video content directly to each multicast group. As the network condition of the client may change dynamically, the evaluation and update of the multicast group are performed in real-time.

To partition multicast groups effectively based on client network conditions, we propose a combination approach of the Self-Organizing Mapping (SOM) network and the k-means clustering algorithm. SOM is an unsupervised learning neural network technique specifically designed to map high-dimensional input data to a lower-dimensional topology (typically a two-dimensional grid) while preserving the topological properties of the input data. It first performs random initialization of the neuron weights, and then finds the neuron closest to the input vector, the best matching unit (BMU), by calculating the Euclidean distance between the input vector and the weights of each neuron. Next, the weights of neurons in the BMU and their neighborhoods are adjusted to better map the input data structure. The weight adjustment formula is:

$$w(t+1) = w(t) + \theta(t, u, v) \cdot \alpha(t) \cdot (x - w(t)) \qquad (2)$$

where $w(t)$ represents the neuron weight at time t, x is the current input vector, $\theta(t, u, v)$ is the neighborhood kernel function based on the BMU position, and $\alpha(t)$ is the learning rate decreasing over time.

To improve the accuracy and stability of clustering, we use the number of clusters and cluster centers derived from the SOM clustering results as input for the k-means algorithm. This helps converge faster. The algorithm outputs two main results: 1) the cluster label for each client, indicating which cluster each client is assigned to, recorded in an array of size n, and 2) the cluster centers, which are the K determined cluster centers, recorded in an array of size K, each representing the bandwidth value of a cluster center. By adjusting the correlation between cluster centers and data points, the proposed method can effectively handle the potential differences in network conditions among multicast group members, thus optimizing the composition of multicast groups.

Algorithm 1. Multicast Group Partitioning

1: **Input:** $\{BW_{c_1}, BW_{c_2}, \ldots, BW_{c_n}\}$ (Bandwidths of n clients)
2: **Initialize:** SOM weights w, number of clusters K
3: **for** $i = 1$ to n **do**
4:   $BMU_{c_i} \leftarrow \arg\min_w \|BW_{c_i} - w\|$
5:   $w \leftarrow w + \theta \cdot \alpha \cdot (BW_{c_i} - w)$ ▷ Update SOM
6: **end for**
7: $\mu \leftarrow$ SOM cluster centers
8: **for** $x \in X$ **do**
9:   $D(x) \leftarrow \min_{\mu_i} \|x - \mu_i\|$
10: **end for**
11: **for** $k = 2$ to K **do**
12:   $P(x) \leftarrow \frac{D(x)^2}{\sum_{x' \in X} D(x')^2}$
13:   $\mu_k \leftarrow$ select x with probability $P(x)$
14: **end for**
15: **Output:** Cluster labels for each client and the K cluster centers

---

Compared with existing technology, our algorithm can better utilize network resources and efficiently meet the video needs of clients. Moreover, our algorithm is not limited to specific hardware devices, exhibits good robustness, and can be deployed on commercial devices. The relevant experimental methods and performance analysis will be explained in detail in Sect. 5.

5 Experiments

In this section, we design a wireless collaborative VR system in a normal education scene, deploy our algorithms within that system, and verify their efficiency.

5.1 Experiment Results and Analysis

In our experiments, we deployed the algorithm to run on commercial devices and collected VR headset processor data, video FPS, and QoE data. In addition, we compared the collected data with the PAVQ [5], FireFly [8] and the Chen method [1], which are state-of-the-art methods for this challenge. The relevant experimental data and detailed performance indexes are shown in Fig. 2.

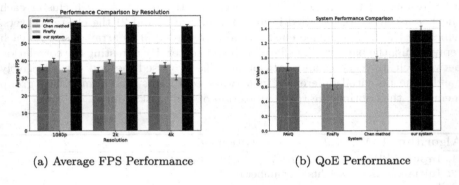

(a) Average FPS Performance   (b) QoE Performance

Fig. 2. Average FPS and QoE Performance

Figure 2(a) shows the advantages of our method in FPS. The frame rate of our system can reach the playback standard of 60 frames per second under different resolutions, which increases by 53.8% compared with the Chen method, 70.2% compared with the PAVQ algorithm, and 78.2% compared with the FireFly algorithm. Figure 2(b) shows the experimental results of the QoE index. Compared with the QoE data obtained by four different methods, our method has the most outstanding performance, with an average QoE value of 1.271, which has certain advantages over the other three methods. Our system has improved performance by approximately 39.33% compared to the Chen method, 57.22% compared to the PAVQ algorithm, and 114.89% compared to the FireFly algorithm. The results of our experiment demonstrate the benefits of our system in improving the viewing experience for users.

5.2 Ablations Experiments

In ablation experiments, we removed the 360-degree video transmission algorithm based on motion prediction and the adaptive multicast algorithm using a hybrid clustering method respectively, and verified the performance improvement brought by each module. The experimental results are shown in Table 2 below. When we used a unicast system to transmit the full video, the average QoE index was 0.41. After adding only the efficient multicast algorithm, the average QoE index was 0.63. After adding only the motion prediction algorithm, the average QoE index was 0.63. After using our complete scheme, the average QoE index was 1.14. The results of the ablation experiments effectively demonstrate the performance improvement brought by our system for this task.

Figure 3 shows the CDF graph before and after adding our algorithms. As can be seen from Fig. 3(a) and 3(b), CPU and GPU utilization have declined to a certain extent, indicating reduced processing demands on the devices. The

Table 2. QoE performance under different scenarios and resolutions in ablation experiments. Red indicates best performance, yellow indicates second-best.

Scenes	Fireworks-2K	Hawaii-2K	Diving-2K	Fireworks-4K	Hawaii-4K	Diving-4K
Unicast system	0.49	0.42	0.47	0.33	0.37	0.38
no motion prediction	0.62	0.59	0.77	0.58	0.52	0.61
no efficient multicast	0.79	0.94	0.69	0.71	0.92	0.58
full system	1.22	0.89	1.29	1.08	0.78	1.12

reason for this phenomenon is that our system reduces the content of a single video through motion prediction and reduces bandwidth consumption and network resource utilization through the multicast algorithm, thus reducing the processing load on VR headset devices and improving overall system efficiency. In Fig. 3(c), the QoE indicator has improved significantly, further proving that adopting our algorithm can significantly improve the user's VR viewing experience. The improvement in QoE directly reflects the optimization of video playback quality, allowing all users in the system to enjoy smoother and higher-quality VR content under lower bandwidth conditions.

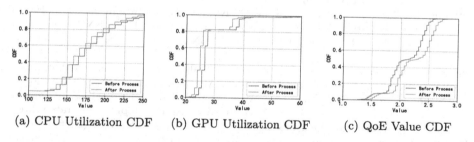

(a) CPU Utilization CDF   (b) GPU Utilization CDF   (c) QoE Value CDF

Fig. 3. Device Performance and QoE Index. The blue line is the performance before adding the optimization algorithm, and the orange line is the performance after adding the optimization algorithm. (Color figure online)

6 Conclusion

In this study, we propose two algorithms to optimize the commercial collaborative VR system. By applying and evaluating our two algorithms, we compare and analyze them against the widely used algorithms today. The experimental results show that our algorithms achieve significant improvements in key indicators such as system performance, FPS, and QoE index.

However, there are some limitations in the study. Regarding motion prediction, the model still has room for improvement in accuracy. Additionally, our algorithm performs well in scenarios with similar network conditions, but there is still room for optimization when there are large differences in bandwidth resources among users.

References

1. Chen, J., Qian, F., Li, B.: Enhancing quality of experience for collaborative virtual reality with commodity mobile devices. In: 2022 IEEE 42nd International Conference on Distributed Computing Systems (ICDCS), pp. 1018–1028 (2022)
2. Chopra, L., Chakraborty, S., Mondal, A., Chakraborty, S.: PARIMA: viewport adaptive 360-degree video streaming. In: Proceedings of the Web Conference 2021, WWW '21. ACM (2021)
3. Feng, X., Swaminathan, V., Wei, S.: Viewport prediction for live 360-degree mobile video streaming using user-content hybrid motion tracking. Proc. ACM Interact. Mob. Wearable Ubiquitous Technol. **3**(2) (2019)
4. Fremerey, S., Singla, A., Meseberg, K., Raake, A.: AVtrack360: an open dataset and software recording people's head rotations watching 360° contents on an HMD, pp. 1–6 (2018)
5. Joseph, V., de Veciana, G.: Jointly optimizing multi-user rate adaptation for video transport over wireless systems: mean-fairness-variability tradeoffs. In: 2012 Proceedings IEEE INFOCOM, pp. 567–575 (2012)
6. Kuzyakov, E.: Next-generation video encoding techniques for 360 video and VR
7. Li, Z., et al.: Probe and adapt: rate adaptation for HTTP video streaming at scale. IEEE J. Sel. Areas Commun. **32**(4), 719–733 (2014)
8. Liu, X., Vlachou, C., Qian, F., Wang, C., Kim, K.H.: Firefly: untethered multi-user VR for commodity mobile devices. In: Proceedings of the 2020 USENIX Conference on Usenix Annual Technical Conference, USENIX ATC'20. USENIX Association (2020)
9. Mintz, R., Litvak, S., Yair, Y.Y.: 3D-virtual reality in science education: an implication for astronomy teaching. J. Comput. Math. Sci. Teach. **20**, 293–305 (2001)
10. Nicholson, D.T., Chalk, C., Funnell, W.R., Daniel, S.J.: Can virtual reality improve anatomy education? A randomized controlled study of a computer-generated three-dimensional anatomical ear model. Med. Educ. **40**(11), 1081–1087 (2006)
11. Ota, D., Loftin, B., Saito, T., Lea, R., Keller, J.: Virtual reality in surgical education. Comput. Biol. Med. **25**(2), 127–137 (1995)
12. Son, J., Ryu, E.S.: Tile-based 360-degree video streaming for mobile virtual reality in cyber physical system. Comput. Electr. Eng. **72**, 361–368 (2018). https://doi.org/10.1016/j.compeleceng.2018.10.002
13. Stockhammer, T.: Dynamic adaptive streaming over HTTP –: standards and design principles. In: Proceedings of the Second Annual ACM Conference on Multimedia Systems, pp. 133–144 (2011)
14. Wei, B., Song, H., Wang, S., Katto, J.: Performance analysis of adaptive bitrate algorithms for multi-user dash video streaming. In: 2021 IEEE Wireless Communications and Networking Conference (WCNC), pp. 1–6 (2021)

MEGKT: Multi-edge Features Enhancement for Graph-Based Knowledge Tracing

Lei Zhang, Linlin Zhao[✉], and Zhenguo Zhang

Department of Computer Science and Technology, Yanbian University,
977 Gongyuan Road, Yanji 133002, China
{2023050078,llzhao,zgzhang}@ybu.edu.cn

Abstract. Knowledge tracing (KT) captures the mastery status and proficiency level of students by modeling the answer information in their historical answering records, and then predicts their future answering situations. It can help teachers better grasp their learning situation and make more reasonable and targeted teaching plans. The previous methods mainly focus on the temporal changes in the learning process of students, ignoring the spatial relationships between students, exercises, and knowledge concepts. Graph is an effective way to model these relationships and several works have been shown it's feasibility. However, existing graph-based KT methods mainly extract features from the predefined simple graphs, which usually is a single type of information in students' answering records. To better utilize the abundant information hidden in multiple correlations, we propose a new model called Multi-Edge Features Enhancement for Graph-Based Knowledge Tracing (MEGKT). In MEGKT, we first model student historical learning records as two heterogeneous graphs, where there are three types of edges between students and exercises, students and students, exercises and exercises. Then, we explore the implicit information of student answering situations from the constructed heterogeneous graphs by employing Edge Aggregated Graph Attention Network (EGAT), to build different student models. On this basis, we integrate these student models into a stronger teacher model through online knowledge distillation, allowing the teacher model to supervise the learning of the student model to achieve better results and make predictions through the teacher model. To evaluate the performance of the proposed MEGKT, we do experiments on two real education datasets and the results show that our MEGKT has achieved state-of-the-art performance in predicting student answering exercises.

Keywords: Knowledge Tracing · Heterogeneous Graph · Multiple Edge Features · Online Knowledge Distillation

1 Introduction

With the continuous development of educational technology, gaining an in-depth understanding of students' learning processes has become increasingly important. Knowledge tracing [7], as one of the key issues in the field of education, aims to reveal students' acquisition and evolution of knowledge during the learning process, identify their weaknesses, and pinpoint the key and difficult points in a particular subject. This helps teachers better understand students' learning status and formulate further learning plans.

Traditional models, such as the BKT [2] model and factor analysis models based on Item Response Theory (IRT), have significant limitations when dealing with large-scale complex data. Many scholars have considered using deep learning techniques to enhance its performance. DKT [11] is the first to apply deep learning to knowledge tracing. As research progressing, models based on attention mechanism, such as SAKT [10] and SAINT [1], and models based on Graph Neural Networks, such as GKT [9], have been proposed. These models outperform traditional models in terms of performance and scalability. Existing graph-based models primarily explore information from single associations in learning data. However, the information from these single associations is limited, which cannot fully capture the overall learning state of students. To address this issue, we consider multiple associations in learning data and explore changes in students' knowledge states from various perspectives in both time and space.

In this paper, we propose a graph-based knowledge tracing model enhanced with multiple edge features. We consider various relationships among students and questions, forming a heterogeneous graph with different types of edges. Then we use the Edge-feature Graph Attention Network (EGAT) [15] to train it. Additionally, we employ online knowledge distillation to integrate knowledge extracted from different edge feature and achieve good performance.

2 Related Works

2.1 Knowledge Tracing

Knowledge tracing aims to observe, represent, and quantify students' knowledge states. The main task is to analyze data generated from students' past learning activities, determining their knowledge mastery, and predicting their performance in future interactions. To address the shortcomings of traditional methods, deep learning approaches have been introduced. *Piech et al.* propose the DKT, which represents student's knowledge state using a single hidden vector. *Zhang et al.* propose the DKVMN [16], which use an external memory matrix to store knowledge concepts and updates student's mastery of corresponding knowledge. *Pandey et al.* propose the Self-Attentive Knowledge Tracing (SAKT) model, which utilize a transformer [12] to identify relevant knowledge concepts (KCs) [14] and assign weights to answered exercises based on student's performance.

2.2 Graph Neural Network

Graph neural networks (GNNs) [3] are widely used models in deep learning, which commonly employ for processing graph-structured data. They effectively capture relationships between different nodes, as well as between nodes and edges, enabling them to tackle various complex tasks. The core of graph neural networks lies in aggregating neighborhood information of nodes through a message passing mechanism and updating node feature representations. Classical models in GNNs, such as GCN [6], GraphSAGE [3], and GAT [13], have stronger ability to capture information within graphs. In this paper, we use a Edge-featured graph attention network (EGAT) [15] to capture the information embedded in multiple relationships among students, exercises, and knowledge concepts.

2.3 Online Knowledge Distillation

Knowledge distillation is first introduced by *Hinton* in "Distilling the Knowledge in a Neural Network" [5]. It's a model compression technique, which transfer knowledge from a complex and large Teacher model to a simpler Student model, enhancing the generalization ability of the Student model. Depending on the transfer method, it can be categorized into target-based distillation and feature-based distillation. Offline distillation method requires pre-training a large network model, which has significant limitations. Therefore, we adopt an online knowledge distillation, where a group of Student models are trained simultaneously and integrated into a stronger Teacher model. And the Teacher model provides predictions as additional supervision to the Student models.

3 Proposed Method

In this section, we explore the relationships between students and exercises, between students, and between exercises during the learning process. Our model can capture students' knowledge states and predict their future responses based on multiple edge features embeddings and online knowledge distillation. Figure 1 illustrates the overall architecture of the model:

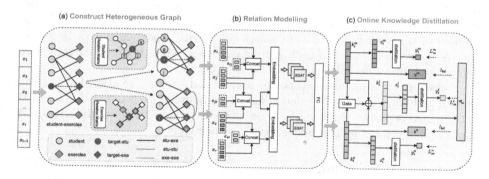

Fig. 1. The architecture of MEGKT

The first part is Construct Heterogeneous Graph, we calculate the student and skill relationship matrix, then we construct two heterogeneous graphs with different edges. Next, in Relation Modelling, we use EGAT to learn node representations and structural information in the edge features, resulting in updated node representations for two graph structures. Finally, in Online Knowledge Distillation, we treat two sub-models as peer Student models, then we integrate them into a Teacher model to enhance the prediction performance.

3.1 Problem Formulation

The knowledge tracing task can be formalized as a supervised sequence learning task. Given the student's past exercise interactions $X = (x_1, x_2, ..., x_t)$, the goal is to predict the response of the next interaction x_{t+1}. Here, $x_t = (e_t, r_t)$, where e_t represents the question that student answers at time t, and r_t indicates whether the answer is correct (1 for correct, 0 for incorrect). Then we predict the probability $p(r_{t+1} = 1|e_{t+1}, X)$ of the student correctly answering question e_{t+1} at time $t + 1$.

3.2 Construct Heterogeneous Graph

In this section, we will introduce how to construct a heterogeneous bipartite graph that integrates various edge features.

First, we need to construct a basic heterogeneous graph based on the interactions between students and exercises. We denote students $s_i \in S$ and exercises $e_i \in E$ as student nodes and exercise nodes in the heterogeneous graph, respectively. The correctness of a student's response $c_i \in C$ is considered as edge features between students and exercises. Next, we will calculate the student and the exercise relationship matrix to obtain the associations between students and between exercises.

Based on each student's answering behavior, we use the Pearson correlation coefficient to calculate the similarity matrix $M_s \in R^{S \times S}$ between students, which represents as $r \in [-1, 1]$. When $r > 0$, it indicates a positive correlation between two students; the larger the value of r, the more similar the two students are. For the exercise relationship matrix, we construct edges between exercises based on the knowledge concepts related to each exercise, Then we calculate the co-occurrence matrix $M_e \in R^{E \times E}$. Each question e_i corresponds to one or more knowledge concepts $\{k_1, k_2, \cdots, k_n\}$. Each value in the matrix represents the number of times each pair of exercises co-occur with a particular KC. Based on the co-occurrence counts of each pair of exercises, edges between exercises are categorized into different types.

Based on the student relationship matrix and exercise relationship matrix, we can add edges between students and edges between exercises to the original heterogeneous graph, resulting in heterogeneous graphs with multiple types of edges centered on students and exercises, respectively.

3.3 Relation Modelling

In this paper, we use the Edge-feature Graph Attention Network (EGAT) to learn the node representations and information from edge features in the graph. The network structure of EGAT is shown in Fig. 2. EGAT aggregates node representations as follows:

$$x_i^{(l)} = \sigma \left(W_c^{(l)} x_i^{(l-1)} + \sum_{j \in R_i} \alpha_{ij} W_b^{(l)} x_j^{(l-1)} \right) \quad (1)$$

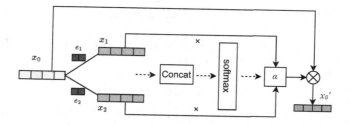

Fig. 2. Massage passing in EGAT

In Eq. (1), $x_i^{(l)}$ denotes the representation of node i at the l-th layer, and $x_i^{(0)}$ is initialized with a one-hot vector representing the initial node. R_i denotes the set of neighbors of node i. $W_a^{(l)}$ and $W_b^{(l)}$ respectively represents the learnable weights for self-loops and receiving messages from neighboring nodes at layer l. σ is the activation function. The calculation of α_{ij} in the formula is:

$$\alpha_{ij} = \frac{\exp\left(\sigma\left(W_\alpha\left[W_k x_i \| W_k x_j \| W_e e_{ij}\right]\right)\right)}{\sum_{k \in R_i} \exp\left(\sigma\left(W_\alpha\left[W_k x_i \| W_k x_k \| W_e e_{ik}\right]\right)\right)} \quad (2)$$

α_{ij} is the attention coefficient between node i and node j. W_α, W_k, and W_e represent the attention weight, node feature weight, and edge feature weight, respectively. These parameters are all learnable. The graph attention coefficient is computed by using the edge feature vector e_{ij}, which allows the incorporation of correctness and timestamp information into the message-passing process. e_{ij} can be represented as $e_{ij} = concat\,(c_{ij}, t_{ij})$. After obtaining the updated node representations through EGAT, they are concatenated along the column direction, as shown in $x_i' = concat\,(x_i^1, x_i^2, \ldots, x_i^N)$. Next, the representations of the target student's nodes x_s' and x_e' are concatenated to be used for predicting the correctness of answering each exercise by the student, which is $x_c' = concat\,(x_s', x_e')$.

Finally, we perform linear transformations and activation functions on the target node representations, and randomly drop out some neurons. We also apply a linear transformation to the output layer. The specific process is:

$$h = RELU\,(x_c' W_1 + b_1) \quad (3)$$

$$k = linear(dropout(h, \alpha)) \quad (4)$$

3.4 Online Knowledge Distillation

We consider the two sub-models as equivalent Student models, denoted as k_i^m and k_i^n, which represent the learned knowledge states from the two types of graphs, respectively. We integrate them into a Teacher model by using a gating mechanism. The computation of the gate θ is as follows:

$$\theta = \sigma(k) = \sigma\left(W_m k_i^m + b_m + W_n k_i^n + b_n\right) \quad (5)$$

where W_m and W_n are weight matrices, b_m and b_n are bias vectors. σ represents the activation function like Sigmoid.

Then we can calculate the knowledge state of the Teacher model by Eq. (6), where $k_i'^m = \theta \cdot k_i{}^m$ and $k_i'^n = (1-\theta) \cdot k_i{}^n$.

$$k_i{}^t = cat\left(k_i'^m, k_i'^n, -1\right) \tag{6}$$

To guide the training of the Student model by the Teacher model, we need to compute both soft-target and hard-target predictions for the classification task. For multi-class classification problems, soft-targets represent the probabilities for each possible class. By adding a temperature T to the original softmax function, the Student model can learn the implicit information in the negative labels. The updated softmax function can be represented as follows:

$$q_i = \frac{\exp\left(z_i/T\right)}{\sum_j \exp\left(z_j/T\right)} \tag{7}$$

where z represents the logit for each class. The logits of each model are calculated according to $z_i = w^T k_i + b$. Then we obtain the logit of two Student models, denoted as z_i^m and z_i^N, and the logits of Teacher model, denoted as z_i^T. Then, we use updated softmax function to compute soft-targets with $y_i' = (1 + e^{-z_i/T})^{-1}$. Thus we obtain the soft-targets for the Student model, denoted as $y_i'^m, y_i'^n, y_i'^t$. We can compute the distillation loss during training process by Eq. (8). Additionally, we compute the cross-entropy loss generated during training for each model based on predicted values y_i and actual response r_{t+1} by Eq. (9).

$$loss_{kd} = \frac{\sum_{i=1}^{n} \|y_i'^t - y_i'^m\|_1 + \|y_i'^t - y_i'^n\|_1}{n} \tag{8}$$

$$loss_{bce} = -\frac{\sum_{i=1}^{N}\left(y_i \log\left(r_{t+1}\right) + (1-y_i) \log\left(1 - r_{t+1}\right)\right)}{N} \tag{9}$$

The sum of cross-entropy losses of Teacher model and Student model is $loss_{train}$. The final total loss of the model is represented by Eq. (10), where μ is used to control the weight between the training loss and the distillation loss.

$$Loss = loss_{train} + \mu loss_{kd} \tag{10}$$

4 Experiment

4.1 Implementation Details

Datasets. To evaluate the model's performance, two real-world datasets are used in our experiments: ASSIST17[1] and EdNet[2]. ASSIST17 is provided by

---
[1] https://sites.google.com/view/assistmentsdatamining/data-mining-competition-2017.
[2] https://github.com/riiid/ednet.

the online education platform ASSISTment. It comprises 942,816 records from 1,709 students answering 3,162 questions. EdNet is collected by Santa, contains over 130 million records. For our experiments, we selected 1,500,000 records from 5,760 students answering 2,394 questions as other research works. Table 1 provides a detailed overview of the data.

Parameter and Evaluation Methodology. In the experiment, we split 80% of the dataset as training set and the remaining 20% as validation set. We set the maximum sequence length to 64, the dimensionality of hidden layer to 32, and the number of hidden layers to 4. The hyperparameters such as epochs, learning rate, dropout rate, and batch size are set to 20, 2e−3, 0.5, and 128. We use the Adam optimizer to optimize the model parameters and perform five-fold cross-validation to obtain the final experimental results. To assess the effectiveness of the methods proposed by the MEGKT model, we will use the Area Under the Curve (AUC) and Accuracy (ACC) to evaluate the model's effectiveness.

Table 1. Datasets statistics

	ASSIST17	Ednet
# Student	1709	5760
# Exercise	3162	2394
# Concept	102	188
# Record	942816	1500000
Concepts per exercise	1.00	2.26
Student per record	551.68	260.42

4.2 Baselines

We compared the proposed MEGKT model to eight baseline models:

DKT [11]: DKT tracks student knowledge states with RNN and LSTM, it assumes a single hidden knowledge concept in the student's knowledge state.

DKVMN [16]: DKVMN enhances the external memory structure, keeps the key matrix static while designing the value matrix as dynamic.

SAKT [10]: SAKT applies attention mechanism to knowledge tracing, which learn the attention weights of questions in a series of interactions.

SAINT [1]: SAINT separates student interaction sequences into question and response embedding sequences, then use them as inputs to the encoder and decoder.

GKT [9]: GKT applies GNN to KT for the first time and rephrases KT problems as a time-series node-level classification problem.

PEBG [8]: PEBG constructs a bipartite graph of exercise-concept relationships to learn pre-trained embeddings of exercises.

IGKT [4]: IGKT introduces a new inductive graph-based KT framework that fully utilizes subgraph extraction methods of the student-exercise bipartite graph.

4.3 Results

Effect of Parameters. To investigate the impact of parameter variations on the experimental results, we conduct further experiments to study the effects of temperature T, hidden layer dimension d, and distillation weight coefficient μ. The specific experimental conditions are shown in Fig. 3, Fig. 4 and Fig. 5. From the experiments we can find that the best performance is $T = 0.5$, $d = 32$, and $\mu = 5e-6$.

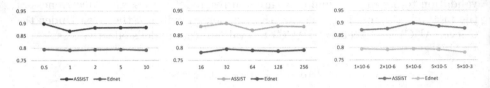

Fig. 3. Sensitivity of T **Fig. 4.** Sensitivity of d **Fig. 5.** Sensitivity of μ

Performance Prediction. The experimental results on the two datasets are shown in Table 2: It can be observed that our model achieved the best performance compared to other baselines. Our model outperformed the second-best model IGKT by 1.5% and 1.9% in terms of AUC and ACC on the ASSIST17 dataset, and by 1.0% and 1.3% on the Ednet dataset.

Table 2. Comparison of prediction performance

Method	ASSIST17		Ednet	
	AUC	ACC	AUC	ACC
DKT	0.703	0.680	0.679	0.688
DKVMN	0.693	0.677	0.671	0.679
SAKT	0.668	0.641	0.646	0.639
SAINT	0.793	0.755	0.766	0.703
GKT	0.762	0.730	0.726	0.698
PEBG	0.829	0.783	0.718	0.696
IGKT	0.884	0.799	0.785	0.742
MEGKT	**0.899**	**0.818**	**0.795**	**0.755**

From the results, we can also find that:

(1) Graph-based models exhibit better performance in predicting student answers than the models based on attention mechanism. Our proposed model outperform the second-best model SAINT in AUC and ACC by 6.75% and 5.75% on average.
(2) Considering multiple edge features in graph-based models is more conducive. Our model showed improvements of 10.30% and 7.25% in AUC and ACC compared to the GKT which only considers single associations.
(3) Incorporating knowledge distillation enhances the performance significantly. Compared to the PEBG which does not use knowledge distillation, MEGKT achieved improvements of 7.35% and 4.7% in AUC and ACC on average.

4.4 Ablation Study

To better validate the rationality and effectiveness of our method, we conducted a series of ablation experiments. We designed the following five variants:

MEGKT-OP: The graph contains only one type of edge between students and questions.

MEGKT-AE: Both input graph structures for the two student models include edges between questions.

MEGKT-AS: Both input graph structures for the two student models include edges between students.

MEGKT-GAT: Replace the Edge Aggregating Graph Attention Network with a general Graph Attention Network for training the model.

MEGKT-ED: Remove knowledge distillation from the model.

The results of the ablation experiments are shown in Table 3. Based on the results, we can get the following conclusions:

(1) Effectiveness of Adding Multiple Edges: Considering multiple edges simultaneously yield better results than considering single type of edge. Removing some edges led to decrease in AUC and ACC by 1.32% and 1.28% on average.
(2) Effectiveness of EGAT: Removing EGAT led to decrease in AUC and ACC by 12.8% and 9.1% on the average. This indicates the importance of handling edge features for graph-based knowledge tracing methods.
(3) Effectiveness of Knowledge Distillation: Models with knowledge distillation outperform those without it. Removing knowledge distillation led to decrease by 1.35% in AUC and by 1.3% in ACC on the average of two datasets.

Table 3. Performance of ablation study

Method	ASSIST17		Ednet	
	AUC	ACC	AUC	ACC
MEGKT-OP	0.881(↓ 1.8%)	0.802(↓ 1.6%)	0.790(↓ 0.5%)	0.752(↓ 0.3%)
MEGKT-AE	0.880(↓ 1.9%)	0.797(↓ 2.1%)	0.788(↓ 0.7%)	0.749(↓ 0.6%)
MEGKT-AS	0.874(↓ 2.5%)	0.796(↓ 2.2%)	0.790(↓ 0.5%)	0.746(↓ 0.9%)
MEGKT-GAT	0.820(↓ 7.9%)	0.736(↓ 8.2%)	0.618(↓ 17.7%)	0.655(↓ 10.0%)
MEGKT-ED	0.881(↓ 1.8%)	0.795(↓ 2.3%)	0.786(↓ 0.9%)	0.752(↓ 0.3%)

5 Conclusion

This paper proposes a knowledge tracing model enhanced with multiple edge features based on a heterogeneous graph, termed MEGKT. We explore changes in students' knowledge states during the learning process in the form of a graph, constructing various edges between students and exercises, between students, and between exercises, thus forming different student models. We also utilize online knowledge distillation to integrate the Student models into a Teacher model. This Teacher model learns the information embedded in the graph from different Student models and supervises the learning of the Student models, achieving better modeling performance. We conducted experiments on two public datasets, and the results demonstrate that our model effectively captures the overall knowledge mastery of students and achieves the best prediction performance. In the future, we will consider the influence of more factors on knowledge state to better improve performance, such as forgetting factor.

Acknowledgment. This work is supported by the Science and Technology Project of Jilin Provincial Education Department [JJKH20240681KJ] and the 14th Five-Year Plan Project of Jilin Province Education Science (GH23513).

References

1. Choi, Y., et al.: Towards an appropriate query, key, and value computation for knowledge tracing. In: Proceedings of the Seventh ACM Conference on Learning@ Scale, pp. 341–344 (2020)
2. Corbett, A.T., Anderson, J.R.: Knowledge tracing: modeling the acquisition of procedural knowledge. User Model. User-Adap. Inter. **4**, 253–278 (1994)
3. Hamilton, W., Ying, Z., Leskovec, J.: Inductive representation learning on large graphs. In: Advances in Neural Information Processing Systems 30 (2017)
4. Han, D., Kim, D., Kim, M., Han, K., Yi, M.Y.: Temporal enhanced inductive graph knowledge tracing. Appl. Intell. **53**(23), 29282–29299 (2023)
5. Hinton, G., Vinyals, O., Dean, J.: Distilling the knowledge in a neural network. arXiv preprint arXiv:1503.02531 (2015)
6. Kipf, T.N., Welling, M.: Semi-supervised classification with graph convolutional networks. arXiv preprint arXiv:1609.02907 (2016)

7. Liu, S., et al.: A hierarchical memory network for knowledge tracing. Expert Syst. Appl. **177**, 114935 (2021)
8. Liu, Y., Yang, Y., Chen, X., Shen, J., Zhang, H., Yu, Y.: Improving knowledge tracing via pre-training question embeddings. arXiv preprint arXiv:2012.05031 (2020)
9. Nakagawa, H., Iwasawa, Y., Matsuo, Y.: Graph-based knowledge tracing: modeling student proficiency using graph neural network. In: IEEE/WIC/ACM International Conference on Web Intelligence, pp. 156–163 (2019)
10. Pandey, S., Karypis, G.: A self-attentive model for knowledge tracing. arXiv preprint arXiv:1907.06837 (2019)
11. Piech, C., et al.: Deep knowledge tracing. In: Advances in Neural Information Processing Systems 28 (2015)
12. Vaswani, A., et al.: Attention is all you need. In: Advances in Neural Information Processing Systems 30 (2017)
13. Velickovic, P., Cucurull, G., Casanova, A., Romero, A., Lio, P., Bengio, Y., et al.: Graph attention networks. Stat **1050**(20), 10–48550 (2017)
14. Wang, W., Ma, H., Zhao, Y., Li, Z., He, X.: Tracking knowledge proficiency of students with calibrated q-matrix. Expert Syst. Appl. **192**, 116454 (2022)
15. Wang, Z., Chen, J., Chen, H.: EGAT: edge-featured graph attention network. In: Artificial Neural Networks and Machine Learning–ICANN 2021: 30th International Conference on Artificial Neural Networks, Bratislava, Slovakia, 14–17 September 2021, Proceedings, Part I 30, pp. 253–264. Springer (2021)
16. Zhang, J., Shi, X., King, I., Yeung, D.Y.: Dynamic key-value memory networks for knowledge tracing. In: Proceedings of the 26th International Conference on World Wide Web, pp. 765–774 (2017)

STBDM Workshop

Collaborative Inference for Adaptive DNN Pipeline-Aware Alignment

Jiankang Ren[1], Lin Liu[1], Ran Bi[2,3(✉)], Simeng Li[1], Shengyu Li[1], and Xian Lv[1]

[1] Dalian University of Technology, Dalian, China
[2] Northeastern University at Qinhuangdao, Qinhuangdao, China
biran@neuq.edu.cn
[3] Hebei Key Laboratory of Marine Perception Network and Data Processing, Qinhuangdao, China

Abstract. Recent advancements in Internet of Things (IoT) and deep neural network (DNN) have enabled spatio-temporal data-based inference tasks with substantial computational demands. Edge computing mitigates these demands by offloading DNN tasks to edge servers. However, existing studies mainly focus on single-task optimization for delay reduction, neglecting system throughput and energy efficiency, which is critical for device lifespan and performance. Our work addresses the high throughput and real-time requirements of DNN inference tasks for spatio-temporal data, while reducing system energy overhead and ensuring task accuracy. We employ early exit techniques and consider inter-task parallelism to optimize energy consumption and enhance system throughput. We refine the original flow graph and achieve efficient pipeline alignment through effective scheduling. When selecting early exit models, we only consider tasks that meet accuracy requirements, thus avoiding models with poor inference accuracy. Experimental results show that compared with the existing state-of-the-art schemes, our method improves the throughput by 14.2% to 60% on models such as VGG19, Googlenet and Resnet-50, and reduces the average inference energy consumption of tasks by about 11.1% to 49.8%.

Keywords: edge inference · real time · DNN partitioning · throughput

1 Introduction

In recent years, the rapid development and integration of the Internet of Things (IoT) and Artificial Intelligence (AI) have revolutionized fields like smart cities and smart transportation. For instance, intelligent transportation systems can enhance risk management and reduce road congestion by leveraging 5G and deep neural network (DNN) [1]. These applications often involve data with spatial and temporal characteristics, such as severe traffic congestion during peak hours and sparse traffic at night [2]. Moreover, these applications are highly sensitive to latency and require real-time processing, which demands more computational power from edge terminal devices than traditional spatio-temporal data-based applications [3]. To address these challenges, edge computing has emerged, bringing computational power closer to the task requests. By offloading DNN inference tasks from edge devices to edge servers for collaborative processing, an edge inference solution has been developed to meet the real-time requirements of AI applications such as autonomous driving.

Many existing studies [5,6,13,14] in edge computing focus on partitioning single tasks to minimize execution time, thus improving the user's service quality in terms of delay. However, for spatio-temporal data inference tasks, the primary goal is to enhance system throughput while satisfying delay constraints. Directly applying these single-task optimization methods can result in performance loss. Additionally, edge inference task requests fluctuate over time due to spatio-temporal characteristics, influenced by factors like bad weather and nighttime conditions [7,8]. Researchers have proposed optimization techniques such as early exit, pruning, and quantization to enable DNNs to run on computation-constrained local device (LD) [4]. It has been shown that computational power needed for different inference tasks (e.g., clear vs. fuzzy images) to achieve the same accuracy can vary [9,10]. Thus, the early exit technique can reduce unnecessary computation and improve throughput by selecting appropriate DNN model exit points that meet accuracy requirements. Moreover, since LD are often battery-powered with limited energy capacity, efficient partitioning of DNN inference tasks and optimal allocation of computational resources are crucial for prolonging device lifespan and enhancing task performance. However, few existing studies on edge inference optimize both throughput and energy consumption.

Our main contributions are as follows: 1. We propose the Rectified Adaptive Graph-like Deep Neural Network Surgeon (RAGDNS) algorithm, which leverages an early exit mechanism to achieve real-time high throughput for spatio-temporal data inference tasks. 2. We design a rectified max-flow min-cut algorithm for balanced partitioning by considering the differences in the start execution times of LD and ES, thus fully utilizing computing resources to improve system throughput. 3. We construct temporal relationships between constraints to optimize system energy consumption. This ensures that resource allocation meets sequential consistency and real-time execution requirements both within and across tasks, while also reducing energy usage.

2 Related Work

Efficient DNN inference on mobile devices requires overcoming significant computational, latency, and energy constraints [11]. We review the relevant literature on DNN partition and task offloading for optimizing inference-tasks execution.

Previous research has focused on reducing inference latency in IoT devices through various model partition and offloading computations. Neurosurgeon [5] found that cut-point selection impacts both end-to-end latency and energy consumption of IoT devices due to differences in data sizes and computational requirements at each layer. This approach splits the DNN model into two parts, offloading the more resource-intensive part to a powerful cloud. Building on this, Edgent [12] refines partition to further reduce latency by training multiple master AI models to adaptively select model size. However, the fragile wireless link between the device and edge server may become a bottleneck. To mitigate this, CLIO [13] and SPINN [14] compress the DNN layers at the partition points, reducing data transmission delays. Unlike these approaches, MoDNN [15] maps different parts of the DNN layers to mobile devices to accelerate computation.

Recent studies focus on minimizing both task processing delay and energy consumption in DNN inference by dynamic task offloading and workload partitioning. Bozorgchenani et al. [16] proposed an evolutionary algorithm to minimize both task processing delay and energy consumption in MEC networks. Zeng et al. [17] introduced multi-edge devices with heterogeneous computing capabilities to collaborate on DNN inference, focusing on dynamic workload partitioning and optimization. However, their strategies are limited to chain topology DNNs. As DNN structures evolve from traditional chain to directed acyclic graphs (DAGs) models like GoogleNet and ResNet [18]. Hu et al. [6] explored DNN partitioning in networks composed of edge servers and clouds, by solving maximum stream minimum cut problem. They further developed the DSH algorithm to enhance throughput demands under high loads [19]. Yang et al. [18] also applied the max-flow min-cut algorithm for DAG-structured DNNs, using a budgeted early exit strategy to compute partitions.

The work by Yang et al. [20] partitions DNN layers into blocks, utilizing edge devices for distributed computation, with each device completing one or more blocks. Liu et al. [21] improves system throughput by offloading all DNN inference tasks to the edge server, scheduling computational power and transmission bandwidth to form a pipeline for task computation and transmission. However, this approach is highly network-dependent and may degrade partitioning performance by combining DAG-like DNN branches into large blocks. Our approach differs from previous studies by addressing the high throughput and real-time demands of DNN inference tasks for spatio-temporal data, reducing system energy overhead while meeting accuracy requirements.

3 System Model and Problem Formulation

3.1 System Model

Inference tasks are assumed to be performed by the device in each time slot. These tasks are generated at the beginning of the time slot and share the same end-to-end inference delay requirement, equal to the time slot length. Tasks are processed sequentially by queuing within the local device (LD) and edge server (ES). Tasks in the queue that exceed the delay constraint are discarded.

The ith task generated by the LD in the time slot is denoted as τ_i. These tasks use a common DNN inference model, such as Googlenet. Similar to [18], the edge inference task uses a budgeted early exit optimization mechanism. The DNN network structure of different early-exit branching models is fixed and known, with pre-trained model parameters stored for use. The DNN model with the jth exit branch for the ith task is represented as $G_{i,j} = (V_{i,j}, E_{i,j})$. For clarity, subscripts are omitted when no confusion, i.e., $G_i = (V_i, E_i)$, where V_i is the set of DNN layers and E_i is the set of layer dependencies. V_i contains $|V_i|$ layers: $v_{i,0}, v_{i,1}, \ldots, v_{i,|V_i|-1}$, where $v_{i,0}$ is a virtual input layer. Node $v_{i,1}$ denotes the first layer of data input processing, and $v_{i,|V_i|-1}$ denotes the last layer, outputting the inference result. Each edge $(v_{i,j}, v_{i,k}) \in E_i$ indicates that the computation of layer $v_{i,j}$ must be completed before computing layer $v_{i,k}$. Considering that a layer's input data may come from multiple layers, the result of one layer's computation serves as input for multiple subsequent layers. Thus, $N^+(v_{i,j}) = \{v_{i,k} \mid (v_{i,j}, v_{i,k}) \in E_i\}$ denotes nodes using the output of $v_{i,j}$ as input, while $N^-(v_{i,k}) = \{v_{i,j} \mid (v_{i,j}, v_{i,k}) \in E_i\}$ denotes the nodes required for the input data for the computation of node $v_{i,k}$ (Fig. 1).

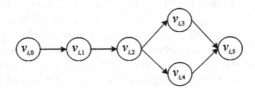

Fig. 1. DNN model graph

To fully utilize ES and LD computing resources for accelerating DNN inference tasks and enhancing system throughput, each DNN inference task τ_i can be divided into two disjoint subsets V_i^l and V_i^e. Here, V_i^l and V_i^e represent layers processed locally and offloaded to ES for collaborative processing, respectively. These subsets satisfy $V_i = V_i^l \cup V_i^e$ and $V_i^l \cap V_i^e = \emptyset$. Additionally, we denote by V_i^{tr} as the layer (nodes) that transmits computation results to ES for further process, ensuring $V_i^{tr} \subseteq V_i^l$. It is assumed that if an inference layer $v_{i,j}$ is processed at ES, any node $v_{i,k}$ requiring its output as input (indicated by an edge $(v_{i,j}, v_{i,k})$) must also be processed at ES. Therefore, when $v_{i,j} \in V_i^e$, the nodes $N^+(v_{i,j}) \subseteq V_i^e$ ensure the graph of the DNN inference task is partitioned into only two parts.

3.2 End-to-End Processing Delay and Energy Consumption Model

Let C_i^l and C_i^e denote the computing resources allocated to τ_i at the LD and ES, respectively, in terms of CPU frequency (cycles/s). These resources are controlled by dynamic voltage and frequency scaling techniques and constrained by $C_{\max}^l$ and $C_{\max}^e$. The required number of floating-point operations for layer $v_{i,j}$ is $f(v_{i,j})$. The transmission power of the LD is P^{tr}, and the transmission rate for the current time slot is R.

End-to-End Processing Delay Model. The end-to-end processing delay of a task depends on the LD start execution time, LD execution duration, transmission delay, ES start execution time, and ES execution duration. The execution duration for layer $v_{i,j}$ is $\frac{f(v_{i,j})}{C}$, where C is the allocated computing resource. Thus, the LD execution duration of τ_i is

$$t\left(V_i^l\right) = \sum\nolimits_{v_{i,j} \in V_i^l} \frac{f(v_{i,j})}{C_i^l} \tag{1}$$

Similarly, the ES execution duration of $task_i$ is

$$t\left(V_i^e\right) = \sum\nolimits_{v_{i,j} \in V_i^e} \frac{f(v_{i,j})}{C_i^e} \tag{2}$$

Let t_i^{ls} and t_i^{es} denote the LD and ES start execution time, respectively. Based on queue execution characteristics, the relationship between the start times of adjacent tasks on the LD is as follows:

$$t_i^{ls} = t_{i-1}^{ls} + t\left(V_{i-1}^l\right) \tag{3}$$

$O_{i,j}$ is the output data of layer $v_{i,j}$. Thus, the transmission delay of $task_i$ is

$$t\left(V_i^{tr}\right) = \sum\nolimits_{v_{i,j} \in V_i^{tr}} \frac{O_{i,j}}{R} \tag{4}$$

The start execution time of ES depends on the task execution logic dependencies and queue characteristics:

$$t_i^{es} = \max\left(t_i^{ls} + t\left(V_i^{tr}\right), t_{i-1}^{es} + t\left(V_{i-1}^e\right)\right) \tag{5}$$

We can obtain the end-to-end processing delay of τ_i as

$$t_i^\lambda = t_i^{es} + t\left(V_i^e\right) \tag{6}$$

where t_i^λ is the end-to-end processing delay of τ_i. Formula (6) holds even if tasks are not offloaded.

End-to-End Processing Energy Consumption Model. The end-to-end processing energy consumption of a task comprises LD execution energy consumption, transmission energy, and ES execution energy consumption. The transmission energy consumption of τ_i is given by $e^{tr} = P^{tr} \times t(V_i^{tr})$. Based on [22], CPU power consumption P^e is a superlinear function of execution frequency, given by $P^e = \epsilon C^3$, where ϵ is a constant dependent on the chip, and C is the allocated computing resource. The execution duration of layer $v_{i,j}$ is $\frac{f(v_{i,j})}{C}$, resulting in an energy consumption of $\epsilon C^2 f(v_{i,j})$ for $v_{i,j}$. Thus, the LD execution energy consumption $e\left(V_i^l\right)$ and the ES execution energy consumption $e\left(V_i^e\right)$ for τ_i are given by:

$$e\left(V_i^l\right) = \sum\nolimits_{v_{i,j} \in V_i^l} \epsilon\left(C_i^l\right)^2 f(v_{i,j}) \tag{7}$$

$$e\left(V_i^e\right) = \sum\nolimits_{v_{i,j} \in V_i^e} \epsilon\left(C_i^e\right)^2 f(v_{i,j}) \tag{8}$$

The total end-to-end processing energy consumption of τ_i is

$$E_i = e\left(V_i^l\right) + e\left(V_i^{tr}\right) + e\left(V_i^e\right) \tag{9}$$

3.3 Problem Formulation

Our goal is to maximize system utility, which consist of throughput, model accuracy gain and computation slack gain. We normalize the computation slack gain to address magnitude differences and to incorporate load-awareness for balancing throughput and model accuracy. Unlike Edgent [12], we express accuracy as a logarithmic function instead of an exponential one. Let $\mathbb{C}$, $\mathbb{P}$, and S represent the allocated computing resources, partition strategies, and selected exit branching models for the inference tasks, respectively. The problem is formulated as follows:

$$(\mathcal{P}) : (\mathbb{C}, \mathbb{P}, S) = \max |S| + 1 - \frac{E_s}{E_f} + \sum_{G_{i,j} \in S} \log\left(1 + p_{i,j} - p_L\right)$$

$$s.t.\ 0 \le c_i^l \le C_l, \forall G_{i,j} \in S \tag{10a}$$

$$0 \le c_i^e \le C_e, \forall G_{i,j} \in S \tag{10b}$$

$$t_i^\lambda \le D, \forall G_{i,j} \in S \tag{10c}$$

$$t_i^{ls} + t\left(V_i^l\right) \le t_i^{es}, \forall G_{i,j} \in S \tag{10d}$$

where E_s and E_f denote the processing energy consumption with and without resource slack, respectively. $p_{i,j}$ is the accuracy of the selected exit branching model, and p_L is the minimum accuracy requirement for task τ_i. c_i^l and c_i^e denote the computing resources allocated to task τ_i on LD and ES, respectively. C_l and C_e are the maximum computing resources of LD and ES, respectively. (10a)–(10c) represent the resource and end-to-end processing delay constraints. Constraint (10d) indicates the dependency of task execution time.

Theorem 1. *Problem $\mathcal{P}$ is NP hard.*

Due to space limitations, the proof is omitted; please refer to [23].

4 Rectified Adaptive Graph-Liked Deep Neural Network Surgeon

4.1 RAGDNS Algorithm Design

For Problem $\mathcal{P}$, there are three main challenges: 1) Selecting the appropriate early exit branching model for each task. 2) Partitioning tasks to minimize delay increase for subsequent tasks while fully utilizing parallel computation on different devices, thus allowing the system to complete more inference tasks. 3) Performing computation relaxation to meet end-to-end processing delay constraints, while addressing the incompatibility of temporal logic dependencies within and between tasks. To address these challenges, we propose an online algorithm called Rectified Adaptive Graph-like Deep Neural Network Surgeon (RAGDNS). This algorithm includes two sub-algorithms to tackle the partitioning and computation relaxation problems. Due to space limitations, we omit the details of the rectified max-flow min-cut algorithm (Algorithm 2) and construction of constraint and relax computing resource algorithm (Algorithm 3). For full details, please refer to [23]. The RAGDNS algorithm is presented in Algorithm 1.

Algorithm 1: RAGDNS algorithm

Require:
  number of device tasks n, channel noise σ, transmission bandwidth B_l, Transmission Power P^{trans}, alternative branch exit model set s, minimum task accuracy $P_{lowbound}$, device computation resources C_l, server computation resources C_e, end-to-end delay constraint D

Ensure:
  allocated computational resource $\mathbb{C}$, the partition strategy $\mathbb{P}$, the set of exit branching models selected for the inference task S

1: Initial the partition strategy $\mathbb{P}$, the set of exit branching models selected for the inference task S, task local start time t^{ls}, task start executing time on edge server t^{es}
2: **for** $i = 1; i \leq n; i \leftarrow i + 1$ **do**
3:    $profit_{temp} = 0$
4:    Initial g_{temp} as arbitrary exit model, initial $partition_{temp}$ as arbitrary
5:    Initial $t^{ls}_{temp}, t^{es}_{temp}$
6:    Initial $flag \leftarrow False$
7:    **for each** $g \in s$ **do**
8:      **if** the accuracy of $g \geq P_{lowbound}$ **then**
9:         Call rectified max flow min cut algorithm to to obtain the partition $partition_{cur}$ of the selected model g, as well as the local start time of the next task t^{ls}_{cur} and the server start time of execution t^{es}_{cur}, end to end delay d^{e2e}_g
10:         **if** $d^{e2e}_g \geq D$ **then**
11:            break
12:         **end if**
13:         Call build constrain and relex computation algorithm to botain the current system profit $profit_{cur}$ and current computation resources $\mathbb{C}_{cur}$
14:         **if** $profit_{temp} \leq profit_{cur}$ **then**
15:            $flag = True$
16:            $profit_{temp} \leftarrow profit_{cur}$
17:            $g_{temp} \leftarrow g$
18:            $partition_{temp} \leftarrow partition_{cur}$
19:            $\mathbb{C} \leftarrow \mathbb{C}_{cur}$
20:            $t^{ls}_{temp} \leftarrow t^{ls}_{cur}$
21:            $t^{es}_{temp} \leftarrow t^{es}_{cur}$
22:         **end if**
23:       **end if**
24:    **end for**
25:    **if** $flag == True$ **then**
26:       The exit branching model g_{temp} with the highest gain is partition $partition_{temp}$ added to the model selection set S as well as the Partition set $\mathbb{P}$.
27:    **end if**
28: **end for**
29: **return** $\mathbb{C}$, $\mathbb{P}$, S

To handle the coupling between these challenges, our algorithm employs an approach similar to alternating optimization, sequentially solving each sub-problem while fixing the conditions of the others.

Selecting the exit branching model for a task to meet accuracy requirements has been shown to be NP-hard. In RAGDNS, we address this by evaluating each exit branch model that satisfies the task's accuracy requirements. We partition the task, construct end-to-end processing delay constraints based on the previously chosen model, and perform computation relaxation to determine the system utility of each model. The model with the largest tility is selected. This approach effectively tackles the model selection problem for each task.

4.2 Rectified Max-Flow Min-Cut Algorithm

To solve the second problem, we determine both the model chosen for the inference task and the allocated computing resources on the LD and ES. The model is determined through enumeration. Initially, we fix the computing resources at their maximum values for the LD and ES. This approach has two advantages: 1) If the partitioned task does not meet the end-to-end delay requirement, the algorithm can exit directly. 2) If the partitioned task meets the end-to-end delay with maximum resources, there is a feasible solution through computation relaxation.

Once the exit branching model is selected, the network structure to be partitioned is fixed, and the computation and I/O data for each node are known. With allocated computing resources, we can determine the execution duration for each network layer on the LD and ES, as well as the transmission delay for offloading inference tasks. Using this information, we construct an auxiliary graph and apply the max-flow min-cut algorithm to find the shortest execution latency for the inference task without considering the logical dependencies between task execution times, which will be discussed later. Here, we briefly describe the construction of an auxiliary flow network and using the max-flow min-cut algorithm to solve inference task partition.

Construct Auxiliary Flow Network. Given a DNN model $G_i = (V_i, E_i)$, we construct an auxiliary network flow graph $G'_i = (V'_i, E'_i)$ with edge capacity $w(\cdot)$. To do this, we add two special node s (source) and t (sink). For G_i, let $V_i^{mo} = \{v_{i,j} \mid d_o(v_{i,j}) \geq 1\}$ denote the set of nodes with multiple out-degrees, where $d_o(v_{i,j})$ is the out-degree of $v_{i,j}$. We then add virtual nodes $v'_{i,j}$ corresponding to each node $v_{i,j} \in V_i^{mo}$. For each virtual node $v'_{i,j}$, we set $N^+(v'_{i,j}) = N^+(v_{i,j})$ and delete the original edges $(v_{i,j}, v_{i,k})$. We then add an edge $(v_{i,j}, v'_{i,j})$. Additionally, we add edges $(s, v_{i,j})$ for all $v_{i,j} \in V_i$ and add edges $(v_{i,j}, t)$ for all $v_{i,j} \in V_i \setminus \{v_0\}$. Thus, the node set of the auxiliary network flow graph is $V'_i = \{s, t\} \cup V_i \cup V_i^{vir}$, where $V_i^{vir} = \{v'_{i,j} \mid \forall v_{i,j} \in V_i^{mo}\}$. The edge sets are defined as follows:

- Outgoing edges of the source node: $E_i^s = \{(s, v_{i,j}) \mid v_{i,j} \in V_i\}$,
- Incoming edges of the sink node: $E_i^t \{(v_{i,j}, t) \mid v_{i,j} \in V_i \setminus v_{i,0}\}$,

- Same edges in DNN model G and auxiliary network flow graph G'_i: $E_i^{sa} = E_i - \{(v_{i,j}, v_{i,k}) \in E_i \mid \forall v_{i,j} \in V_i^{mo}\}$,
- Outgoing edges of virtual nodes: $E_i^{viro} = \{(v'_{i,j}, v_{i,k}) \mid \forall (v_{i,j}, v_{i,k}) \in E_i\}$.
- Incoming edges of virtual nodes: $E_i^{viri} = \{(v_{i,j}, v'_{i,j}) \mid \forall v_{i,j} \in V_i^{mo}\}$.

Thus the edge set of the auxiliary network flow graph is $E'_i = E_i^s \cup E_i^t \cup E_i^{same} \cup E_i^{viro} \cup E_i^{viri}$. In summary, the auxiliary network flow graph is constructed by adding a source node s connected to all original nodes, and a sink node t connected from all nodes except $v_{i,0}$. Virtual nodes are introduced to replace connections for nodes with multiple out-degrees. Figure 2(a) illustrates the transformation of the original DNN model into an auxiliary network flow graph.

Initial Flow Network Capacity. The construction of the auxiliary network flow graph ensures accurate partitioning of a DNN model into local and edge server execution, taking into account node execution and transmission times. This process, facilitated by the max-flow min-cut algorithm, is crucial for optimizing task execution. The capacity of edge $w(s, v_{i,0})$ is infinite, i.e. $w(s, v_{i,0}) = \infty$. For each edge $(s, v_{i,j}) \in E'_i$, where $v_{i,j} \in V_i \backslash v_{i,0}$ the capacity is execution latency of node $v_{i,j}$ executing on ES. For each edge $(v_{i,j}, t)$, where $v_{i,j} \in V_i \backslash v_{i,0}$, the capacity is the execution delay of node $v_{i,j}$ on LD. For each edge in E_i^{same} and E_i^{viri}, the capacity is the transmission latency of the node. The remaining edges in the auxiliary flow network G'_i have infinite capacities.

Introducing virtual nodes and adjusting connections and capacities addresses nodes with multiple out-degrees, ensuring transmission edges are calculated only once. This setup prevents redundant data transmission when multiple nodes are executed on the edge server. Infinite capacities for edges connected to virtual and multi-outdegree nodes ensure correct partitioning. The same applies to edges between the source node and output layer nodes. This construction enables splitting the graph into two subgraphs, distinguishing local execution times from edge server execution times and transmission times. The max-flow min-cut algorithm then identifies the minimum cut set by computing the network's maximum flow. The equivalence between partitioning model G_i cutting the auxiliary network flow graph G'_i will be demonstrated in the algorithm analysis.

When the approach from [19] is used to segment and form a pipeline as in Fig. 3(a), it creates different bottleneck segments, resulting in wasted computing resources. Using our proposed algorithm 2 and 3, we can obtain the pipeline-aware alignment as shown in Fig. 3(b).

Inspired by the max-flow min-cut algorithm [24], which continuously finds augmenting paths and modifies flow capacity, we propose modifying the capacities of related edges in the network flow graph after constructing the auxiliary network flow graph. The max-flow min-cut algorithm is then used to obtain the optimal partition, leading to the design of the rectified max-flow min-cut algorithm.

First, perform a topological sort of the nodes in the graph to determine their execution order. Then, calculate the surplus time by finding the difference between the start execution time at ES and LD. Next, examine the sorted nodes

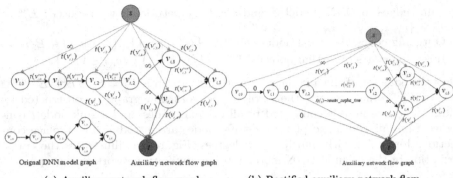

(a) Auxiliary network flow graph   (b) Rectified auxiliary network flow

Fig. 2. The illusion of auxiliary network flow graph

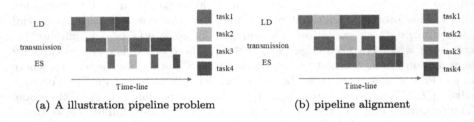

(a) A illustration pipeline problem   (b) pipeline alignment

Fig. 3. The illusion of pipeline

sequentially. If the surplus time is greater than or equal to a node's local execution time, set the capacity of the edge connecting that node to the sink node in the auxiliary network flow graph to 0, thereby assigning it to local execution. Update the surplus time accordingly and compare the surplus time and transmission delay, adjusting the edge capacity if necessary. Note that the surplus time is not updated at this point since no transmission delay occurs if the node is not offloaded. The remaining surplus time is then use to subsequent nodes. The modified auxiliary network flow graph is shown in Fig. 2(b).

4.3 Construction of Constraint and Relax Computing Resource Algorithm

After partitioning, the next challenge is slack scheduling of computing resources. An intuitive approach is to calculate the end-to-end processing delay of each inference task using the rectified max-flow min-cut algorithm and then solve an optimization problem that minimizes energy consumption. However, this approach encounters issues with temporal dependencies for intra-task and inter-task processing. To address this, we design a constraint-construction and resource-relaxation algorithm to establish related constraints and solve for the computing resource allocation scheme for all selected reference tasks, as well as the current system utility. These utilities are then used in Algorithm 1 to determine which exit branching model to choose for the inference tasks.

The execution relationship constraint graph is constructed by traversing the identified inference task models and their partitions, as well as the partitions of the current inference models under examination. This constraint graph addresses the dependencies between local execution and offloading of the same task to the ES, as well as the execution dependencies of successive tasks on the same devices (LD/ES). For this purpose, each inference task model is divided into two nodes based on the two connectivity graphs it partitions: one for local execution (connected to the source node) and one for execution at the ES (connected to the sink node). By traversing the connectivity graph, we obtain the local execution time, ES execution time, and the amount of data transmitted, then calculate the transmission delay. These operations are performed sequentially for both the partitioned inference tasks with a determined selection model and the current inference task model under examination, resulting in a constraint graph based on execution dependencies.

The critical path of the constructed constraint graph is identified and used to calculate start time for all nodes, ensuring compliance with delay constraints. An optimization algorithm is then employed to determine computing resource allocation and system utility. The critical path of the constraint graph is identified and used to calculate start times for all nodes, ensuring compliance with delay constraints. Start times for critical nodes are determined by traversing the critical path by traversing this path forward and backward, while start times for ES-executed nodes not on the critical path are relaxed. The graph is then traversed forward from nodes with zero in-degrees to calculate start times and construct constraints based on start time, node execution latency, and transmission delay for cross-device edges. Additional constraints ensure the completion time of the last node on each device meets delay requirements. Finally, an optimization algorithm determines the computing resource allocation for each task on the devices, and the system's utility is calculated based on the optimization objective and related information.

4.4 Algorithm Analysis

In the following, we theoretically analyze the performance of the constructed auxiliary flow network graph G'_i and show the correctness of the proposed algorithm. We also analyze the time complexity of the RAGDNS algorithm. Detailed proofs are available in [23].

Theorem 2. *Any partition of the DNN graph G_i corresponds to a set of cut edges in the auxiliary flow network G'_i, dividing it into two connected graphs that include the source and sink nodes, respectively.*

Theorem 3. *The set of cut edges in the auxiliary flow network G'_i corresponds to the partition in the DNN model G_i.*

Proof. This proof follows the reverse process of the constructive proof of Theorem 2, hence it is omitted.

The constructive proof not only maps the partition of the DNN model G_i to the cut in the auxiliary flow network G'_i, but also outlines how to transform the auxiliary network flow graph obtained via the max-flow min-cut algorithm back into the original partition. This involves checking whether the weights of the constructed edges in the auxiliary network flow graph are zero and then dividing the set accordingly, ignoring the residual edges introduced during the execution of the max-flow min-cut algorithm.

Theorem 4. *The modified max-flow min-cut algorithm obtains the partition that can obtain the smallest end-to-end delay increment.*

Theorem 5. *The time complexity of RAGDNS is $O(NT(N(|V|+|E|)+F|V|+N^3))$ where N is the number of reference task, T is the number of types of exit branching models for inference tasks, $|V|$ and $|E|$ are numbers of nodes and edges in graph, respectively, and F is the maximum flow.*

5 Performance Evaluation

In this section, we evaluate the performance of the proposed algorithms through experimental simulations.

5.1 Experimental Settings

According to the Raspberry Pi 4 chip parameters, LD is set to 4 cores at 1.5 GHz and ES is set to 64 cores at 1.5 GHz, similar to a GPU. The energy factor ϵ for both devices are set to 10^{-26}. The transmission power is 1 W. Our experiments involve three DNN models: one chained model (VGG) and two DAG models (ResNet-50 and GoogLeNet). GoogLeNet's exit branches are configured according to Ref. [18], VGG19 has three exit points after each pooling layer, and ResNet-50 has exits at the last, penultimate, and middle residual blocks. Experiments were conducted on these DNN models under conditions of fixed bandwidth, with transmission rates ranging from 1 MBps to 14 MBps. Later experiments show that the computing power and strategy will ultimately affect the algorithm performance in this range, so the transmission range of network bandwidth is reasonable and robust. To evaluate the algorithm's performance, we set up three baselines as follows:

1. Neurosurgeon [5]: The algorithm divides the chained DNNs to minimise the task execution time.
2. Edgent [12]: This algorithm improves on Neurosurgeon by introducing early exit branching models. After partitioning all branch models, he chooses the model with the highest accuracy that can satisfy the delay constraint.

3. DADS [19]: This algorithm partition the DAG DNN model using the max-flow min-cut algorithm to minimise the execution time of the task and to cope with high throughput scenarios by forming a simple flow.

In the above algorithms, Neurosurgeon and Edgent do not design a solution for high throughput scenarios, we use a similar approach to DADS to form simple pipeline for the partition. The line chart and scatter plot in sub-picture (a) of the experimental results show the system utility and throughput, respectively, where the ordinate of system revenue is on the left and the ordinate of throughput is on the right.

5.2 Evaluation Result

Due to the limited space, we omit the experiments and analysis on VGG, for details, please see [23].

As shown in Fig. 4, GoogLeNet's performance varies across transmission rates, with RAGDNS outperforming other algorithms due to its flexible partitioning and efficient utilization of computing power, despite a slight decrease in accuracy caused by the early exit mechanism. When network conditions are suitable, all algorithms exhibit similar partitioning schemes and performance. DADS, being more flexible, outperforms Neurosurgeon under poor network conditions, while Edgent's early exit mechanism makes its performance similar to DADS. Figure 4 indicates that beyond a 9 MBps transmission rate, computing power limits system's performance. RAGDNS outperforms all other algorithms in terms of system revenue, throughput, and average energy consumption due to its flexible partitioning, early exit mechanism, alignment pipeline mechanism, and computing resource relaxation.

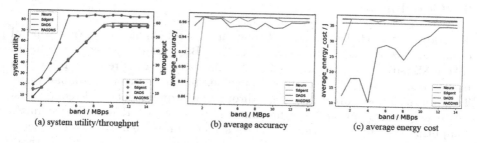

Fig. 4. Performance on Googlenet over different transmission rate

ResNet-50, despite its DAG model, closely resembles a chain model, leading to similar performance across Neurosurgeon, Edgent, and DADS. However, Edgent's early exit mechanism shows a slight performance advantage over DADS' flexible division in chain models, as depicted in Fig. 5. RAGDNS is superior to these algorithms in system revenue, throughput, and energy efficiency by integrating both early exit and flexible partitioning mechanisms. By fully leveraging computing resource RAGDNS enhances throughput even under performance bottlenecks. Moreover, its computing resource relaxation mechanism significantly reduces average task energy consumption by nearly 50% compared to other algorithms.

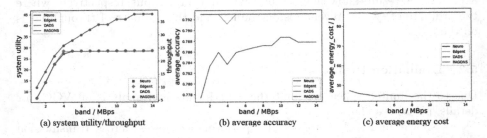

Fig. 5. Performance on Resnet-50 over different transmission rate

Combining the above experiments, we can obtain that on VGG19, Googlenet and Resnet-50 models, compared with the existing advanced methods, our method improves the throughput by 14.2% to 60%, and reduces the average inference energy consumption of tasks by about 11.1% to 49.8%. Experiments show that the computing power and strategy will ultimately affect the algorithm performance in above transmission of network bandwidth, so the proposed RAGDN is robust.

6 Conclusion

To address the challenge of real-time high throughput in spatio-temporal data inference tasks, we proposed RAGDNS. By considering task start times, RAGDNS partitions inference tasks and utilizes a computing resource relaxation mechanism to dynamically align the pipeline, minimizing computational waste. This approach meets real-time high throughput requirements and reduces system energy consumption. Experimental results show that RAGDNS maximizes computing resource utilization, enhances system throughput, and lowers energy consumption, with minimal performance loss compared to algorithms without early exit mechanisms.

References

1. Ma, C., et al.: Attention based multi-unit spatial-temporal network for traffic flow forecasting. In: 2021 8th IEEE CSCloud/2021 7th IEEE EdgeCom (2021)
2. Si, J., et al.: STEGNN: spatial-temporal embedding graph neural networks for road network forecasting. In: 2022 IEEE 28th ICPADS (2023)

3. Yao, Y., et al.: MVSTGN: a multi-view spatial-temporal graph network for cellular traffic prediction. In: IEEE TMC (2023)
4. Shuvo, M.M.H., et al.: Efficient acceleration of deep learning inference on resource-constrained edge devices: a review. In: Proceedings of the IEEE (2023)
5. Kang, et al.: Neurosurgeon: collaborative intelligence between the cloud and mobile edge. ACM SIGARCH Comput. Archit. News (2017)
6. Hu, C., et al.: Dynamic adaptive DNN surgery for inference acceleration on the edge. In: IEEE INFOCOM 2019 - Conference on Computer Communications (2019)
7. Zhao, N., et al.: Spatial-temporal aggregation graph convolution network for efficient mobile cellular traffic prediction. IEEE Commun. Lett. (2022)
8. Deng, Y., et al.: Robust spatial-temporal correlation model for background initialization in severe scene. In: ICASSP 2021 (2021)
9. Teerapittayanon, S., et al.: BranchyNet: fast inference via early exiting from deep neural networks. In: 2016 23rd ICPR (2016)
10. Kaya, Y., et al.: Shallow-deep networks: understanding and mitigating network overthinking. In: International Conference on Machine Learning. PMLR (2019)
11. Dong, F., et al.: Multi-exit DNN inference acceleration based on multi-dimensional optimization for edge intelligence. In: IEEE TMC (2023)
12. Li, E., et al.: Edge intelligence: on-demand deep learning model co-inference with device-edge synergy. In: Proceedings of the 2018 Workshop on Mobile Edge Communications (2018)
13. Huang, J., et al.: Clio: enabling automatic compilation of deep learning pipelines across IoT and cloud. In: Proceedings of the 26th Annual International Conference on Mobile Computing and Networking (2020)
14. Laskaridis, et al.: SPINN: synergistic progressive inference of neural networks over device and cloud. In: Proceedings of the 26th Annual International Conference on Mobile Computing and Networking (2020)
15. Mao, J., et al.: MoDNN: local distributed mobile computing system for deep neural network. DATE (2017)
16. Bozorgchenani, et al.: Multi-objective computation sharing in energy and delay constrained mobile edge computing environments. IEEE TMC (2020)
17. Zeng, et al.: Coedge: cooperative DNN inference with adaptive workload partitioning over heterogeneous edge devices. IEEE/ACM Trans. Netw. (2020)
18. Yang, et al.: On-demand inference acceleration for directed acyclic graph neural networks over edge-cloud collaboration. JPDC (2023)
19. Liang, H., et al.: DNN surgery: accelerating DNN inference on the edge through layer partitioning. In: IEEE TCC (2023)
20. Yang, X., et al.: Towards efficient inference: adaptively cooperate in heterogeneous IoT edge cluster. In: 2021 IEEE 41st ICDCS (2021)
21. Liu, Z., et al.: Resource allocation for multiuser edge inference with batching and early exiting. IEEE J. Sel. Areas Commun. (2023)
22. Zhang, X., et al.: EFFECT-DNN: energy-efficient edge framework for real-time DNN inference. In: 2023 IEEE 24th WoWMoM (2023)
23. Ren, J., et al.: Collaborative Inference for Adaptive DNN Pipeline-aware Alignment (2024). https://pan.baidu.com/s/1s1mlyXEnmhJNeo6JMpaOEA?pwd=q5fx
24. Cormen, T.H., et al.: Introduction to Algorithms. MIT Press (2022)

A Unified Framework for Link Prediction on Heterogeneous Temporal Graph

Chongjian Yue[1], Qiao Mi[2], and Lun Du[1(✉)]

[1] Peking University, Beijing, China
chongjian.yue@stu.pku.edu.cn, dulun@pku.edu.cn
[2] Qinghai Normal University, Xining, China

Abstract. Link prediction is a critical research frontier in graph data analytics. In real-world scenarios, graph data often exhibit heterogeneity and temporal dynamism, adding layers of complexity to analytical models. Existing work typically considers either heterogeneity or temporal dynamism, with very few studies modeling both properties simultaneously. In this study, we propose a novel link prediction framework designed to integrate both heterogeneity and temporal variations within graph data, applicable across a wide range of domains. Our framework includes a Dual-Window Strategy and a Mix Information Graph Neural Network (MIGNN) model. The MIGNN synthesizes temporal node representations by aggregating heterogeneous temporal information, while the Dual-Window Strategy enhances the model's ability to capture long-term distributional characteristics inherent in graph data. We also incorporate additional experiments using a more equitable data-splitting approach and perform comparisons with an expanded range of baseline methods.

Keywords: link prediction · heterogeneous temporal networks · graph neural networks · deep learning

1 Introduction

Link prediction constitutes a foundational challenge in graph analysis research, aiming to discern potential interactions or relations between entities. It has gained increasing attention in the research area of networks and has various applications [13], such as predicting co-authorship collaborations [8], recommending friends in social networks [2], and identifying molecular interactions in protein networks [24].

However, in practical application scenarios, data often presents complexity, such as dynamics (evolves over time) and heterogeneity (diversity of entities and interactions) [1,11,17]. For example, a dynamic network may have new links appearing at each time point, while a heterogeneous network may have different types of events and links that need to be distinguished. Once we take

---

C. Yue and Q. Mi—Co-First Author, Equal Contribution.

these temporal and heterogeneous characteristics into consideration, the problem becomes extraordinarily complex, and existing models are often ill-suited to deal with such issues. A common link prediction framework is needed to better discover potential links in networks. Besides, link prediction methods have been used to predict the links that may appear in the future in temporal networks [4]. Most of these works focus on predicting whether a link will appear in the future. Nevertheless, a more valuable question in real life is when links are most likely to appear. In general, we summarize three challenges that we need to solve: **1. How to efficiently process the voluminous historical events present in temporal information; 2. How to learn the long-term distribution of links; 3. How to extract temporal and heterogeneous information from historical data.** For the above reasons, we propose a new model design that specifically targets such complex scenarios to more effectively tackle the complex data challenges in the link prediction task.

The three main contributions of this paper are:

1 **Efficient Processing of Voluminous Historical Events in Temporal Information:** We introduce a graph-time processing approach where historical signals (events) are treated as a stream. As time progresses, the number of signals in this stream increases. Once a critical mass of signals is reached, or after a designated time has passed, the model extracts signals from the stream for training. This process resembles a window sliding over the event stream, which we term the "past window".
2 **Understanding the Long-term Distribution of Links:** We have adopted a "future window" following the past window to emulate the long-term distribution of links. This future window, sliding along the event stream after the past window, encapsulates both authentic positive samples and certain negative samples, serving to enhance the model's predictive capabilities. Together, the past and future windows constitute our innovative "Dual-Window Strategy", enabling the model to understand the data's extended distribution.
3 **Extraction of Temporal and Heterogeneous Information from Historical Data:** We incorporate the Memory Module and Temporal Embedding Generator with the Dual-Window Strategy as the "Graph-aware Operator". Temporal and heterogeneous data are extracted using distinct feature extraction functions and are subsequently input into the Memory Module to enhance computational efficiency and reduce operational costs. During the prediction phase, the Temporal Embedding Generator retrieves the curated historical data, creating node representations at specific times for prediction.

To evaluate the effectiveness of our model, we participated in the Link Prediction Task and achieved top performance on both datasets. To further validate the time sensitivity, we proposed a dataset-splitting method that is more sensitive to time. The results indicated that compared to state-of-the-art GNN baselines, our model improved results by 4.45%–5.01%.

2 Related Work

Most of the existing link prediction methods, which are based on deep learning, are under the framework of Embedding-Classifier. The existing methods can be divided into two categories: Network Embedding methods [29] and Graph Neural Networks methods [27].

Network Embedding methods generate corresponding representations for nodes by retaining the structure information of the graph. DeepWalk [15], Node2vec [5], and LINE [20] all rely on the skip-gram model to generate node representations. However, these methods do not work well on heterogeneous temporal networks since they ignore both the heterogeneous and temporal information of the network.

Graphical Neural Networks are the other mainstream method. According to their network settings, they can be divided into four categories: static graph neural networks, heterogeneous graph neural networks, temporal graph neural networks, and heterogeneous temporal graph neural networks.

Static Graph Neural Networks generate representations for nodes through the message passing mechanism. GCN [9] averages neighbors' information with a linear projection. GraphSAGE [6] selects only a part of its neighbors for message passing, which makes it better to learn large-scale graphs. GAT [23] uses self-attention to distinguish the importance of different neighbors. By stacking multiple layers, this kind of method can retain the global structure information, but also lack the heterogeneous and temporal information of the network.

Heterogeneous Graph Neural Networks consider the type of link or node in message passing. RGCN [19] designs different linear projections for different types of links for information aggregation. CompGCN [22] also learns relation embeddings of heterogeneous information networks. HAN [25] proposes hierarchical attention to aggregate information from the neighbors. HetGNN [28] adopts different RNNs for different node types to integrate multi-modal features. Although these methods fully learn the heterogeneous information in the graph, they ignore the temporal information of the network.

Temporal Graph Neural Networks learn node embeddings on the temporal networks. CTDNE [14] proposes to use temporal walks to learning temporally valid embeddings with temporally valid information. DySAT [18] uses self-attention to learn node representations on temporal networks. TGN [16], Jodie [10], and DyRep [21] store historical information in a memory network.

Heterogeneous Temporal Graph Neural Networks are new methods proposed in recent years. Change2vec [3] leverages meta-path-based node embedding and models the development of networks to preserve both heterogeneous and temporal features of networks. HDGAN [12] uses three levels of attention to account for the heterogeneity and dynamicity of the network at the same time. THINE [8] not only use attention mechanism and meta-path to preserve structures and semantics but also combines the Hawkes Process [26] to simulate the evolution of the temporal network. These models have explored the dynamicity and heterogeneity of networks from different perspectives.

3 Preliminary and Problem Definition

3.1 Heterogeneous Temporal Networks

A heterogeneous temporal network is defined as a graph $G = \{V, E, \gamma, \delta\}$, where V is a set of nodes and each node $v \in V$ represents an entity in the graph. E is a set of links, and $e \in E$ represents a link between two entities in the graph. $\gamma : E \rightarrow R$ is a link type mapping function, where each link $e \in E$ corresponds to a specific type $r \in R$. $\delta : E \rightarrow T$ is a link timestamp mapping function, where each link $e \in E$ corresponds to a specific timestamp $t \in T$. Thus, each link in the graph can be represented by a quadruple (u, v, r, t), which means node u and node v have a link of type r at time t.

3.2 Heterogeneous Temporal Link Prediction

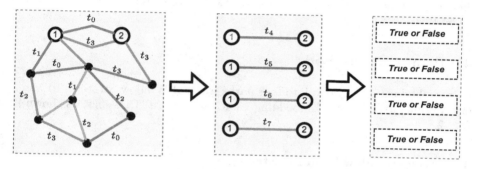

Fig. 1. A toy example of Heterogeneous Temporal Link Prediction. Given all links before t_3, predict whether a link of a given type will occur at a specific time $t(t > t_3)$. The color of the link represents the type of link.

In a heterogeneous temporal network $G = \{V, E, \gamma, \delta\}$, given all links (u, v, r, t) before time T_k. We aim to predict whether a link (u, v, r, t) is true, where $T_k \leq t \leq T_p$. The period $[T_k, T_p]$ is called "the prediction period".

As a toy example shown in Fig. 1, the link color represents the link type. And the timestamp represents when the link occurred. We have knowledge of all links before time t_3. We should predict the links after time t_3.

In terms of application, heterogeneous temporal link prediction can be used in social networks, user-item networks, etc. In social networks, this new link prediction can determine whether two users will have a specified interaction at a specified time, such as adding friends, chatting, etc. In the user-item network, this link prediction can predict whether the user will purchase or unsubscribe the product at a specified time based on the user's past behavior.

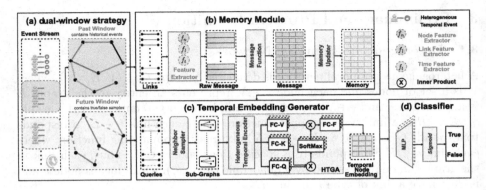

Fig. 2. Architecture of Dual-Window Strategy and Mix Information Graph Neural Network (MIGNN). (a) The Dual-Window strategy generates two windows. (b) The memory module extracts heterogeneous and temporal information and stores them in the memory vector. (c) The temporal embedding generator generates temporal embedding for the target nodes. (d) The classifier calculates the probability of a link at the target time.

4 Proposed Method

Our method can be divided into four parts: a) Dual-Window Strategy, b) Memory Module, c) Temporal Embedding Generator, and d) Classifier, as shown in Fig. 2.

4.1 Dual-Window Training Strategy

To tackle the long-term prediction problem, we deviate from the standard training strategy, which typically uses information from the previous step to predict the current step. Instead, we utilize information from multiple steps prior to the current step to predict the link in the current step during the training phase. This approach helps us avoid the out-of-distribution problem.

Specifically, we designate the previous steps as the Past Window (PW) and the current step as the Future Window (FW). We collect and model all the events within the PW to predict both positive and negative link samples that are sampled within the FW. The positive samples include all the links present within the FW. For the negative samples, we generate them by sampling along four dimensions: source node, destination node, relation, and time, while keeping the other dimensions unchanged. We then combine these four types of negative samples to construct the final negative link sample set.

4.2 Memory Module

The Memory module is designed to memorize historical information. Each node v is associated with a memory vector m_v. When an interaction involving a source

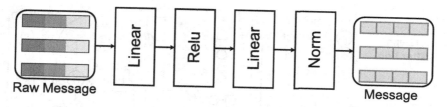

Fig. 3. Message Function.

node s, a destination node d, a relation type r, and a timestamp t is input, we update the memory of node s and the node d. For each historical link in the Past Window, the feature extractor extracts all related information to construct a raw message vector $rmsg$, as shown in Eq. 1:

$$rmsg = f_n(u,v,r,t)\|f_l(u,v,r,t)\|f_t(u,v,r,t) \qquad (1)$$

where the node feature extractor f_n concatenates the memory vectors m_u and m_v; the link feature extractor f_l takes the type information and generates type embedding E_r for each link type r. E_r is a learnable embedding that is initialized randomly. For the dataset without link-type features, E_r is a learnable embedding vector; otherwise, E_r will be a learnable function with the link-type feature as input. The time feature extractor $f_t(\cdot)$ is used to extract features about the timestamp. Given that temporal data usually exhibit periodicity, we designed a period time encoder:

$$f_t(u,v,r,t) = MLP(g_1(t)\|g_2(t)\|g_3(t)\|g_4(t)), \qquad (2)$$

where $g_1(t)$ outputs t is which day in a week, $g_2(t)$ outputs which week in a month, $g_3(t)$ outputs which day in a month, $g_4(t)$ outputs which month in a year, and MLP is a Multilayer Perceptron. The $rmsg$ only performs a simple concatenation operation on the features of links, which may result in a high-dimensional raw message. Therefore, a message function is used to project raw message into a low-dimensional space, as shown in Fig. 3. Finally, we use a GRU module to update the memory of each node. Since there are many links related to a node in each batch from the Past Window, we use the mean aggregation of these links to update the old memory with the GRU module.

4.3 Temporal Embedding Generator

Within the memory module, we transform the historical information from the past window into a memory vector. Following that, using the data queries from the future window and the temporal embedding generator, we produce the representation vector required for classification. This module can be divided into two parts: the neighbor sampler and the Mix Graph Attention (MGA, as shown in Fig. 4). For each node in the queries, the neighbor sampler samples a fixed number of nodes with the most recent connection times from its neighborhood to

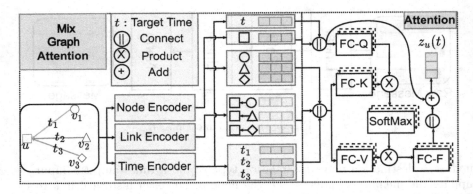

Fig. 4. Mix Graph Attention (MGA). The MGA module uses an attention encoder to obtain node, link, and time representations, uses attention to weigh the importance of each neighbor node, and aggregates these weighted representations to generate a temporal embedding $z_u(t)$ for a node u at a target time t.

form a subgraph, and this subgraph is used for message aggregation. We believe that neighbors with closer temporal intervals have stronger importance, and the impact of a link gradually diminishes over time. After that, the Heterogeneous-Temporal Encoder in MGA is used to generate three kinds of vectors as shown in Eq. 3 4 5:

$$f(u) = w_{ne} \cdot ne_u + w_m \cdot m_u \quad (3)$$
$$g(u,v) = E_{r_{u,v}} \quad (4)$$
$$h(t_{u,v}) = \phi(t_{u,v}) \quad (5)$$

The node encoder $f(\cdot)$ balances the importance between the memory vector and the node embedding by taking the self-adaptive weight sum as the node encoding result. The link encoder $g(\cdot)$ is an identity function that provides the embedding of the link type $r_{u,v}$ as the link encoding result. For the time encoder $h(\cdot)$, we use the output of the period time encoder $\phi(\cdot)$ as the time encoding result. After obtaining the three vectors, we employ a multi-head attention mechanism to weigh the impact on nodes caused by links of different types and times, as shown in Eqs. 6 7 8:

$$q_u = (f(u)||h(t)) \cdot W_q \quad (6)$$
$$k_v = (f(v)||h(t_{u,v})||g(u,v)) \cdot W_k \quad (7)$$
$$v_v = (f(v)||h(t_{u,v})||g(u,v)) \cdot W_v \quad (8)$$

5 Experiment

5.1 Dataset and Setup

We use the datasets from the WSDM2022 Challenge - Large Scale Temporal Graph Link Prediction to evaluate our methods. The challenge includes two large heterogeneous temporal datasets. Dataset A: a dynamic event graph with entities as nodes and different types of events as links; Dataset B: a user-item graph with users and items as nodes and different types of interactions as links. In Dataset A, both nodes and link types have features, while in Dataset B, only the links have features.

For baselines, we use the implementation provided in DGL. For the link prediction task, AUC (Area Under the Curve) is a commonly used evaluation metric. For network embedding methods and static graph neural networks, we ignore link types and timestamps in the datasets. For heterogeneous graph neural networks, we only remove the timestamps. For temporal graph neural networks, we only ignore the link types. Unless specified otherwise, all hyperparameters are consistent with the original implementation.

Static graph neural networks (excluding GraphSAGE) and heterogeneous graph neural networks require aggregating all neighbors' information at each step, which incurs significant time and space costs. Therefore, we randomly select 50 neighbors each time we perform message passing.

Because the number of nodes in Dataset B is much larger than that in Dataset A, we set the embedding size for Dataset A to 128 and for Dataset B to 64. Our model is optimized using the AdamW optimizer.

5.2 Splitting Methods

Table 1. AUC in the time-sensitive scenario

Category	Model	AUC A	AUC B
Network Embedding Methods	DeepWalk	0.5764(±0.0010)	0.5314(±0.0008)
	AntGraph	0.5865(±0.0004)	0.5387(±0.0013)
Static Graph Neural Networks	GCN [9]	0.5915(±0.0294)	0.5633(±0.0062)
	GraphSAGE [6]	0.5950(±0.0288)	0.5511(±0.0222)
	GAT [23]	0.5993(±0.0190)	0.5535(±0.0137)
Heterogeneous Graph Neural Networks	RGCN [19]	0.6026(±0.0752)	0.5710(±0.0116)
	CompGCN [22]	0.6047(±0.0060)	0.5788(±0.0072)
Temporal Graph Neural Networks	DyRep [21]	0.6099(±0.0011)	0.5197(±0.0013)
	Jodie [10]	0.6104(±0.0005)	0.5208(±0.0006)
	TGN [16]	0.6127(±0.0018)	0.5382(±0.0005)
Heterogeneous Temporal Graph Neural Networks	HGT [7]	0.6164(±0.0063)	0.5825(±0.0014)
	MIGNN (Ours)	**0.6438(±0.0002)**	**0.6117(±0.0034)**
-	%Improv.	4.45%~11.69%	5.01%~15.11%

We add a new time-sensitive splitting method because the original splitting method, as mentioned in the final report of the organizers, is not time-sensitive.

The original method requires predicting if a link will occur regardless of the timestamp, which is not equitable for evaluation.

Specifically, the new splitting method divides the dataset into several time windows $\{TW_1, TW_2, ..., TW_n\}$. Each time window has a length of 7 days. For each time window $D_i \in TW_i$ contains all the links $\{(s,d,r)\}$ that occurred on the i-th day of the time window TW_i. We take all links in the window as positive samples. If a link (s,d,r) only occurs on certain days $\{D_i\}$ in this time window, the link (s,d,r) on the remaining days $\{D_j\}$ is set as a negative sample.

The experimental results of the proposed method and baselines on the new splitting method are summarized in Table 1. **Bold** indicates the best result, and Underline indicates the sub-optimal result.

5.3 Ablation Study

Table 2. Ablation Study

Models	Dataset A	Dataset B
MIGNN	0.6438	0.6117
MIGNN-TE	0.6010 (7.12%)	0.5316 (15.07%)
MIGNN-MEM	0.6335 (1.62%)	0.5485 (11.52%)
MIGNN-ATTN_TIME	0.6076 (5.95%)	0.5451 (12.22%)
MIGNN-ATTN_TYPE	0.6435 (0.04%)	0.6081 (0.59%)
MIGNN-ATTN_TIME_TYPE	0.6056 (6.30%)	0.5052 (21.08%)

The core components in MIGNN are the period time encoder, memory module, and the heterogeneous temporal graph attention. To further analyze their effectiveness, we conducted an ablation study by removing these components from MIGNN. The ablation results are shown in Table 2. The names of the models are explained as follows:

- **MIGNN-TE**. The period time encoder is replaced with a digits time encoder, which splits a timestamp into a digit vector.
- **MIGNN-MEM**. The memory module is removed from MIGNN.
- **MIGNN-ATTN_TIME**. The time factor is removed from MIGNN.
- **MIGNN-ATTN_TYPE**. The link type factor is removed from MIGNN.
- **MIGNN-ATTN_TIME_TYPE**. Both the time and link type factors are removed from MIGNN.

The importance of periodicity is illustrated by MIGNN-TE. When a digits time encoder is used instead of a period time encoder, the AUC value decreases by 7.12% on Dataset A and 15.07% on Dataset B. This indicates that periodic characteristics are more important for Dataset B.

From MIGNN-MEM, we observe that historical events are meaningful for both datasets, though they contribute differently. When the memory module is removed, the AUC value on Dataset A decreases slightly, but on Dataset B, the AUC value decreases by 11.52%.

Experiments with MIGNN-ATTN_TIME, MIGNN-ATTN_TYPE, and MIGNN-ATTN_TIME_TYPE show the varying importance of time and type factors in Mix Graph Attention (MGA). The time factor is the most significant. When the time factor is removed (MIGNN-ATTN_TIME), the AUC values on Datasets A and B decrease significantly, by 5.95% and 12.22%, respectively. However, when the link type factor is removed (MIGNN-ATTN_TYPE), the AUC values decrease only slightly. When both factors are removed (MIGNN-ATTN_TIME_TYPE), AUC decreases more on both datasets, demonstrating the importance of considering both factors simultaneously.

5.4 Influence of Dynamicity and Heterogeneity

Dynamicity and heterogeneity are key factors in heterogeneous temporal networks. To evaluate the robustness of MIGNN, we assess its performance with varying time window sizes and link type proportions. As shown in Fig. 5, he x-axis represents the size of the time window, and the y-axis represents the AUC score. On both datasets, MIGNN consistently outperforms other models across all window sizes, demonstrating its adaptability to different prediction periods.

While Dataset A shows an overall improvement in effectiveness with increasing window sizes, Dataset B does not exhibit a clear trend. Additionally, we evaluate the influence of link type proportions, where x on the x-axis indicates the proportion of link types preserved. MIGNN performs better than other models under any link type proportion, indicating its strong ability to learn heterogeneous information.

These experiments illustrate that MIGNN is robust to both time window size and link type variations, making it a versatile choice for link prediction in heterogeneous temporal networks.

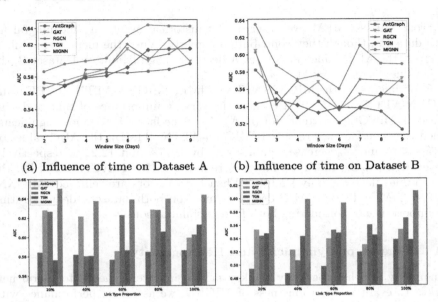

(a) Influence of time on Dataset A (b) Influence of time on Dataset B

(c) Influence of link type on Dataset A (d) Influence of link type on Dataset B

Fig. 5. Influence of Dynamicity and Heterogeneity

6 Conclusion

In this paper, we introduce a task for heterogeneous temporal link prediction in heterogeneous temporal networks. This task addresses the complexities of the time factor while incorporating heterogeneous information. To tackle this challenge, we design a Mix Information Graph Neural Network (MIGNN) equipped with a Dual-Window Strategy. We evaluate our approach using two real-world datasets, and the results demonstrate that our method surpasses existing techniques. Additionally, we introduce a time-sensitive splitting method and compare its performance with the baseline.

References

1. Aggarwal, C.C., Xie, Y., Yu, P.S.: A framework for dynamic link prediction in heterogeneous networks. Stat. Anal. Data Min. **7**(1), 14–33 (2014)
2. Aiello, L.M., Barrat, A., Schifanella, R., Cattuto, C., Markines, B., Menczer, F.: Friendship prediction and homophily in social media. ACM Trans. Web **6**(2), 9:1–9:33 (2012)
3. Bian, R., Koh, Y.S., Dobbie, G., Divoli, A.: Network embedding and change modeling in dynamic heterogeneous networks. In: SIGIR, pp. 861–864. ACM (2019)
4. Dunlavy, D.M., Kolda, T.G., Acar, E.: Temporal link prediction using matrix and tensor factorizations. ACM Trans. Knowl. Discov. Data **5**(2), 10:1–10:27 (2011)

5. Grover, A., Leskovec, J.: node2vec: scalable feature learning for networks. In: KDD, pp. 855–864. ACM (2016)
6. Hamilton, W.L., Ying, Z., Leskovec, J.: Inductive representation learning on large graphs. In: NIPS, pp. 1024–1034 (2017)
7. Hu, Z., Dong, Y., Wang, K., Sun, Y.: Heterogeneous graph transformer. In: WWW, pp. 2704–2710. ACM/IW3C2 (2020)
8. Huang, H., Shi, R., Zhou, W., Wang, X., Jin, H., Fu, X.: Temporal heterogeneous information network embedding. In: IJCAI, pp. 1470–1476. ijcai.org (2021)
9. Kipf, T.N., Welling, M.: Semi-supervised classification with graph convolutional networks. In: ICLR (Poster). OpenReview.net (2017)
10. Kumar, S., Zhang, X., Leskovec, J.: Predicting dynamic embedding trajectory in temporal interaction networks. In: KDD, pp. 1269–1278. ACM (2019)
11. Jaya Lakshmi, T., Durga Bhavani, S.: Link prediction in temporal heterogeneous networks. In: Wang, G.A., Chau, M., Chen, H. (eds.) PAISI 2017. LNCS, vol. 10241, pp. 83–98. Springer, Cham (2017). https://doi.org/10.1007/978-3-319-57463-9_6
12. Li, Q., Shang, Y., Qiao, X., Dai, W.: Heterogeneous dynamic graph attention network. In: ICKG, pp. 404–411. IEEE (2020)
13. Lu, L., Zhou, T.: Link prediction in complex networks: a survey. CoRR abs/1010.0725 (2010)
14. Nguyen, G.H., Lee, J.B., Rossi, R.A., Ahmed, N.K., Koh, E., Kim, S.: Continuous-time dynamic network embeddings. In: WWW (Companion Volume), pp. 969–976. ACM (2018)
15. Perozzi, B., Al-Rfou, R., Skiena, S.: Deepwalk: on learning of social representations. In: KDD, pp. 701–710. ACM (2014)
16. Rossi, E., Chamberlain, B., Frasca, F., Eynard, D., Monti, F., Bronstein, M.M.: Temporal graph networks for deep learning on dynamic graphs. CoRR abs/2006.10637 (2020)
17. Rümmele, N., Ichise, R., Werthner, H.: Exploring supervised methods for temporal link prediction in heterogeneous social networks. In: Gangemi, A., Leonardi, S., Panconesi, A. (eds.) Proceedings of the 24th International Conference on World Wide Web Companion, WWW 2015, Florence, Italy, 18–22 May 2015 - Companion Volume, pp. 1363–1368. ACM (2015)
18. Sankar, A., Wu, Y., Gou, L., Zhang, W., Yang, H.: Dysat: deep neural representation learning on dynamic graphs via self-attention networks. In: WSDM, pp. 519–527. ACM (2020)
19. Schlichtkrull, M., Kipf, T.N., Bloem, P., van den Berg, R., Titov, I., Welling, M.: Modeling relational data with graph convolutional networks. In: Gangemi, A., et al. (eds.) ESWC 2018. LNCS, vol. 10843, pp. 593–607. Springer, Cham (2018). https://doi.org/10.1007/978-3-319-93417-4_38
20. Tang, J., Qu, M., Wang, M., Zhang, M., Yan, J., Mei, Q.: LINE: large-scale information network embedding. In: WWW, pp. 1067–1077. ACM (2015)
21. Trivedi, R., Farajtabar, M., Biswal, P., Zha, H.: Dyrep: learning representations over dynamic graphs. In: ICLR (Poster). OpenReview.net (2019)
22. Vashishth, S., Sanyal, S., Nitin, V., Talukdar, P.P.: Composition-based multi-relational graph convolutional networks. In: ICLR. OpenReview.net (2020)
23. Velickovic, P., Cucurull, G., Casanova, A., Romero, A., Liò, P., Bengio, Y.: Graph attention networks. In: ICLR (Poster). OpenReview.net (2018)
24. Wan, X., et al.: An inductive graph neural network model for compound-protein interaction prediction based on a homogeneous graph. Briefings Bioinform. **23**(3) (2022)

25. Wang, X., et al.: Heterogeneous graph attention network. In: WWW, pp. 2022–2032. ACM (2019)
26. Wheatley, S., Filimonov, V., Sornette, D.: The hawkes process with renewal immigration & its estimation with an EM algorithm. Comput. Stat. Data Anal. **94**, 120–135 (2016)
27. Wu, Z., Pan, S., Chen, F., Long, G., Zhang, C., Yu, P.S.: A comprehensive survey on graph neural networks. IEEE Trans. Neural Netw. Learn. Syst. **32**(1), 4–24 (2021)
28. Zhang, C., Song, D., Huang, C., Swami, A., Chawla, N.V.: Heterogeneous graph neural network. In: KDD, pp. 793–803. ACM (2019)
29. Zhang, D., Yin, J., Zhu, X., Zhang, C.: Network representation learning: a survey. IEEE Trans. Big Data **6**(1), 3–28 (2020)

Massive-Parallel Game of Politicians

Andrew Schumann[1(✉)] and Krzysztof Pancerz[2]

[1] Department of Cognitive Science and Mathematical Modelling, University of Information Technology and Management in Rzeszow, Sucharskiego 2, 35-225 Rzeszow, Poland
andrew.schumann@gmail.com

[2] The John Paul II Catholic University of Lublin, Al. Racławickie 14, 20-950 Lublin, Poland

Abstract. Mould or fungal propagation can serve as a universal model for a massive-parallel game of many players, moving concurrently. Based on this model, a game of politicians is introduced to occupy as many political keywords as possible in the media. If a politician begins to be mentioned every time with some word or group of words, then this means that he or she has occupied this word or group of words. In the election campaign, the winner is the politician who has occupied the most words. The study was carried out on the example of Polish politicians. In addition, a space for keywords related to mentions in the messages of 780 politicians of the Sejm and Senate of the last two terms of office was created. Thus, the reflection game of politicians for keywords was implemented. In this game, you can explain the role of politicians in their activities for their party or for themselves. The period of the game is for three years. This game shows the dynamics of political discourse in Poland during this period. In different media, this game is realized in different ways. For example, party-related keywords in the media correlate differently with party program texts. This means that journalists from various portals influence this game of politicians. Data modeling as a game of politicians for keywords makes it possible to identify hidden players and their degree of reflexivity.

Keywords: Concurrent Game · Massive-Parallel Game · Fungi

1 Introduction

Cellular automata are well suited for simulating the propagation of various colonies of organisms: slime mould [1, 2], colonies of bacteria [3, 4], colonies of fungi [5–7], and swarms as such [8]. A cell in such automata can be defined in different ways. However, the successive occupation of these cells by organisms can be interpreted as their propagation. In particular, Markus Margenstern defines a cellular automaton on the heptagrid (see Fig. 1) by a local transition function in form of a table, where each row defines a rule and the table has nine columns numbered from 0 to 8 so that each entry of the table containing a state of the automaton. For more details, see [3, 4].

However, this model proposed by M. Margenstern can also simulate resource colonization by mould and fungal colonies. Such experiments are available to everyone. If you do not wash the cup after coffee, then after some time the damn mould and fungi will appear on it, Fig. 2. They will grow in the form of multicellular filaments called hyphae, conditionally called by us here the 'green' and 'white' mould. They are in a competition for the heptagrid space and at the discovering stage they are trying to sketch main directions of their propagation. At the logistic stage they build networks, called a mycelium, to product spores (conidia) formed by differentiation at the ends of hyphae.

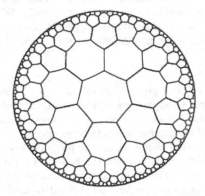

Fig. 1. The heptagrid as a space for cellular automata simulating the growth of bacterial or fungal colonies, see [3].

In this paper, fungal propagation will be considered as a basic model for understanding the phenomenon of propagation of many players moving concurrently. We have some cells with the resources of a cellular automaton, and occupation captures these cells according to the rules of this automaton as shown in the picture of Fig. 2. Our task is to show that the election race of politicians can be represented by the model of fungal occupation of space. Politicians are interested in their mention in the media. Therefore, we can say that the purpose of the political race is to promote their mention in different cells (clusters) of keywords.

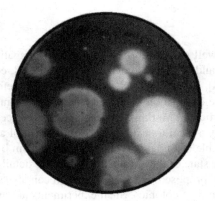

Fig. 2. The competition between the 'green' and 'white' fungi in the same unwashed cup of coffee.

In our research, we studied the mention of Polish politicians over three years (2017–2019) in two major media portals:

- https://www.gazetaprawna.pl/;
- https://www.gazeta.pl/.

We have constructed our model on these data, because they had no external outrage due to COVID-19 in 2020 and the war in Ukraine in 2022. During the period of time from 2017 to 2019, the following political parties and alliances were active in Poland:

- *PiS* – "Prawo i Sprawiedliwość" (Law and Justice);
- *PSL* – "Polskie Stronnictwo Ludowe" (Polish People's Party);
- *PO* – "Platforma Obywatelska" (Civic Platform);
- *Kukiz* – "Kukiz'15";
- *NL* – "Nowa Lewica" (New Left);
- *Conf.* – "Konfederacja Wolność i Niepodległość" (Confederation Liberty and Independence).

We considered these parties as multi-agent players in the game of occupying clusters of political words. For each major news portal there is a strong correlation of mentioning political parties and their success in elections. For example, the frequency of mentions for the Polish parties in the news from 2017 to 2019 is given in Table 1. The correlation with their success in the elections in the Sejm of the 9th codification is 0.947892. That is why politicians strive to occupy as many clusters of political words in the media as possible. This means that they tend to be more associated with these clusters – to be mentioned strictly together with the keywords of the cluster. In this regard, they behave like a fungus of Fig. 2 that seeks to occupy as many heptagrid cells of a cellular automaton as possible. Only instead of cells here are clusters of keywords.

Table 1. Frequency of mentions of Polish political parties and alliances from 2017 to 2019 in https://www.gazetaprawna.pl/.

Party/Alliance	Its Frequency in Media
PiS	5888
PO	1462
Conf.	64
NL	183
PSL-Kukiz	374

For collecting keywords from the Polish media we used natural language processing and text mining methods to analyze the relevance of keywords related to politicians' mentions in the news. The process of pre-processing text data included the following operations: (a) removing punctuation marks; (b) removing numbers; (c) removing the names and surnames of politicians; (d) converting all letters in texts to lower case, removing words from the stop list (so-called stop-words), i.e. words of little importance (the stop list for the Polish language was used); (e) reducing words to their basic forms (the service available in the form of REST API on the website of the CLARIN-PL consortium http://clarin-pl.eu/ was used); (f) removing words denoting state and local government institutions and positions (e.g. president, prime minister, minister, marshal, etc.), names of political parties, names of media, names of days of the week, names of months and the most frequently used verbs in press reports (e.g. say, speak, mark, announce, etc.) – an own list of words was used.

Determination of the importance of keywords was based on the calculation of the frequency of words appearing in the texts. The proposed approaches to keyword relevance analysis were implemented in the form of scripts for the R environment. Packages that were used:

- *tm* – providing a wide range of text mining operations;
- *worldcloud* – providing a tool for creating a word cloud.

Additionally, using the *httr* and XML packages, a client application for the REST API provided by CLARIN-PL was created. This application is designed to create an HTTP POST request and parse the REST response. The body of the HTTP POST request is saved in JSON format and contains, among others: text to reduce words. REST responses are sent in the XML format. Basic forms of words were extracted from the XML structure. In addition, the keyword space associated with the mentions of politicians in the news was grouped. The cases subjected to grouping were keywords related to the portals. Cases were grouped in a two-dimensional space, whose variables were the frequency of occurrence of words in two nationwide media, https://www.gazetaprawna.pl/ and https://www.gazeta.pl/. The grouping was carried out for the total values (for 3 years) and for the values for individual months in these years. The k-means algorithm was used for grouping, requiring the declaration of the number of groups into which the set of cases is divided. The number k was determined on the basis of the Gap statistic method algorithm, which allows estimating the number of groups in the data. Its implementation is available in the *factoextra* package for the R environment.

We received 40 keyword clusters. They served for us as cells of the cellular automaton of politicians to occupy these clusters. The cellular automaton for occupying places in clusters is defined in Sect. 1. The visualization of this automaton is given in Sect. 2. A strategy of this game is defined in Sect. 3.

This game is, firstly, bio-inspired – it is based on the natural behavior of colonies, in particular colonies of bacteria, colonies of mould or fungal microparticles, etc. by occupying as many cells of space as possible. Such bio-inspired games have only recently begun to be studied [9–12]. In our case, we presented the parties as big multi-agent players, and individual politicians as their microparticles. The goal of parties and politicians is to occupy as much space as possible in clusters of political words. Secondly, these games are concurrent – in them, the players do not take steps sequentially (as in chess, for example), but simultaneously. The theory of concurrent games has also been developing only recently [13, 14]. In the case of politicians, we are not even dealing with a concurrent game, but with a massively parallel game, when the number of players reaches several hundred or more. And they all go in the game in parallel.

This paper is organized as follows. In Sect. 2 and 3, we describe our model for collecting empirical data. In Sect. 4, we define a reflexive game among Polish politicians, based on empirical data of mentioning some terms in media.

2 Cellular Automaton of Political Keyword Occupation Game

In this game we focus on political keywords. These words have the following main characteristics: (1) they appear in the public media at the same time as mentioning a politician or his/her political party, group, coalition, etc.; (2) they show the relevant rigorous political concept, used as one of the keywords for political reflection (discourse), such as "strategy," "program," etc.; (3) they express certain achievements of politicians, such as "insisting," "proclaiming," "explaining," etc.; (4) they express objects of political reflection (discourse), such as "family," "religion," "state," etc.

Thanks to the developed program by us, all the keywords that appeared when talking about a Polish politician or his/her political party were collected. Then, six classes of the 300 most frequent political words were selected for each of the media: $P(X, PiS)$, $P(X, PSL), P(X, PO), P(X, Kukiz), P(X, NL), P(X, Conf.)$, where X is the information portal. Let X be https://www.gazetaprawna.pl/ (gp). Then we have classes $P(gp, PiS)$, $P(gp, PSL)$, $P(gp, PO)$, $P(gp, Kukiz)$, $P(gp, NL)$, $P(gp, Conf.)$. The class $P(gp, PiS)$ means the set of political words mentioned in gp along with mentioning the party PiS, the class $P(gp, PSL)$ means the set of political words mentioned in gp along with mentioning PSL, etc. Let X be https://www.gazeta.pl/ ($gazeta$). Then we have classes $P(gazeta, PiS)$, $P(gazeta, PSL), P(gazeta, PO), P(gazeta, Kukiz), P(gazeta, NL), P(gazeta, Conf.)$ and so on. The class $P(gazeta, PiS)$ means the set of political words mentioned in $gazeta$ along with mentioning the party PiS, the class $P(gazeta, PSL)$ means the set of political words mentioned in $gazeta$ along with mentioning PSL, etc. We have different political keywords for each portal, e.g. $P(gp, PiS) \neq P(gazeta, PiS)$. Nevertheless, for almost every pair of classes with the same political party (PiS) and different portals (gp and $gazeta$), the intersection of both classes is not empty: $P(gp, PiS) \cap P(gazeta, PiS) \neq \emptyset$.

The game simulates politicians fighting about success in elections by occupying as many clusters of political keywords as possible. Consider the simple case that we only have two large news portals: gp and $gazeta$. Let's define a coordinate for each word a of $P(gp, Y)$ and for each word b of $P(gazeta, Y)$, where Y is one of the six Polish political parties or alliances. It is defined as vector $\langle n_{a_{gp}}, m_{a_{gazeta}} \rangle$ for a of $P(gp, Y)$ and as vector $\langle n_{b_{gazeta}}, m_{b_{gp}} \rangle$ for b of $P(gazeta, Y)$, where $n_{a_{gp}}$ is the frequency of word a in gp for period T, $m_{a_{gazeta}}$ is the frequency of word a in $gazeta$ for period T, $n_{b_{gazeta}}$ is the frequency of word b in $gazeta$ for period T, $m_{b_{gp}}$ is the frequency of word b in gp for period T. Let it be a period of 3 years.

Now let's define two spaces for our game:

$S_{gp} = P(gp, PiS) \cup P(gp, PSL) \cup P(gp, PO) \cup P(gp, Kukiz) \cup P(gp, NL) \cup P(gp, Conf.)$.

$S_{gazeta} = P(gazeta, PiS) \cup P(gazeta, PSL) \cup P(gazeta, PO) \cup P(gazeta, Kukiz) \cup P(gazeta, NL) \cup P(gazeta, Conf.)$.

Then we can define one of the distances for each point $x \in S_{gp}$ with coordinate $\langle x_1, x_2 \rangle$ and for each point $y \in S_{gazeta}$ with coordinate $\langle y_1, y_2 \rangle$, such as the Euclidean distance. In this way, we obtain 40 clusters of political words that are occupied by politicians and, as a result, by their parties or alliances.

3 Visualization of Game

The tool for simulating the game of keywords related to the mentions of politicians in the news was created as a web application in the R environment using the *Shiny* package. The simulator allows for the observation of the relative degree of occupation of individual groups of keywords resulting from grouping by individual players (parties) in different months of the considered period. The user can select a specific month using the slider. Thanks to the use of the reactive mechanism available in applications based on the *Shiny* package, images for individual clusters are automatically refreshed. Images are presented using area charts. The *ggplot2* package available for the R environment was used to generate such plots. The layout of the charts in the main panel was obtained thanks to the capabilities of the *gridExtra* package also available for the R environment. In Figs. 3, 4 and 5 there are given some example images of the degree of occupation of individual clusters, presented for two selected months. Words that are occupied by one or another party will be indicated by the corresponding color: if occupied by *PiS*, then by blue, if by *Conf.*, then by brown, if by *NL*, then by red, if by *PO*, then by orange, if by *PSL*, then by light green, if by *Kukiz*, then by gray.

Let us remember that each term j has a coordinate $\langle n_{j_{gp}}, m_{j_{gazeta}} \rangle$ in the space S_{gp} and a coordinate $\langle m_{j_{gazeta}}, n_{j_{gp}} \rangle$ in the space S_{gazeta}. The goal of every politician (his or her political party or alliance) is to occupy words and then appropriate clusters. They are mentioned along with political keywords and respective 40 clusters step by step. Suppose the game time is $t = 0, 1, 2, \ldots, 35$. Each word j has the value $n_{j_{X,t,gp}}/(n_{j_{gp}} * n_i)$ as the probability of it being occupied by political party X in gp in step t, where.

- $n_{j_{X,t,gp}}$ is the frequency of word j in gp with a mention of the player X at time t,
- $n_{j_{gp}}$ is the frequency of word j in gp for the entire period T,
- n_i is the number of words in the cluster where j is located.

At the end of the game, the payoff of X is defined as

- Words j with the probability of their occupation $\mathbf{P}_{X,gp}(j) = \sum_t (n_{j_{X,t,gp}}/(n_{j_{gp}} * n_i))$,
- Clusters i with occupation probability $\mathbf{P}_{X,gp}(i) = \sum_{n_i} P_{X,gp}(j)$.

To win, the player X should occupy words and clusters with a higher probability. For example, for *gazeta* (i.e. S_{gazeta}) in January 2017, PiS wins. The absence of mentions of a political party will mean its withdrawal from the political struggle.

4 Cellular Automaton of Political Keyword Occupation Game

The goal of politicians is to occupy words and then their clusters. Politicians have strategies to occupy as many words and clusters as possible. These strategies are reflected in political programs. Suppose Pr_X is the priority set of political words for player X, where X is one of the 6 political parties. For example, Pr_X could be a political program for X. Then we say that X occupies the words of the set $P(m, X)$, where $m \in \{gp, gazeta\}$, with reflection level ρ, where ρ is the correlation between Pr_X and $P(m, X)$. Therefore, we assume that the level of reflection runs over the range $[-1,1]$. We calculated that PiS deals with words of $P(gp, PiS)$ with a reflection level of 0.470101 (that is, the correlation ρ between the political program and $P(gp, PiS)$ is equal to 0.470101). Meanwhile, PiS deals with words of $P(gazeta, PiS)$ with a reflection level of 0.44252. The reflection level of other parties is closer to 0.

Let $Expert(Pr_X)$ be a small set of some political keywords selected by a political expert from the set Pr_X. For example, consider the set $Expert(Pr_{PiS})$ consisting of the following words: "chrześcijański" (Christian), "dekomunizacja" (decommunization), "dobrobyt" (welfare), "energia" (energy), "fundusz" (fund), "historia" (history), "katolicki" (Catholic), "kościół," (church), "LGBT," "lustracja" (lustration), "mama" (mom), "modernizacja" (modernization), "narodowy" (national), "naród" (nation), "naturalny" (natural), "ochrona" (protection), "państwo" (noun: state), "państwowy" (adj.: state), "płaca" (salary), "polski" (polish), "prawo" (law), "prawica" (noun: right wing), "prawicowy" (adj.: right-wing), "propaganda," "przedsiąbiorstwo" (enterprise), "remont" (renovation), "rodzina" (family), "ryczałt" (flat rate), "służba" (service), "solidarność" (solidarity), "solidarny" (solidary), "społecznościowy" (community-based), "społeczny" (social), "sprawiedliwy" (fair), "szkoła" (school), "tradycja" (tradition), "tradycyjny" (traditional), "wiara" (faith), "wierzyć" (believe), "wrogi" (enemies), "wspólnie" (together), "wspólnota" (community), "wspólny" (common), "zdrowie" (health), "życie" (life). According to this $Expert(Pr_{PiS})$, PiS occupies $P(gp, PiS)$ with a reflection level of 0.868432 and occupies $P(gazeta, PiS)$ with a reflection level of 0.877105. On the other hand, $Expert(Pr_{PiS})$ is well represented in gp and $gazeta$ (correlation is equal to 0.987055).

In this way, we can consider $Expert(Pr_{PiS})$, $Expert(Pr_{PO})$, $Expert(Pr_{PSL})$, $Expert(Pr_{NL})$, $Expert(Pr_{Kukiz})$, $Expert(Pr_{Conf.})$ as a test group of political words considered important for understanding the political program of the corresponding political party and the degree of their implementation in the politicians' game of political words. For each word j of $Expert(Pr_{PiS})$, $Expert(Pr_{PO})$, $Expert(Pr_{PSL})$, $Expert(Pr_{NL})$, $Expert(Pr_{Kukiz})$, $Expert(Pr_{Conf.})$ there is a coordinate $\langle n_{j_{gp}}, m_{j_{gazeta}} \rangle$ in S_{gp} and a coordinate $\langle n_{j_{gazeta}}, m_{j_{gp}} \rangle$ in S_{gazeta}. The goal of each politician (his or her political party) is to promote his or her political program, and we can see how these programs are represented as clusters related to the actual dynamics of political words.

As we remember, at the end of the game, the payoff of party X in its test set of words is defined as

- Words j with the probability of their set $\mathbf{P}_{X,gp}(j) = \sum_t (n_{j_{X,t,gp}}/(n_{j_{gp}} * n_i))$;
- Clusters i are with occupation probability $\mathbf{P}_{X,gp}(i) = \sum_{n_i} P_{X,gp}(j)$.

But additionally, we calculate a frequency $n_{j_{Expert(Pr_X)}}$ of words j in the expert sets $Expert(Pr_X)$ of the player X. Then we define $\rho_{t,X,gp}(n_{j_{X,t,gp}}, n_{j_{Expert(Pr_X)}})$ to obtain the reflective level of X in gp at time t, and we define $\rho_{t,X,gazeta}(n_{j_{X,t,gazeta}}, n_{j_{Expert(Pr_X)}})$ to obtain the reflective level of X in $gazeta$ at time t.

To win at t, the party X should have a higher level of reflection and better implementation of its political program than all other parties. To win in the whole game, the party X should have a higher level of reflection more often (for most t).

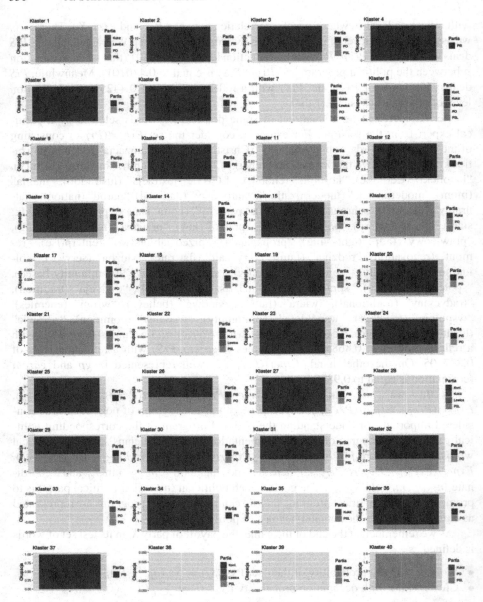

Fig. 3. The occupation of 40 clusters of keywords by Polish parties in the space S_{gazeta} in January 2017. The blue color is to denote the places occupied by *PiS*, while the orange color is to denote the places occupied by *PO*. We see that *PiS* and *PO* are two political leaders by their mention in the media. (Colour figure online)

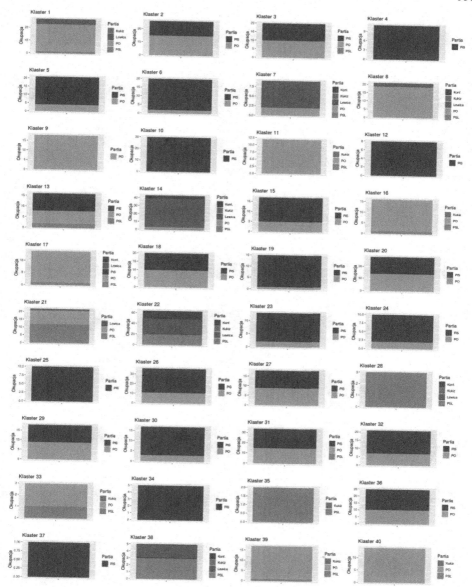

Fig. 4. The occupation of 40 clusters of keywords by Polish parties in the space S_{gazeta} in December 2019. Some clusters are partly occupied by *NL* (red) and *PSL* (green). (Colour figure online)

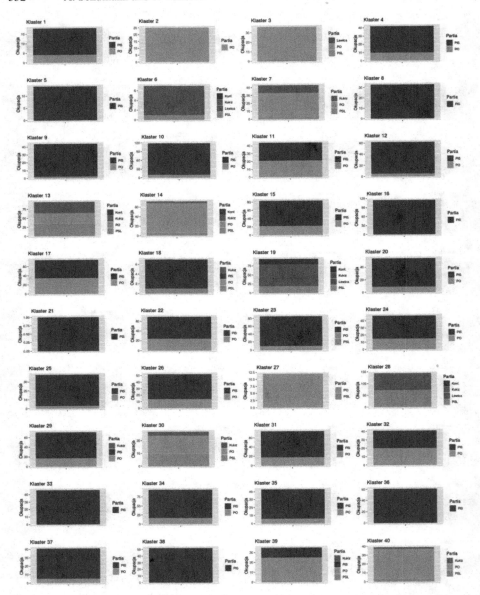

Fig. 5. The occupation of 40 clusters of keywords by Polish parties in the space S_{gp} in January 2017. *PiS* surely wins. (Colour figure online)

5 Conclusions

In this research, components programmed in the R environment for the analysis of political discourse in Poland were verified; in other words, information waves were analyzed in the context of 780 politicians of the Sejm and Senate of the two terms of office and 6 political parties.

The main results of the project are: (i) constant monitoring for the analysis of Polish public resources using an RSS feed reader from news portals, an HTML parser for news posted on national media news portals, an HTML parser for news posted on local media news portals, (ii) inference from large data sets from Polish public resources as part of a designed platform in the R environment for automatic data analysis in real time so as to discover correlations for data grouping and generating conclusions as part of massively parallel games.

As a result of this analysis, we can find a correlation between the success of politicians in the reflexive game in the occupation of political words and their success in political elections. The winner is the one who occupied the largest percentage of the clusters. This behavior corresponds to the standard behavior of mould and fungi in terms of competitive occupation of living space (Fig. 2). Then we can define a reflective level of politicians – how they can implement their political programs in the media.

Acknowledgments. This research was funded by the the Podkarpackie Regional Operational Program (Podkarpackie Innovation Center, $F3\_40$) as the project entitled Massive-Parallel Games and Modeling of Information Warfare by Means of Web Mining and Big Data Analysis.

Disclosure of Interests. The authors declare no conflicts of interests.

References

1. Shirakawa, T., Adamatzky, A., Gunji, Y.P., Miyake, Y.: On simultaneous construction of Voronoi diagram and Delaunay triangulation by Physarum polycephalum. Int. J. Bifurcation Chaos **19**(09), 3109–3117 (2009)
2. Schumann, A., Pancerz, K.: High-Level Models of Unconventional Computations. Springer, Berlin (2019)
3. Margenstern, M.: An algorithmic approach to tilings of hyperbolic spaces: Universality results. Fund. Inform. **138**(1–2), 113–125 (2015)
4. Margenstern, M.: Bacteria inspired patterns grown with hyperbolic cellular automata. In: International Conference on High Performance Computing & Simulation, pp. 757–763. IEEE (2011)
5. Halley, J. M., Comins, H. N., Lawton, J. H., Hassell, M. P.: Competition, succession and pattern in fungal communities: towards a cellular automaton model. Oikos, pp. 435–442 (1994)
6. Mayne, R., Roberts, N., Phillips, N., Weerasekera, R., Adamatzky, A.: Propagation of electrical signals by fungi. Biosystems **229**, 104933 (2023)
7. Adamatzky, A.: Fungal Machines – Sensing and Computing with Fungi. Springer, Berlin (2023)

8. Schumann, A.: Behaviourism in studying swarms: logical models of sensing and motoring. Springer International Publishing (2019)
9. Schumann, A., Pancerz, K., Adamatzky, A., Grube, M.: Bio-inspired game theory: The case of Physarum polycephalum. In: Proceedings of the 8th International Conference on Bioinspired Information and Communications Technologies, pp. 9–16 (2014)
10. Schumann, A., Pancerz, K.: Physarumsoft – a software tool for programming Physarum machines and simulating Physarum games. In: 2015 Federated Conference on Computer Science and Information Systems (FedCSIS), pp. 607–614. IEEE (2015)
11. Pancerz, K., Schumann, A.: Slime mould games based on rough set theory. Int. J. Appl. Math. Comput. Sci. **28**(3), 531–544 (2018)
12. Theocharopoulou, G., Giannakis, K., Papalitsas, C., Fanarioti, S., Andronikos, T.: Elements of game theory in a bio-inspired model of computation. In: 10th International Conference on Information, Intelligence, Systems and Applications (IISA), pp. 1–4. IEEE (2019)
13. Abramsky, S., Mellies, P. A.: Concurrent games and full completeness. In: Proceedings. 14th Symposium on Logic in Computer Science (Cat. No. PR00158), pp. 431–442. IEEE (1999)
14. Maubert, B., Pinchinat, S., Schwarzentruber, F., Stranieri, S.: Concurrent games in dynamic epistemic logic. In: Proceedings of the Twenty-Ninth International Conference on International Joint Conferences on Artificial Intelligence, pp. 1877–1883 (2021)

Multiscale Convolutional Feature Aggregation for Fine-Grained Image Retrieval

Caolin Yang, Kaifeng Ding, and Chengzhuan Yang(✉)

Zhejiang Normal University, Jinhua 321004, Zhejiang, China
{caolinyang,010711358,czyang}@zjnu.edu.cn

Abstract. Fine-grained image retrieval aims to retrieve images from the same subclass as the query images from a database containing a specific class object. To solve the problem of large inter-class differences and small intra-class differences, we propose the Multiscale Convolutional Feature Aggregation (MsCFA) method accomplished in an unsupervised environment. This method comprises an object localization submodule and a multiscale feature fusion module. In the object localization module, we perform a coarse localization of the object based on the response levels within the feature map. Specifically, we crop the high-response regions of the feature map to ensure that the resulting cropped feature map contains more object features and less background noise. Subsequently, we employ the regional maximum activation of convolutions method to extract features from this cropped region. The cropped object feature is fused with the uncropped original feature to obtain the feature representation on a single scale. In the feature fusion module, we input images of different scales into the network to obtain feature representations at various scales and then fuse these different scale features to obtain our multiscale convolutional feature aggregation descriptor. In experiments on five classical fine-grained datasets, our method outperforms other fine-grained retrieval tasks in the same type of unsupervised environment and is comparable to those supervised tasks with end-to-end training.

Keywords: Fine-grained Image Retrieval · Convolutional Feature · Feature Extraction · Feature Fusion · Multiscale Images

1 Introduction

Image retrieval aims to find images from a database that are related to or similar to the query image. In traditional image retrieval, global features (such as color histogram, texture features, shape descriptors, etc.) are usually used to represent images, and similar images are retrieved by calculating the similarity between images. These methods have achieved satisfactory results in certain scenarios, but there are still some limitations when dealing with fine-grained image retrieval. Fine-grained image retrieval poses a more challenging task compared to traditional image retrieval tasks [1–3, 32]. It requires the system to comprehend the subtle differences and features within images and to differentiate between highly similar images. This type of retrieval usually involves the utilization and recognition of local features in an image, such as parts of an object, details, or specific visual features. Compared with fine-grained image classification tasks [4, 5, 19], fine-grained image retrieval mainly examines the generalization ability of the model on unknown categories. Whether it is image classification or image retrieval tasks, a high-quality image representation is indispensable, and effective feature extraction methods are the key to obtaining high-quality image representation.

The initial feature extraction process relied on hand-crafted algorithms to capture local image attributes like colors, textures, shapes, etc. With the development of neural networks, this process is automated, eliminating the need for manual feature design. Neural networks learn advanced features directly from images, enhancing performance and efficiency in fine-grained image retrieval. However, manual extraction was resource-intensive, while neural network-based extraction was constrained by data volume, training duration, and iterations, leading to resource waste. Moreover, subtle differences in fine-grained image retrieval tasks further impede research development.

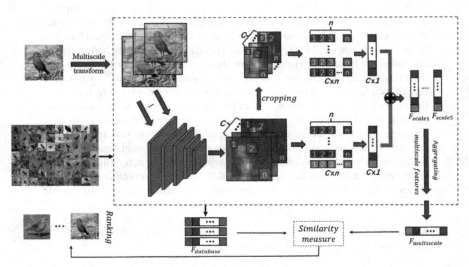

Fig. 1. Overall structure diagram of our method. The 'dotted' line box represents the main process of feature extraction in our method. The 'green' arrow branch within the box denotes the feature extraction process of our object localization submodule, while the 'red' arrow branch represents the feature extraction process of our original feature map. $F_{multiscale}$ represents multiscale features across five different scales, $F_{database}$ represents the feature set of database images, and *Ranking* indicates the ranking results of database images retrieved based on the query image.

To solve these problems, we propose the MsCFA method, whose overall framework is illustrated in Fig. 1. To extract even more discriminative features, we perform a rough object localization by cropping the high-response regions from the original feature maps, as demonstrated by the 'cropping' step in Fig. 1. Simultaneously, features are extracted from both the cropped feature map and the original feature map using the R-MAC method. Subsequently, the features from these two parallel branches are effectively fused to obtain our feature representation on a single scale. As depicted by the 'green' and 'red' arrows in Fig. 1, the 'green' arrow signifies the feature extraction process specific to the cropped feature map, whereas the 'red' arrow represents the process of feature extraction conducted on the original, uncropped feature map. Notably, the entire feature extraction process solely relies on a pre-trained network without the need for end-to-end training. Given the diversity of features, we transform the original image into multiple scales and input them into the network to extract features from different scales. Ultimately, these multiscale features are fused into a feature vector, which serves as the ultimate feature representation of the image. For both the query image and database images, this method is used to extract features, measure the distance between the query image and each database image, and then use it for subsequent retrieval tasks. Our main contributions are as follows:

- We propose an improved submodule of R-MAC based on high-response region object localization.
- We propose a multiscale feature descriptor that describes image features with greater comprehensiveness and accuracy.
- Extensive experiments conducted on five benchmark fine-grained image datasets demonstrate the effectiveness and robust performance of our proposed method.

The rest of this paper is organized as follows: Sect. 2 introduces traditional image retrieval task and fine-grained image retrieval task. Section 3 describes the MsCFA method in detail. In Sect. 4, the content of the experiment is introduced in detail. Secction 5 summarizes the work of this paper and puts forward some suggestions for future research.

2 Related Work

In this section, we will introduce related work from two aspects: traditional image retrieval and fine-grained image retrieval.

2.1 Traditional Image Retrieval

Traditional image retrieval primarily involves retrieving images belonging to the same broad category as the query image from multiple distinct classes, and current extensive research in this field has achieved quite remarkable results. [6] first utilized the activation of the top layer in CNN as a descriptor for image retrieval, demonstrating its excellent performance even when the model is trained on unrelated data [7]. [32] proposed a masking scheme that selects representative local convolution features through embedding and aggregation methods. [1] directly employed the deep features of pre-trained CNN for image retrieval, primarily extracting additional features from the last convolutional layer to enhance feature representation. The SPoC [8] also involves weighted sum-pooling on the feature map. In contrast, the CroW [9] applies weighting to the feature map in spatial and channel dimensions before sum-pooling. The R-MAC [10] method utilizes sliding windows of different sizes on the feature map, extracting local features using the results of max-pooling, and ultimately summing up this series of local features to obtain the global representation of the image.

2.2 Fine-Grained Image Retrieval

Traditional image retrieval methods often fail to achieve ideal results when processing fine-grained images due to inadequate local feature description, loss of semantic information, and limitation of feature expression ability. Therefore, more advanced methods and techniques are needed to solve these problems. The emergence of SCDA [11] has brought phased progress to FGIR tasks [12, 33, 34]. [11] selected relevant deep descriptors by performing object region localization, which was then aggregated into a compact feature vector to serve as the comprehensive image representation, and the entire process of object localization was implemented using an unsupervised approach.

Fig. 2. The raw images depicted in this figure originate from several open-source datasets and serve as the input for our feature extraction and analysis procedures.

In our proposed object localization submodule, when performing the feature map's clipping, the clipping region's determination is likewise inspired by the methodology above. Also inspired is the weakly supervised localization module proposed in [12]. To enhance model generalization and improve retrieval performance, [12] and [13] respectively proposed the centralized ranking loss (CRL) function and the segmented cross-entropy loss function. Whether it is the improvement and optimization of loss functions by [12–14] or the optimization of feature embedding by [17, 18, 31] previous work, they all employ metric learning techniques to extract discriminative features for the object, thereby achieving the task of fine-grained image retrieval.

3 Feature Extraction and Multiscale Feature Fusion

In this section, we detail our proposed MsCFA method for fine-grained image retrieval. Initially, we crop the object region based on the submodule's high activation response. Following this, we acquire local descriptors. Finally, we fuse multiscale features to achieve optimal retrieval performance.

3.1 Object Localization

In publicly available datasets, most images are randomly captured in natural environments, thus frequently contaminated by background noise, which blends with vital features within the images, as depicted in Fig. 2. When utilizing neural networks to extract features, such noise is also incorporated into the feature representations, potentially obscuring or masking the intrinsically important feature information in the images. Therefore, like most research efforts, we perform a coarse localization of the object regions. Our approach to localization is inspired by the methodology in [20] but differs in that instead of directly returning the feature map of the localized region as the image descriptor, our submodule crops the feature map and utilizes the cropped version to extract the local features.

Obtain the Mask Map. Given an image of size $H \times W \times C$, where H and W represent the height and width of the input image, and C is typically 3 for a color image. When this image is fed into a pre-trained convolutional neural network, it produces a feature map X of size $h \times w \times c$, where h and w denote the height and width of the resulting feature map, and c represents the number of channels in the feature map. For instance, in this paper, the pre-trained network used is VGG16. If the input image is a 224×224 color image, the output from this network's last convolutional layer, namely $pool5$, would result in a feature map of size $7 \times 7 \times 512$. Subsequently, summing along the third dimension yields a 3D tensor A of size $h \times w \times 1$, which satisfies the following relationship with the feature map X,

$$A = \sum_{n=1}^{c} X_n \qquad (1)$$

where X_n represents the 2D tensor (or matrix) of the response of the n-th feature channel output by the network at a given spatial position, and the pixel value at each position of the $h * w$ pixels in A is the result of summing the channel dimensions.

For the obtained A, typically, the higher the activation response value of a particular location (i,j), the more likely it is that the location belongs to the object region, where $i \in \{1,...,h\}, j \in \{1,...,w\}$. Therefore, we average the activation values across the $h * w$ positions of A, as shown in formula (2),

$$\hat{a} = \frac{1}{h * w} \sum_{i=1}^{h} \sum_{j=1}^{w} A(i,j) \qquad (2)$$

where $A(i,j)$ represents the activation value at the current position (i,j) in A.

The obtained value $\hat{a}$ is used as a threshold to determine whether a location is a high-response area (i.e., the object area). If the activation value at a location exceeds the threshold, it is marked as 1; otherwise, it is marked as 0. This process yields the mask diagram M of the object area, as illustrated in formula (3),

$$M(i,j) = \begin{cases} 1 & \text{if } A_{i,j} > \hat{a} \\ 0 & \text{otherwise} \end{cases} \qquad (3)$$

where M is also a 3D tensor with a shape of $h \times w \times 1$, the same as that of A.

Crop Object Region. First, we employ the '8-connected' criterion to determine the connectivity between pixels, upon which we identify and extract all connected regions within M. According to the definition of '8-connected', two pixels are considered connected when their edges or corners are directly adjacent. Specifically, if two adjacent pixels both have a value of '1' and are adjacent in horizontal, vertical, or diagonal directions, we classify these two pixels as belonging to the same object or region. Assign different labels sequentially to different connected regions, while identical labels are allocated to the same connected region. Subsequently, the connected region with the largest area is selected as our object region. Instead of performing edge thinning on the object region, we crop the smallest rectangle encompassing the largest connected region to facilitate subsequent local feature extraction. This cropped region is then mapped to the original feature map. The resulting feature map, X_{crop}, has dimensions of $h_c \times w_c \times 1$, where h_c and w_c represent the height and width of the cropped feature graph, respectively, while the number of channels, c, remains unchanged. Figure 3 illustrates the main steps of our cropping process.

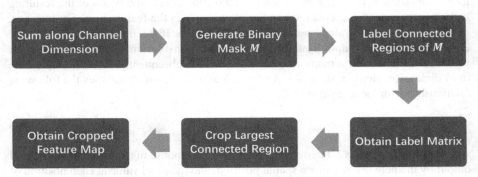

Fig. 3. The figure shows the process of cropping the feature map.

3.2 Local Feature Extraction

In fine-grained tasks, slight differences between categories often manifest in local features of the image, such as the bird's beak, wings, feet, and tail. Precisely because the extraction of effective local information is crucial for FGIR tasks, the concept of local regions has emerged. Initially, [21] introduced a region structure implemented on the original image, and later, [10] proposed a region structure applied to the CNN response map, known as the R-MAC [10] method. This method has proven effective for traditional image retrieval tasks. Our local feature extraction is also inspired by this approach. To fully leverage the strengths of the R-MAC [10] method, we devise a dual-path strategy for local feature extraction: part of the features is directly derived from the raw CNN response maps, while the other part stems from the cropped response maps. Specifically, an image is input into a CNN, which then undergoes a series of convolution and pooling operations to produce its corresponding feature map. For this feature map, we adopt a dual-path strategy: one path directly extracts R-MAC features to capture global information; the other path crops the image based on object localization described in Sec 3.1 and then extracts R-MAC features to focus on the object region. Finally, these two sets of features are fused to obtain a single-scale integrated feature representation. The detailed feature extraction process is elaborated below.

Determine the Local Region Size. The first is the determination of the size of the local region, analogous to the scale definition proposed in [10], based on the scale on the short side $min\{h_c, w_c\}$. We aggregate the feature vectors computed at different scales l, where $l \in \{1, 2, 3, \ldots\}$, to form the feature representation of the entire image. Different scales, l, where $l \in \{1, 2, 3 \ldots\}$, are superimposed to obtain the feature representation of the entire image. If the scale on the long side $max\{h_c, w_c\}$ is s, then the number of local regions n under this scale l is considered to be $l * s$. (Note: the scale here is not the same as the scale in the title, the scale here is related to the size of the local region, and the title refers to the size of the input image.) The various scales, l, can be interpreted as the size of sliding windows on the feature map. The value of l corresponds to the number of sliding windows, where $l-1$ represents the number of slides of the sliding windows on the short side $min\{h_c, w_c\}$. When $l = 1$, the local region is the largest. To maintain consistency with the scale definitions in [10] while minimizing the loss of detail information, we set the overlap ratio between consecutive regions along the longer edge to 40% and 50% along the shorter edge. Consequently, the height h_r and width w_r of the local region are $2min\{h_c, w_c\}/(l + 1)$. Figure 4 shows the sampling information at different scales.

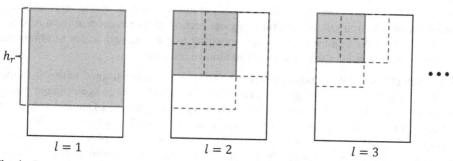

Fig. 4. Sample regions were extracted at several different scales ($l = 1,2,3$). hr is the size of the local region; we show the first and adjacent local regions under different scales; the green region is the first local region, and the red border is its adjacent local region.

Obtain the Local Descriptor. Each local region is represented as a short feature vector, and then the feature representation fr of the image at the current scale is shown in formula (4),

$$fr = \sum_{i=1}^{n} r_i \qquad (4)$$

where $i \in \{1,\ldots,n\}$, n is the number of local regions on the current feature map, and r_i is a C-dimensional vector.

Let's define L, If L = 1 take l =1; If L = 2 take l =1, l =2; If L = 3, take l =1, l = 2, l =3…Then, the feature of the image represents Fr, as shown in formula (5).

$$Fr = [fr_1, fr_2, fr_3 \ldots\ldots, fr_z], z \in L \tag{5}$$

We stack Fr along the channel dimension to obtain a C-dimensional vector, which serves as the local feature descriptor.

The above outlines our entire feature extraction process. As previously mentioned, the cropping in our submodule is based on the approximate shape of the object. Since the cropped height h_c and width w_c may not be equal, the number of local regions during local feature extraction is not fixed. This characteristic renders the extracted features more discriminative and robust, enhancing the image's feature representation capability. Therefore, for both branches of feature extraction, we employ the abovementioned methods to extract features and subsequently fuse these two sets of features through a weighted combination, resulting in a single-scale feature representation of the image.

3.3 Multiscale Global Feature Fusion

Feature Fusion. Feature fusion is a common method of feature extraction, which can integrate features from different sources or levels to enhance the ability to express the internal structure and semantic information of images. Especially in FGIR tasks, it has a wide range of applications [8, 22–24]. Due to the limitation of feature expression from a single-scale image, we fuse multiple scale features of the image and then make full use of the complementarity of features at different scales to achieve more accurate and robust image feature expression. Specifically, we input multiple scales of the original image into the network simultaneously. Following our preliminary extraction of local features, we concatenate the feature vectors corresponding to the obtained scales to obtain the feature representation of the image for final image retrieval.

Scale Selection. IIn this study, we selected VGG16 as the pre-trained network framework, whose pre-trained weights are based on ImageNet, a large-scale image classification dataset. Compared to complex networks like ResNet and DenseNet, VGG16 achieves a more intuitive and stable feature extraction capability with its concise architecture and streamlined parameters. Adopting the standard setting of VGG16, we set the input image size at 224 × 224 pixels, which serves as the baseline minimum scale for our research. Given the wide range of image sizes encountered in actual datasets, spanning from 206 × 240 to 1500 × 1000, we devised a multiscale strategy. By incrementally expanding the scale by half of the baseline each time, we constructed five distinct scales: 224 × 224, 336 × 336,…, up to 672 × 672. Subsequently, these images at various scales were concurrently fed into VGG16 for feature extraction, and the features were then fused to create our multiscale feature representation.

4 Experiments

In this section, we present our approach's experimental results. First, we introduce some experimental settings, including the datasets, evaluation metrics, and certain experimental details. Then, we compare our method's retrieval performance with other methods. Finally, to verify our method's effectiveness, we conducted an ablation experiment.

4.1 Experimental Setting

Datasets. To ensure the credibility of our experiments, we employed five classic publicly available datasets in fine-grained tasks. These include CUB200–2011 [25] with 200 classes and 11,788 images (5,994 for training, 5,794 for testing); Stanford Dogs [26] with 120 classes and 20,580 images (100 per class for training, remainder for testing); Oxford Flower [27] with 8,189 images across 102 classes; Aircraft [28] with 10,000 images from 100 classes (100 per class, evenly split into training, validation, and testing sets, with the first two subsets used for training); and Cars [29] with 196 classes and 16,185 images (8,144 for training, 8,041 for testing).

Evaluation Metrics. We adopted the commonly used evaluation index — mean Average Precision(mAP) to evaluate the performance [16]. Precisely, we first calculate each query's average accuracy (AP) score and then take their average as the mAP score. This evaluation method can more accurately measure the performance of our model on the query task. The relevant formula is as follows,

$$AP = \frac{1}{N_r} \sum_{k=1}^{K} (P(k).R(k)) \quad (6)$$

where N_r is the number of relevant images in the current query, $P(k)$ is the accuracy of the current location k, and $R(k)$ is the correlation of the current location k, with the correlation being 1, otherwise 0.

$$mAP = \frac{1}{Q} \sum_{q=1}^{Q} AP(q) \quad (7)$$

where Q is the number of query images.

Implementation Details. Our experiments were conducted using MATLAB R2023b on a 64-bit Ubuntu 22.04.4 LTS system. Since the proposed method is training-free, there is no need to split the data into training and testing sets. To verify the accuracy of our method, we experimented with the maximum amount of data possible, treating each image in the dataset as a query image and comparing it with the database images. The top-6 most relevant images were then returned. While calculating the mAP, we excluded the original query image that ranked first and evaluated the remaining five images as the top-5 results.

4.2 Comparative Experiment

Methods for Comparison. R-MAC [10] extracts the features by encoding several image regions into low-dimensional vectors. SCDA [11] maps the locations of high-response activations in the feature map back to the original image, utilizing this region as the primary feature descriptor. CRL [12] introduces a novel centralized ranking loss aimed at efficiently learning discriminative features. Similarly, PCE [13] proposes a variant of cross-entropy loss known as the piecewise cross-entropy loss function. DGCRL [14] eliminated the gap between the inner product and the Euclidean distance in the training and test stages by adding a Normalize-Scale layer to enhance the intra-class separability and inter-class compactness with their Decorrelated Global-aware Centralized Ranking Loss [15, 16]. Additionally, FPM [30] introduces data mining techniques into FGIR tasks by leveraging frequent pattern mining to explore correlations between different image components.

Table 1. Comparison of the mAP accuracy (%) of the top-5 returned over five datasets, with the best results for each column shown in bold and the second-best results underlined.

Method	CUB	Dog	Flower	Aircraft	Car
R-MAC [10]	59.02	66.28	78.19	54.94	52.98
SCDA [11]	65.79	79.24	77.70	58.64	45.16
CRL [12]	67.23	76.43	78.65	60.21	48.16
PCE [14]	66.79	78.43	78.23	60.64	47.16
DGCRL [13]	67.97	78.25	79.21	59.34	49.67
FPM [30]	68.32	**79.68**	79.62	**62.45**	48.76
OURS	**68.68**	79.52	**93.92**	61.29	**56.10**

Results and Causes. We have conducted a comparison of MsCFA with existing methods. As can be observed from Table 1, our method exhibits significant improvements over the data mining approach of FPM [30], specifically achieving gains of 0.36%, 14.3%, and 7.34% for the CUB, Flower, and Car datasets, respectively. Notably, the retrieval performance of our method is comparable to or even surpasses, those of end-to-end supervised methods leveraging loss functions, achieving superior performance in most of the five datasets. When compared to the original R-MAC method, our method achieves significant gains in retrieval performance, with improvements of 9.66%, 13.24%, 15.73%, 6.35%, and 3.12%, respectively, across the five datasets. Notably, although most methods exhibit low retrieval performance on Car datasets, our method still maintains the advantages of the R-MAC method on Car datasets. However, our results on the Dog and Aircraft datasets fall slightly short of the state-of-the-art FPM [30] method by 0.16% and 1.16%, respectively. A possible explanation for this is that, unlike FPM, which extracts features by frequently mining the intrinsic relationships between different parts, our model may still have limitations in capturing these intricate correlations, leading to a slightly inferior performance on these two datasets.

4.3 Ablation Experiment

To validate the effectiveness of the various components of our method, we conducted ablation experiments. We refer to single-scale features without the object localization submodules as "non-localization single-scale" features (NL_SS), features with the submodules as "localization single-scale" features (L_SS), multiscale features without the localization submodules as "non-localization multiscale" features (NL_MS), and finally our "localization multiscale"(L_MS) method. The specific experimental results are presented in Table 2.

Table 2. Listed from top to bottom are the mAP results for "non-localization single-scale", "localization single-scale", "non-localization multi-scale", and "localization multi-scale".

Method	CUB	Dog	Flower	Aircraft	Car
NL_SS	58.53	72.69	85.63	48.20	41.24
L_SS	58.76	72.85	85.70	48.21	42.10
NL_MS	68.46	77.98	93.60	60.99	55.03
L_MS	68.68	79.52	93.92	61.29	56.10

As can be observed from Table 2, after introducing our localization submodule, the retrieval performance of the five datasets under the single-scale condition achieved slight improvements, but none of the improvements exceeded 1%. However, after introducing the feature fusion module, the performance enhancements on these datasets became particularly significant, with specific improvement rates of 9.93%, 5.29%, 7.97%, 12.79%, and 13.79%, respectively. Notably, when both the localization submodule and the feature fusion module were introduced simultaneously, the performance improvements were even more pronounced, reaching 10.15%, 6.83%, 8.29%, 13.09%, and 14.86%, respectively. This result shows that the introduction of multiscale features has a more significant effect on improving the retrieval performance compared to our localization submodule, and when the two are combined, the optimal effect can be achieved.

5 Conclusions

In this paper, we introduce a multiscale convolutional feature aggregation method for fine-grained image retrieval. This multiscale convolutional feature aggregation method demonstrates remarkable advantages in fine-grained image retrieval, owing to its exceptional ability to capture image features and its unsupervised feature extraction process. We evaluated our approach on five classic fine-grained datasets, and the extensive experimental results indicate that our method enhances the performance of fine-grained image retrieval. However, the adoption of a multi-input approach for our model's inputs has, to some extent, increased the computational complexity and processing time of the model, posing challenges for real-time applications.

In future work, to optimize the model's efficiency, we plan to apply transformation techniques to a single feature map to generate feature representations at multiple scales, thereby replacing the multi-input approach. This not only effectively reduces computational costs but also has the potential to further improve feature discriminability and model robustness. Furthermore, given the inherent similarity between fine-grained classification and fine-grained retrieval tasks, we are confident that this method also holds vast potential for application in fine-grained classification tasks.

Acknowledgments. This work was supported by the National Natural Science Foundation of China (Grant No. 62106227), the Key Projects of Jinhua Science and Technology Bureau (Project no. 2024-2-015), the China Postdoctoral Science Foundation (Grant No. 2023M743132), and the

"Teacher Professional Development Project" for Domestic Visiting Scholars in 2023 (Project No. FX2023007).

References

1. Forcen, J.I., Pagola, M., Barrenechea, E., Bustince, H.: Co-occurrence of deep convolutional features for image search. Image Vis. Comput. **97**, 103909 (2020)
2. Razavian, A.S., Sullivan, J., Carlsson, S., Maki, A.: Visual instance retrieval with deep convolutional networks. ITE Trans. Media Technol. Appl. **4**(3), 251–258 (2016)
3. Fadaei, S., Dehghani, A., Ravaei, B.: Content-based image retrieval using multi-scale averaging local binary patterns. Digit. Sig. Process. **146**, 104391 (2024)
4. Gao, Y., Han, X., Wang, X., Huang, W., Scott, M.: Channel interaction networks for fine-grained image categorization. In: Proceedings of the AAAI Conference on Artificial Intelligence, vol. 34, no. 07, pp. 10818–10825, April 2020
5. Wei, Q., Feng, L., Sun, H., Wang, R., Guo, C., Yin, Y.: Fine-grained classification with noisy labels. In: Proceedings of the IEEE/CVF Conference on Computer Vision and Pattern Recognition, pp. 11651–11660 (2023)
6. Babenko, A., Slesarev, A., Chigorin, A., Lempitsky, V.: Neural codes for image retrieval. In: Computer Vision–ECCV 2014: 13th European Conference, Zurich, Switzerland, 6–12 September 2014, Proceedings, Part I 13, pp. 584–599 (2014)
7. Dubey, S.R.: A decade survey of content-based image retrieval using deep learning. IEEE Trans. Circuits Syst. Video Technol. **32**(5), 2687–2704 (2021)
8. Babenko, A., Lempitsky, V.: Aggregating local deep features for image retrieval. In: Proceedings of the IEEE International Conference on Computer Vision, pp. 1269–1277 (2015)
9. Kalantidis, Y., Mellina, C., Osindero, S.: Cross-dimensional weighting for aggregated deep convolutional features. In: Computer Vision–ECCV 2016 Workshops: Amsterdam, The Netherlands, October 8–10 and 15–16, 2016, Proceedings, Part I 14, pp. 685–701 (2016)
10. Tolias, G., Sicre, R., Jégou, H.: Particular object retrieval with integral max-pooling of CNN activations [EB/OL]. (2015-11-18) [2022-04-05]. https://arxiv.org/abs/1511.05879
11. Wei, X.S., Luo, J.H., Wu, J., Zhou, Z.H.: Selective convolutional descriptor aggregation for fine-grained image retrieval. IEEE Trans. Image Process. **26**(6), 2868–2881 (2017)
12. Zheng, X., Ji, R., Sun, X., Wu, Y., Yang, Y.: Centralized ranking loss with weakly supervised localization for fine-grained object retrieval. In: IJCAI, pp. 1226–1233, July 2018
13. Zeng, X., Zhang, Y., Wang, X., Chen, K., Li, D., Yang, W.: Fine-grained image retrieval via piecewise cross entropy loss. Image Vis. Comput. **93**, 103820 (2020)
14. Zheng, X., Ji, R., Sun, X., Zhang, B., Wu, Y., Huang, F.: Towards optimal fine-grained retrieval via decorrelated centralized loss with normalize-scale layer. In: Proceedings of the AAAI Conference on Artificial Intelligence, vol. 33, no. 01, pp. 9291–9298, July 2019
15. Wei, X.S., et al.: Fine-grained image analysis with deep learning: a survey. IEEE Trans. Pattern Anal. Mach. Intell. **44**(12), 8927–8948 (2021)
16. Wei, X. S., Wu, J., Cui, Q.: Deep learning for fine-grained image analysis: a survey (2019). arXiv preprint arXiv:1907.03069
17. Zhang, X., Zhou, F., Lin, Y., Zhang, S.: Embedding label structures for fine-grained feature representation. In: Proceedings of the IEEE Conference on Computer Vision and Pattern Recognition, pp. 1114–1123 (2016)
18. Oh Song, H., Xiang, Y., Jegelka, S., Savarese, S.: Deep metric learning via lifted structured feature embedding. In: Proceedings of the IEEE Conference on Computer Vision and Pattern Recognition, pp. 4004–4012 (2016)

19. Sun, M., Yuan, Y., Zhou, F., Ding, E.: Multi-attention multi-class constraint for fine-grained image recognition. In: Proceedings of the European Conference on Computer Vision (ECCV), pp. 805–821 (2018)
20. Musgrave, K., Belongie, S., Lim, S.N.: A metric learning reality check. In: Computer Vision–ECCV 2020: 16th European Conference, Glasgow, UK, 23–28 August 2020, Proceedings, Part XXV 16, pp. 681–699. Springer (2020)
21. Ali, S.R., Sullivan, J., Maki, A., Carlsson, S.: A baseline for visual instance retrieval with deep convolutional networks. In: Proceedings of International Conference on Learning Representations (2015)
22. Liu, J., Fan, X., Jiang, J., Liu, R., Luo, Z.: Learning a deep multi-scale feature ensemble and an edge-attention guidance for image fusion. IEEE Trans. Circuits Syst. Video Technol. **32**(1), 105–119 (2021)
23. Zhang, Z., Xie, Y., Zhang, W., Tian, Q.: Effective image retrieval via multilinear multi-index fusion. IEEE Trans. Multimedia **21**(11), 2878–2890 (2019)
24. Imbriaco, R., Sebastian, C., Bondarev, E., de With, P.H.: Aggregated deep local features for remote sensing image retrieval. Remote Sens. **11**(5), 493 (2019)
25. Wah, C., Branson, S., Welinder, P., Perona, P., Belongie, S.: The Caltech-UCSD Birds-200-2011 Dataset. Technical Report, CNS-TR-2011-001, California Institute of Technology (2011)
26. Khosla, A., Jayadevaprakash, N., Yao, B., Li, F.F.: Novel dataset for fine-grained image categorization: Stanford dogs. In: Proceedings of CVPR Workshop on Fine-grained visual Categorization (FGVC), vol. 2, no. 1. Citeseer, June 2011
27. Nilsback, M.E., Zisserman, A.: Automated flower classification over a large number of classes. In: 2008 Sixth Indian Conference on Computer Vision, Graphics & Image Processing, pp. 722–729. IEEE, December 2008
28. Maji, S., Rahtu, E., Kannala, J., Blaschko, M., Vedaldi, A.: Fine-grained visual classification of aircraft (2013). arXiv preprint arXiv:1306.5151
29. Krause, J., Stark, M., Deng, J., Fei-Fei, L.: 3d object representations for fine-grained categorization. In: Proceedings of the IEEE International Conference on Computer Vision Workshops, pp. 554–561 (2013)
30. Zheng, M., Geng, Y., Li, Q.: Revisiting local descriptors via frequent pattern mining for fine-grained image retrieval. Entropy **24**(2), 156 (2022)
31. Huang, C., Loy, C.C., Tang, X.: Local similarity-aware deep feature embedding. In: Advances in Neural Information Processing Systems, vol. 29 (2016)
32. Hoang, T., Do, T.T., Le Tan, D.K., Cheung, N.M.: Selective deep convolutional features for image retrieval. In: Proceedings of the 25th ACM international Conference on Multimedia, pp. 1600–1608, October 2017
33. Wei, X.S., Shen, Y., Sun, X., Wang, P., Peng, Y.: Attribute-aware deep hashing with self-consistency for large-scale fine-grained image retrieval. IEEE Trans. Pattern Anal. Mach. Intell. (2023)
34. Zhu, Y., Cao, G., Yang, Z., Lu, X.: Learning relation-based features for fine-grained image retrieval. Pattern Recogn. **140**, 109543 (2023)

Author Index

B
Bi, Ran 295
Bi, Yuanqiao 200

C
Chai, Yanfeng 30
Chen, Chong 163
Chen, Jili 249
Chen, Yizhou 260
Chen, Zhangze 239
Chi, Kaikai 271
Cui, Haixia 212

D
De Meo, Pasquale 229
Ding, Kaifeng 335
Ding, Linlin 115, 123
Du, Dong 175
Du, Lun 310
Du, Shumei 81
Du, Yuefeng 81, 115, 123
Duan, Xiaodong 45, 57, 69

G
Gan, Lin 45, 57
Gao, Ying 45, 57
Ge, Jiake 30
Guo, Hanghui 239

H
Huang, Jin 260
Huang, Liang 271
Huang, Qionghao 249
Huang, Zixuan 260

K
Kou, Yue 135

L
Li, Changlong 151
Li, Dong 81, 135
Li, Jiaming 69
Li, Jiashu 30
Li, Shengyu 295
Li, Simeng 295
Li, Tianrui 123
Li, Weijian 200
Li, Ying 17
Li, Yuqi 271
Liang, Guiyuan 69
Lin, Youfang 163
Ling, Jianxia 229
Liu, Lin 295
Liu, Tingting 115, 123, 135
Liu, Zhicong 135
Lu, Bo 45, 57, 69
Luo, Qi 3
Lv, Xian 295

M
Mi, Qiao 310

P
Pancerz, Krzysztof 323

Q
Qin, Dao-Ju 103
Qin, Lei 81
Qu, Haoran 135

R
Ran, Wen 151
Ren, Jiankang 295

S
Schumann, Andrew 323
Sha, Edwin H.-M. 151
Shang, Pei-Pei 103

Shestakevych, Tetiana 239
Shi, Jianyang 229
Shuai, Jiaqi 212
Song, Baoyan 123

T
Tan, Zhen 17

U
Ukey, Nimish 3

W
Wang, Bo 163
Wang, Ke 271
Wang, Xin 30
Wang, Yilong 175
Wei, Wei 94
Wei, Yong 81
Wen, Dong 3
Wen, Weichang 212
Wu, Weijun 260
Wu, Xiaojin 57

X
Xiao, Weidong 17
Xie, Guiyuan 200

Y
Yan, Haonan 3
Yang, Caolin 335
Yang, Chengzhuan 335
Yang, Hongxin 115, 123
Yang, Xi 239
Yang, Zhengyi 3
Yang, Zhenyu 188
Yu, Weihao 260
Yu, Yang 175
Yuan, Xia 45
Yue, Chongjian 310

Z
Zhang, Lei 281
Zhang, Qiang 30
Zhang, Qingyu 135
Zhang, Tianming 3
Zhang, Yuanwei 212
Zhang, Zhenguo 281
Zhao, Linlin 281
Zhao, Tianbao 45, 57, 69
Zhong, Tianyuan 57
Zhu, Haibin 45
Zhu, Jia 229

Printed in the United States
by Baker & Taylor Publisher Services